S0-AID-262

Transforming Undergraduate
Science Teaching

Studies in the
Postmodern Theory of Education

Joe L. Kincheloe and Shirley R. Steinberg
General Editors

Vol. 189

PETER LANG
New York • Washington, D.C./Baltimore • Bern
Frankfurt am Main • Berlin • Brussels • Vienna • Oxford

Transforming Undergraduate Science Teaching

Social Constructivist Perspectives

EDITED BY
Peter C. Taylor, Penny J. Gilmer,
& Kenneth Tobin

PETER LANG
New York • Washington, D.C./Baltimore • Bern
Frankfurt am Main • Berlin • Brussels • Vienna • Oxford

CABRINI COLLEGE LIBRARY
610 KING OF PRUSSIA ROAD
RADNOR, PA 19087

Q
181
.T814
2002

#45861969

Library of Congress Cataloging-in-Publication Data

Transforming undergraduate science teaching: social constructivist perspectives /
Edited by Peter C. Taylor, Penny J. Gilmer, and Kenneth Tobin.
p. cm. — (Counterpoints: Studies in the postmodern theory of education; v. 189)
Includes bibliographical references and index.
1. Science—Study and teaching (Higher). 2. Constructivism (Philosophy).
I. Taylor, Peter. II. Gilmer, Penny J. III. Tobin, Kenneth George.
IV. Counterpoints (New York, N.Y.); vol. 189.
Q181 .T814   507'.1'1—dc 21   2001029072
ISBN 0-8204-5293-9
ISSN 1058-1634

Die Deutsche Bibliothek-CIP-Einheitsaufnahme

Transforming undergraduate science teaching:
social constructivist perspectives / ed. by: Peter C. Taylor....
–New York; Washington, D.C./Baltimore; Bern;
Frankfurt am Main; Berlin; Brussels; Vienna; Oxford: Lang.
(Counterpoints; Vol. 189)
ISBN 0-8204-5293-9

Cover design by Lisa Dillon

The paper in this book meets the guidelines for permanence and durability
of the Committee on Production Guidelines for Book Longevity
of the Council of Library Resources.

© 2002 Peter Lang Publishing, Inc., New York

All rights reserved.
Reprint or reproduction, even partially, in all forms such as microfilm,
xerography, microfiche, microcard, and offset strictly prohibited.

Printed in the United States of America

CABRINI COLLEGE LIBRARY
610 KING OF PRUSSIA ROAD
RADNOR, PA 19087

Table of Contents

Acknowledgements

This book project has been underway for over five years. It started with a dinner with the three co-editors in Tallahassee in December 1995, just as Peter Taylor was finishing his sabbatical with Kenneth Tobin. The book has had a long gestation period, and it is about to be born. It has been a wonderful opportunity for learning between and among the authors and co-editors, and with the people with whom we have worked.

We particularly want to acknowledge the competent and dedicated service of Ms. Lisa Upham for her help in the final stages of editing, doing the graphics, and putting together of the book.

We also appreciate the patience of our families throughout this effort.

PREFACE

 Introducing Transformations

Peter C. Taylor, Penny J. Gilmer,
and Kenneth Tobin

This book comes at a time when epistemological reform is sweeping through the global community of science education. Since the 1970s, the epistemologies of practice (or theories of knowing embodied in the teaching activities) of school science teachers have been undergoing a major transformation toward more learner-sensitive standpoints. But undergraduate science teaching, the breeding ground of teachers of school science, has remained largely unaffected; teacher-centered teaching remains the dominant epistemology of practice in many lecture theatres and laboratories. Little wonder then that newly graduated science teachers arrive in school science classrooms ill-prepared to practice learner-sensitive teaching.

Thus, the motivating force behind this book is our wish to expedite the process of epistemological reform of undergraduate science teaching, to align it with the reform goals of the science education community. We hope that what is learned from the unique collection of studies presented here will be extended to science courses for all students, those going on directly into science, those planning to become K-12 teachers of science, and those in allied fields as well.

As editors of this book (and as science educators and a professional scientist), our epistemological standpoint is that of social constructivism, a powerful and multifaceted perspective. It focuses attention on the way learners make sense of their experiences in terms of their extant worldviews and the way others, including peers and teachers, mediate the sense-making process.

Social constructivism is a theory of knowing-in-action (i.e., a theory of knowing embodied in our actions) that serves as a powerful referent for thinking about teaching, learning, and relationships between teachers and students. Significantly, it is a powerful referent for contesting the dominance of the traditional worldview referent of objectivism, which tends to construe teaching in terms of the transmission metaphor of teaching as delivery of disciplinary knowledge (through "lecturing" and "professing") and learning as reception of knowledge (through listening and recording). Beliefs that accord with objectivism can de-emphasize the role of learners' prior knowledge in building deep understandings of science. In such cases, teachers regard science as being learned in isolation from the contexts of the learner's lifeworld.

In contrast, a social constructivist perspective recognizes science as a social construction and regards what is known already as a foundation for the learning of science. What individuals know, feel, and value is a critical part of what they will come to know about science. Social constructivism promotes the joint metaphors of *knowing as constructing* (contingent and viable understandings) and *knowing as participating* (actively and collaboratively) in a specific discourse community. Any science that is known by anyone is constructed within a sociocultural context and is re-presented by the language of the knower (whether professional scientist or student). Approaches to the teaching of science that are based on social constructivism tend to value the participative actions of learners as they use language to make sense of experiences and build a new discourse that is science-like.

If students are to build deep learning of science they must be able to *co-participate*, that is, to access and appropriate a discourse to give them power over the subject matter they are endeavoring to learn. The essence of the challenge is to create a community where students can solve problems in which they see and experience science. For this to be the case they need to create a science-like discourse and have opportunities to apply it and, as necessary, to make adjustments to what they know and can do. Co-participation implies access to others who already know and can do, including teachers, tutors, students from previous years, and peers. Co-participation requires knowledge to be put to the test in the presence of others and implies self- and peer-evaluation as part of the learning process, and an active teacher who pays attention to the evolution of what his/her students know and are able to do.

No matter its size, any classroom can be regarded as a discourse community. If learning of science is to occur within that community it seems essential that learners must access (linguistic) resources to enable them to make sense of what it is they are to learn. To paraphrase Ausubel's (1968) entreaty to teachers, on entry to a course find out what students know and can do and teach them accordingly. This premise entails assisting them to find resources to connect from what they know and can do to what they need to know and be able to do. How to facilitate this is a matter of ongoing creative design mediated by reflective thinking. Learner-sensitive teaching is not easy; it does not follow a recipe or a master narrative. The details of how to accomplish learner-sensitive teaching must be worked out for every possible context. Every learner needs structure, a scaffold, to get from where s/he is to some place else. Some learners can provide their own scaffolds by reading books and accessing resources that they know about from their own lifeworlds. But others cannot provide the necessary structure and are more dependent on teachers in class, tutorials, and individual assistance during office hours. If teachers are sensitive to the needs for all learners to have structure to support learning, then they can find opportunities to mediate learning in ways that make sense to them.

The struggle to transform undergraduate science teaching and learning is not, however, simply a matter of replacing the transmission metaphor with constructivist metaphors. We recognize that science teaching is a cultural (institutional) activity and that cultures are inherently conservative. A much more fruitful approach, we believe, is to engage professors of science in a gradual transformation characterized by their development of a repertoire of referents, including those associated with both objectivism and constructivism. A transformative science teacher is one who can make a well-justified professional judgment about which teaching-learning metaphors are most appropriate for a particular pedagogical purpose and context. This judgment involves a dialectical rationality: keeping one eye on the pragmatic requirements of the extant culture and the other (visionary eye) on possibilities for cultural trans-formation. Nevertheless, although we champion epistemological pluralism for undergraduate science teaching, our focus in this book is largely on transformative possibilities associated with enacting the referent of social constructivism and the joint metaphors of knowing as constructing and participating.

We hasten to add that this paradigm change is not just the dream-child of reform-minded science educators. It is being driven by a well-recognized need to better prepare young people for a rapidly changing society, a society in which advances in technology are as breathtaking as the new relativism that is contesting the comfortable certainties of our traditional (mono-) cultural values and standards. At the beginning of a new millennium, we find a dawning (postmodern?) age of uncertainty, pluralism and difference, an age that for some time now has been shaping professional practices in the fields of film, art, architecture, multimedia design, literary criticism, journalism, history, philosophy, anthropology, psychology, and the social sciences. If science education is to remain relevant to society, it too must change, expand, and diversify. Science education cannot afford to offer young people anything less than a sense of dynamism and empowerment as agents of both cultural reproduction and cultural transformation.

In this book, we draw attention to the beginnings of a paradigm change by portraying, through various innovative research "lenses," the current and future "state of play" of undergraduate science teaching. Table 1 summarizes the major features of an extensive set of research studies of undergraduate science teaching. As readers of this book, you may want to follow threads of common themes, such as *language, power,* or *co-participation.* For instance, the reader can easily identify that the chapters by Bowen, Griffiths, Moscovici, Roth and Tobin, and Gilmer include issues of language in the science classroom. Alternately, the reader may wish to focus on science or science education courses, and the table will allow you to find those chapters and identify the research settings. If you are interested in following a particular research methodology, such as action research, hermeneutic, phenomenology or interpretive ethnography, this is indicated for each chapter as well.

Table 1: Overview of Book

	Author	Research Style	Issues	Course	Students
1	Taylor	• Impressionistic tales • Hermeneutic phenomenology • Critical reflexivity	• Cultural archetypes • Jung's collective unconscious • Pedagogical thoughtfulness • Existential	Health Science & Mathematics	Community college students

Continued on next page

Table 1 (Continued)

	Author	Research Style	Issues	Course	Students
			awareness & emotionality		
2	Bowen	• Interpretive ethnography • Fictionalized narrative	• Student account • Technical language of teacher • Educative rapport	Chemistry	Liberal studies university students
3	Griffiths	• Interpretive ethnography	• Ue of biology content language • Border crossing	Biology	Prospective elementary school teachers
4	Moscovici	• Interpretive ethnography	• Uneven power relationships • Language of power and control • Critical thinking	Biology	Prospective elementary school teachers
5	Scantlebury	• Literature review and autobiography	• Feminist pedagogy • Gender equity strategies • Gender and power • Cooperative learning	Science education	Prospective middle/high school teachers
6	Roth & Tobin	• Interpretive ethnography • Autobiographical tale	• Philosophy of language • Border pedagogy • Cultural myths Problem solving • Reflective discourse	Physical science	Prospective elementary teachers
7	Abbas, Goldsby, & Gilmer	• Interpretive ethnography • Participant-Observer • Critical autobiography	• Journey-related metaphors • Active learning	Physical Science	Prospective elementary teachers
8	Humerick	• Critical action research • Interpretive	• Guided inquiry • Active learning • Caring	Chemistry	Community college science

Continued on next page

Table 1 (Continued)

	Author	Research Style	Issues	Course	Students
		• ethnography	• Caring		majors
9	White	• Critical action research	• Problem-based learning • Autonomous learning • Non-dualistic way of knowing • Critical thinking	Biochem-istry	University-level science majors
10	Mattson	• Fictionalized narrative • Participant-observer	• Alternative assessments • Student-centered learning	Biology	Prospective elementary teachers
11	Brush	• Interpretive ethnography	• Collaboration • Team teaching	Physical Science	Prospective elementary teachers
12	Briscoe	• Interpretive ethnography	• Social linguistics • Communicative interactions	Biology	Prospective elementary teachers
13	Tobin	• Critical autobiography	• Communities of learners • Co-participation • Discourse communities	Graduate Science Education	Practicing elementary & middle school teachers; science education graduate students
14	Cunningham	• Critical autobiography • Qualitative research	• Practical spirituality • Journey toward wholeness • Legitimacy, representation, emotionality & morality	Educational Action Research	Community college professors
15	Cuervo	• Autobio-graphy	• Spirituality • Impressionistic story • Moral virtues	Mathematics (Bilingual)	Community college students

Continued on next page

Table 1 (Continued)

	Author	Research Style	Issues	Course	Students
16	Williams	Critical action research	• Teacher-researcher • Open discourse • Self-discourse • Jungian dream analysis	Business Computing	University business professors
17	Gilmer	• Critical action research • Critical autobiography • Teacher-learner-researcher	• Self-empowering metaphor • Dream analysis: Self-transformations • Chemistry discourse community • Using language of science	Chemistry	Honors general chemistry students

We hope that this unique collection of studies and accompanying commentary will give insight and inspiration to science teacher educators interested in working as agents of epistemological change with professors of science. We offer a diversity of accounts, ranging across science disciplines (and sometimes beyond) and including both frustrations and successes. We confess to not having pitched the book at science professors per se; for the most part, the chapter authors belong to an educational discourse community. This is not to say that science and mathematical professors should necessarily feel excluded; faculty in college science departments have written eight of our seventeen chapters (see Craig Bowen, Sabitra Brush, Margarita Cuervo, Kate Scantlebury, Kenneth Goldsby, Rosalind Humerick, Harold White, Penny Gilmer).

One of our intended audiences is the science education graduate student wishing to undertake dissertation research on undergraduate science teaching. We have included descriptive accounts of research methodologies, ranging from the well-established canons of *interpretive ethnography* to more innovative approaches such as *hermeneutic phenomenology* and *critical autobiography*. We hear the voices of college teachers speaking to us through their own personal narratives in writing styles that foreground deliberately the author's voice. A number of authors report on their experiences as teacher-researchers engaged in professional development programs in which they conducted *action research* studies. We read compelling *tales of the field* based on a newly emerging impressionistic literary genre that (sometimes) fictionalizes characters and allows the authors' lived experiences to serve as an embodied lens for the understanding of research phenomena. We learn how some college teachers stimulate their creative thinking by tapping into their unconscious minds via the keeping of *dream journals*, and we learn how *metaphor* can be used to articulate the (otherwise) ineffable characteristics of learner-sensitive teaching. Most importantly, we have given prominence to discussions about alternative standards of judgment for the knowledge claims emanating from these more radically subjective research approaches.

One of the defining characteristics of the book is the way in which we, as editors, have engaged in a discourse about the educational significance of each of the studies. From our diverse social constructivist standpoints, we have generated a series of *metalogues* in which we identify and discuss issues of greatest salience to us individually and collectively. Gregory Bateson (1972) defined meta-logues as

> ...a conversation about some problematic subject. This conversation should be such that not only do the participants discuss the problem but also the structure of the conversation as a whole is also relevant to the same subject. (p. 1)

Bateson (1972) has a series of metalogues in which he discusses various problematic issues with his young daughter, Catherine Mary Bateson (also daughter of Margaret Mead). The metalogue, "How much do you know" is especially illuminating, both in its content and in how it says what it says. Catherine Mary Bateson (1984) later writes a memoir of her parents, and she reflects on these metalogues.

To construct the metalogues, each of us took responsibility for initiating a commentary on a number of chapters, which we distributed to the others for their considered response. At times, we found that we were in agreement with each other; at other times our viewpoints diverged. The metalogues permit us to give voice to our criticisms of learner-insensitive teaching; to encourage (perhaps chide) reform-minded science professors to continue to challenge the local cultural (institutional) restraints and extend (self-critically) their newly constructed (transitional) epistemologies of practice. Within the metalogues, we celebrate visionary educators, especially those who are pushing the envelope of pragmatic possibilities. At one point (Chapter 15), we include the voice of an outside critic to help us consider the research validity of a study about a contentious teaching issue. The reader of our metalogues is thus afforded the opportunity of comparing his or her own interpretation of any of the studies with those of the editors. We hope that the powerful discourse of a metalogue engenders in the reader a deeper and more reflective appreciation of the efficacy of social constructivist perspectives on undergraduate science teaching reform.

Identifying the Barriers

The first section of the book provides a fresh understanding of the nature of barriers to effective science teaching. From a range of research perspectives, we witness the (imagined) plight of students who are unable to gain a firm foothold on the precipitous learning curves created by professors whose interests lie almost exclusively in the explication of their disciplinary knowledge rather than with the process of knowledge construction. The barriers identified are cultural in nature and political in practice. Their identification involves an emotional sensitivity by researchers to the quality of educative relationships between professors and their students. It involves also a consideration of alternative epistemologies of teaching and research practice whose theoretical referents are outlined below.

In Chapter 1, **Peter C. Taylor** presents *impressionistic tales* of contrasting opportunities to learn within the classrooms of two professors: Dr. Stern, an authoritarian and monological transmitter of factual information, and Dr. Mary Buenos, a caring teacher who endeavors to connect with her students' personal aspirations. These characters are based on extensive fieldwork in college classrooms. They are suggestive of *cultural archetypes* dwelling in Jung's *collective unconscious*. In subsequent chapters, we see how these archetypes, when unbalanced (and invisible), restrain the epistemologies of practice of professors and the scope of their well-intended (and well-funded) curriculum reforms. Peter explains his *artistic* use of *hermeneutic phenomenology* in creating dialogical texts designed to engage his readers in acts of pedagogical thoughtfulness. Importantly, he explains also the standards of his inquiry, which legitimate amplifying the significance of the researcher's critical reflexivity, existential awareness, and emotionality. These key aspects of researcher subjectivity are addressed in unique and (somewhat) provocative ways in Section III (Chapters 13 to 16).

In Chapter 2, by **Craig Bowen**, we learn about Diane in the form of a fictionalized narrative, a mature-age student with rich life experiences who is preparing for a new career as an elementary school teacher. She enters the chemistry course keen to learn science. As the course progresses, Diane speaks through her letters to a friend of the gradual extinction of her initial eagerness to learn chemistry. In her poignant accounts we read about her struggles to make sense of the technical language of the chemistry professor and her fruitless efforts to break through the rhetoric to establish an educative rapport.

In Chapter 3, **Noelle Griffiths** reports an interpretive study that is concerned primarily with the use of language in the science classroom for prospective elementary school teachers, particularly the way that it acts as a barrier to effective teaching. She monitors two prospective teachers' learning experiences and documents how, in the absence of a shared language, the professor is unable to engage students in a successful *border crossing* into the discourse community of science.

In Chapter 4, **Hedy Moscovici's** participant-observation account of two prospective elementary school teachers of biology helps us to understand how the dynamics of uneven *power relationships* between students and professors can be detrimental to learning. We learn how a professor's predilection for a distancing language of power and control compromised students' development of both conceptually sound understandings of science and of themselves as learners capable of thinking critically.

In Chapter 5, **Kathryn Scantlebury**, a feminist science educator, argues for the transformation of the unjust and prevalent masculinist image of science, which acts as a major barrier for many female students. Drawing on the referents of *gender* and *power,* Kate outlines the characteristics of a *feminist pedagogy*. This pedagogy involves students in cooperative learning and an ethic of care, linking knowledge and personal experience, and contesting critically masculinist stereotypes, which portray science (and students of science) as exclusively rational, logical, unemotional, and positivistic. Kate illustrates feminist pedagogy with gender equity strategies drawn from her own teaching practice as a professor of chemistry.

In Chapter 6, **Wolff-Michael Roth** and **Kenneth Tobin** provide an account of two professors of physical science courses designed specifically for prospective elementary school teachers. An interpretive study of Miller, a physicist *par excellence*, reveals why he was unable to engage students as co-participants in a community of learning. Miller's classroom dialogue and whiteboard diagrams are deconstructed from a variety of perspectives associated with a *philosophy of language*. The study identifies "possible points of departure for transformation in teaching practices" of the Millers, Sterns (see Chapter 1), and Mendelsons (see Chapter 4) of the world of college science teaching. Such professors are held in the thrall of powerful cultural myths, which legitimate foundational views of scientific knowledge and monological classroom discourses. Michael and Ken argue for a *border pedagogy* to empower students as *co-participants* with their teachers in a common discourse from which they can develop *canonical physics talk*. To illustrate the feasibility of border pedagogy, an *autobiographical tale* is presented of Ashmore, a physics professor and a reflective teacher-researcher who has transformed his epistemology of practice. Characteristic of Ashmore's teaching are collaborative small group investigations and problem solving designed to engage students in reflective discourses that bridge from their current worldviews to those of the discipline of physics.

Pushing the Envelope

In the second section, we are concerned primarily with illustrating recent transformative endeavors—promising and problematic—of science professors. The opening chapters (Chapters 7–9) serve as examples of professors who are transforming their science teaching in ways that are placing student learning at the heart of their teaching interests. Subsequent chapters (Chapters 10–12) remind us of the not inconsiderable difficulties likely to face reform-minded/hearted professors as they embark on journeys that involve confrontation with long-standing and institutionally enshrined beliefs and values.

In Chapter 7, **Abdullah O. Abbas** (a graduate student at the time of the study and now professor of science education), **Kenneth A. Goldsby,** and **Penny J. Gilmer** (both professors of chemistry) collaborate in this study. They report an exemplary teacher (the second author) of chemistry who designs science teaching for *active learning*. The teacher conceives of his classroom role in terms of multiple journey-related metaphors. He changes roles and metaphors in order to provide a range of opportunities for students to engage actively in the learning process.

In Chapter 8, **Rosalind Humerick** (a professor of chemistry) outlines her *action research* inquiry aimed at improving the way she teaches chemistry. Rosalind draws on her autobiography to identify long-standing values about teaching science: content mastery and a caring attitude toward students. She explains her *guided inquiry* teaching strategies for engaging students in active learning within the lecture and laboratory, and she presents compelling evidence of their positive impact on student learning.

In Chapter 9, **Harold B. White III** (a professor of biochemistry) explains that his teaching is shaped by his vision of students as autonomous learners who can think critically and reflectively about the quality of their understanding. Harold explains his philosophy of *problem-based learning*, which emphasizes the ambiguity and uncertainty of a non-dualistic way of knowing, and gives examples of student assignments. Problem-based learning provides an educationally powerful rationale for reducing content in order to enhance student thinking and expand their ways of knowing scientifically.

But becoming a learner-sensitive science teacher is not necessarily an easy task, even with good intentions and support. The next three chapters are accounts by researchers who collaborated with science professors in federally funded teaching development programs aimed specifically at enhancing prospective elementary school teachers' learning of science. We learn about these professors' (somewhat frustrated) struggles to transform their epistemologies of practice.

In Chapter 10, **Susan Mattson** presents a participant-observer's account of an innovative biology course designed for prospective elementary school teachers. Susan focuses on collaborative course development, specifically the problematic issue of developing an *alternative assessment* system congruent with the team's curriculum goal of student-centered learning. She uses *fictionalized narratives* to illustrate the team's struggle to reconcile their conflicting beliefs in the nature and purpose of assessment and to adopt a multiple perspectives standpoint. Sue's interpretive commentary is designed to engage the reader in reflective thinking about these issues.

In Chapter 11, **Sabitra Brush** examines three professors who attempted to *team teach* a specially designed *hands-on, minds-on* physical science course for prospective elementary school teachers. The course combines physics and chemistry and integrates the laboratory with the lecture. The chapter presents a series of interpretive assertions, which highlight the difficulties that can be experienced by collaborative-minded professors when their conflicting philosophies of teaching and learning are not resolved.

In Chapter 12, **Carol Briscoe** describes a study in which she worked collaboratively with a professor of biology over a period of two years, facilitating his development as a reflective practitioner. The context of the study is an innovative introductory biology course for prospective elementary school teachers designed to engage students in discourse about scientific ideas. The study utilizes a *social linguistic* perspective and focuses on the teacher's *communicative interactions* with students. Carol provides an account of the teacher struggling to change his teacher-centered instructional beliefs, and of the changes resulting from his successful attempts to listen actively to students.

Potentialities...

The third section is concerned with the potential of alternative (radical) means of constructing new (institutional) cultures of learner-sensitive science teaching. The significance of this section is the longer-term vision that is painted of ways in which professors of science can transcend restrictive (personal, cultural) barriers to epistemological reform. These studies, which are situated beyond the borders of most science professors' familiar discourse communities, are intended to illustrate how new information technologies can be harnessed to promote learner-sensitive teaching and how an ongoing commitment to transformative teaching can be fuelled by our inner energies.

In Chapter 13, **Kenneth Tobin** (professor of science education) presents an impressionistic account of his development and use of the Internet to *connect communities of learners*, notably, geographically dispersed science teachers wishing to study graduate level degree courses in science and science education. At the heart of this innovative teaching is the referent of learners as *co-participants* in discourse communities using the language of the discipline. A number of pedagogical features of this interactive tool are outlined, with special attention paid to the unique types of discourse available to students.

Ben Cunningham (a professional educational action researcher) in chapter 14 provides a critical reflexive *autobiographical* account of his journey toward improving his own professional practice. The focus of Ben's writing is the emergent form of his *practical spirituality* and its role in shaping his inter-personal relationships. He starts with a critical event in his life, and follows with a highly personal narrative account of his research journey toward wholeness. Importantly for science educators, Ben argues for a set of scholarly criteria against which his (and others') knowledge claims can be validated. In doing so, he extends some of the central issues of qualitative research, including issues of legitimacy, representation, emotionality, and morality.

In Chapter 15, **Margarita Cuervo** (professor of mathematics) reflects on how she allows her *spirituality* to shape her professional practice. Margarita draws on her autobiography to provide an *impressionistic* account of the values that shape her teaching. In the context of her bilingual mathematics class, Margarita explains how *moral virtues* are key referents for shaping her teaching, particularly her educative relationships with students who are striving to realize the goal of a well-paid and meaningful job or to obtain entry into a graduate program. For Margarita, being a good teacher means being sensitive to the student as a human being.

In Chapter 16, **Mark Campbell Williams** (a professor of information systems) believes in educating the *whole person* in his business computing lectures and tutorials or other venues where technical rationalism tends to prevail. In the role of teacher-researcher, Mark documents his attempts over four years to improve the quality of learning in his classes. His pedagogical innovations, especially *open discourse*, are designed to enable his students to become more reflective, communicative, and collaborative learners. Mark describes his critical reflexive *heuristic research* in which his own *self-discourse* was developed, largely through Jungian dream analysis. Mark takes us on a journey of the researcher exploring his unconscious self, learning from his dreams how to become a more mature and balanced person. As does Ben Cunningham (Chapter 14), Mark dwells helpfully on the validity and ethics of this research.

...Being Realized

The final section bears witness to the power of transformative ideas expressed in earlier chapters, but especially in Chapters 13 to 16. The significance of the final chapter lies in the status of the author, a science professor, and in her radical journey of transformative teaching.

In Chapter 17, **Penny J. Gilmer** (professor of chemistry) presents an account of how, as a *teacher-learner-researcher* intent on engaging her students as co-participants in the canonical discourse of chemistry, she has moved well down the (career-long) path of reconstructing her epistemology of teaching practice. Penny is a transformative teacher inasmuch as her explicit endeavors to reconstruct student learning both arise from and constitute her own *self-transformative* endeavors. Penny explains how, in opening herself to new learning opportunities, she began to reconstruct her epistemology of practice in accordance with her dream-inspired "triple point" metaphor. The triple point is a self-empowering (chemical) metaphor that serves to integrate and enrich her various professional roles. She attributes her inspiration to Mark Williams's dream-based self-transformative research on teaching (Chapter 16) and to her experience as a co-participant in Ken Tobin's Internet-based communities of learners program (Chapter 13). We learn about Penny's use of electronic mail as a vehicle for engaging her students in reflective thinking about their learning and as a source of evidence about the nature and quality of that learning. It opens up a dialogue with her students, one that is designed to enable her to learn about their goals, dreams, and aspirations, as well as their interest in chemistry and science in general. It is apposite to this book that one of the students who benefits from becoming a co-participant in a rich discourse community of learners (including professional chemists) is a prospective teacher of school science.

BIBLIOGRAPHY

Ausubel, D. P. (1968). *Educational psychology: A cognitive view.* New York: Holt, Rinehart & Winston.

Bateson, C. M. (1984). *With a daughter's eye: A memoir of Margaret Mead and Gregory Bateson/Mary Catherine Bateson.* New York: William Morrow.

Bateson, G. (1972). *Steps to an ecology of the mind: Collected essays in anthropology, psychiatry, evolution, and epistemology.* San Francisco: Chandler.

SECTION I

 **Identifying Barriers**

CHAPTER 1

 ## On Being Impressed by College Teaching

Peter C. Taylor

The firmament in the positivist sky twinkles with precision and rigor. However, the spaces between stars and those hidden by clouds recede and disappear. Phenomenology seeks to name those spaces, their relation to the stars and to us. The unity of the epistemological whole resides in ourselves.

—William Pinar & William Reynolds, *Understanding Curriculum...*

8:55 A.M.

Carefully, I lift the video camera out of its case and balance it precariously on the sloping writing tray attached to a nearby student chair. I reach down and grasp the smooth rectangular black battery and click its leaden weight into place. The camera beeps and whirrs into life.

At the same time, Dr. Lilly's activity catches my eye. She is busy focusing her overhead transparencies onto the classroom wall. "There's no pull-down screen in this room. Why not in such an apparently well-resourced college?" The thought disappears as quickly as it arrived.

I notice how alert I'm feeling. That's a good sign with five minutes to go! No sleep deprivation hangover today to cloud my senses. Last night I slept well. Somewhat surprising for the first night in a strange bed in a strange room in a strange part of the world!

Students are slowly occupying the empty seats arranged in scattered rows. The room is gradually taking on the aura of any one of a thousand mathematics classes that have populated my life, stretching back beyond the horizon of my memory. The familiarity threatens to stifle my ability to engage a part of myself, the reflective-inquirer.

8:57 A.M.

Standing to one side of the room in a position that will enable me, without being too obtrusive, to capture the expressions on (some) students' faces, I feel the hard weight of the camera pressing into my right shoulder. Later, I will lean against the wall to relieve the nagging burden. I have forsaken the use of a tripod in favor of mobility, preferring to roam the side and back spaces of the large carpeted room that easily accommodates the twenty-five students now seating themselves in those ubiquitous plastic chairs that one finds in educational institutions around the world.

I have Dr. Lilly's image in sharp focus as she readies herself at the front of the room. The fingers of my right hand wriggle through the tight plastic restraining strap and feel for the out-of-sight main controls: the toggle switch on top for the zoom lens and the on/off button at the rear.

Brilliant white sunlight streams through several windows and the yellow tungsten glow of the overhead projector bulb reflects from the front wall. How much of a problem this backlight might be is hard to gauge through the camera's tiny black-and-white viewfinder. I contort my neck and note the position of the "backlight" switch on the left side of the camera. Later I will need to find this by touch.

I test the camera's ability to focus close up, by zooming in (quite sluggishly) on the whiteboard's remaining scraps of writing that echo the activity of a previous teacher.

9:00 A.M.

Dr. Lilly smiles at the class, coughs gently, and welcomes the students.

Providing a Context

Several years ago during a delightful (but quite challenging) period of my life while on sabbatical at Southern American University, I visited seven community colleges across the state in order to find out something about the quality of college science-related teaching. My role could be considered as a preliminary field reconnaissance in which a minimally resourced expeditionary force-of-one sallied forth into largely unchartered cultural terrain. But the romance suggested by this metaphor is mostly retrospective.

The construction of my professional role took place within the broader context of a rich and sometimes scary set of whole-of-life experiences. As I traversed the state, my solo tour took me into many cities, towns, and rural areas, and into touch with a surprising diversity of cultures. Amongst my most vivid memories are those of the journey itself. Of driving rental cars on unbelievably fast and furious highways, on the "wrong" side of the road, while desperately trying to read road maps and decipher (culturally) strange signs in search of key exits (which I sometimes missed).

I sometimes felt a little vulnerable as I tried to get un-lost on various occasions. (I wonder to what extent the experience of the journey influenced my construction of Drs. Stern and Buenos?) During the journey, the basic essentials of my everyday living—finding a place to eat, sleep, relax, and plug in a computer modem to communicate with those who populate my "normal" life—took on as great a significance as the official fieldwork activities. And the brief duration of my visits meant that I had to seize every moment and make it work as best I could. I expended an enormous amount of energy!

And so it makes good sense for me to tell tales of my experiences, tales that tell as much about me—my situatedness in the inquiry—as they do about the teaching that I witnessed. Thus, this chapter addresses two key issues. I write about the quality of science-related college teaching, raising a voice of concern and of celebration. At most, I claim to present a *prima facie* case that serves to raise questions for other researchers to pursue; questions that many authors in this book have, indeed, pursued. I write also about the way my tales have been written. This is a methodological issue about new writing possibilities for science education research, particularly the metaphors of *research as writing* and *research as reading*.

The opening (extract of a) tale, which I now[1] entitle *Getting Ready*, does, I believe, illustrate (something of) my situatedness in the research. I wrote that tale in order to provide insight into how I went about producing a series of video recordings of college teaching. The tale is meant to indicate (among other things, which I leave to my readers to consider) how I *constructed* an important source of data, data that did not "speak for themselves." The videotapes (documentary evidence and audiotaped interviews) required interpretation. These data served me as a rich source of stimulation for recounting retrospectively my personal experiences, which I represented in the narrative form of impressionistic tales.

Or, am I misrepresenting the process of writing when I say this? Was it not in the act of writing that I recalled my past experiences, and only then did I feel compelled to refer to the "external" data sources in order to confirm the accuracy (or otherwise) of my recollections? But that seems to imply that I wrote factual tales, and that is not (really) the case. Fact and fiction were woven together (purposefully) to create (hopefully) compelling and reflective reading experiences.

For whom? Apart from other researchers, my intended audience comprises college teachers who may be interested in reflecting critically on their own pedagogies but who, first, might need a means of representing (at least to themselves) the essence of their pedagogical values and beliefs. Telling one's own tales (to oneself), I have found, is an empowering and inspirational process. And, of course, there are graduate students interested in methodological possibilities. So, in this chapter I attempt to explain (and legitimate) the epistemological relationship between my fieldwork, my writing, and my readers.

The Official Mandate

My inquiry was designed to serve as an introduction to a longer-term research study planned to illuminate the learning experiences of prospective teachers of science and mathematics during their pre-teacher-certification college[2] degree courses. Several of the chapters in this book are the fruits of that subsequent research. Why the need for such a study? Many in teacher education are concerned about the unpreparedness of young teachers graduating from teacher certification courses for adopting contemporary teaching and learning practices that are becoming an increasingly important part of elementary and high school curricula.

Epistemological reforms being enacted in school science and mathematics curricula worldwide are envisaging teachers as reflective practitioners involved firsthand in the reconstruction of their own teaching practices. Among reform-minded science and mathematics educators, the metaphor of *learning as socially mediated knowledge construction* is displacing the traditionally entrenched metaphor of *learning as absorbing objective facts*. Historical views of science and mathematics as activities that yield privileged truths of Nature are giving way to more relativistic views that recognize the cultural contextualization of scientific and mathematical knowledge, the provisional status of scientific theories, and the experiential basis of mathematical theorems. Metaphors of *hard control* that underpin teachers' traditional images of their classroom roles are being deconstructed in favor of metaphors of student *empowerment* and *co-*

participation in the design of learning activities and in the evaluation of student learning.

However, in the experience of my science education colleagues at Southern American University, intern teachers experience great difficulty in viewing school students as active constructors of contingent knowledge or in initiating classroom social roles that engage students in rich reflective knowledge-building discourse. Rather, they tend to remain wedded to their traditional views of students as absorbent sponges of universal truths and of teachers as dispensers of privileged knowledge. Why is this so? We suspected that the answer lay, in no small way, in the impoverished nature of the learning experiences provided by college degree programs, experiences legitimated by the daily didactic practices of college professors. Based on student teachers' contemporary anecdotes, our own residual images of ourselves as learners in college courses, and a small amount of research on college teaching, we felt that a largely traditional culture of didactic teaching practice continues within the lecture theatres, tutorial classes, and laboratories of many college teachers.[3] These formative experiences are likely to serve as a source of powerful images of teaching and learning that interns draw upon to govern their own teaching practices.

In recent research on epistemological reform in school science and mathematics, we had found that the process of transforming classroom-based social roles of teachers and students often is blocked by the resilience of established patterns of social behavior (Dawson & Taylor, 1998; Taylor, 1996; Tobin & McRobbie, 1996). Social roles are underpinned by extensive networks of beliefs, values, and images that together constitute a robust culture sustained by ritual, taboos, and unwritten protocols and codes of behavior. What makes the enculturating influence of these networks so strong is the invisibility of much of the beliefs and values and the apparent naturalness of the accompanying social patterns of behavior. So, from this epistemological standpoint, what could we learn about the culture of college teaching, especially in relation to the dynamics of enculturation that conserve and regenerate traditional teaching practices?

The Challenge of How

The challenge that faced me as I prepared to embark on my statewide journey was methodological in nature; indeed, it was an epistemological question that I grappled with. How, I wondered, would I warrant my claims to know something significant about the quality of teaching that I was about to observe in a relatively fleeting manner? My experience in interpretive research had been, until then, largely ethnographic in nature. In previous classroom-based studies I had spent considerable time living in classrooms, studying carefully and critically the relationships between participants' (epistemic) beliefs and actions, including my own (Taylor, 1993b). The epistemological standards I had used required me to spend considerable time generating data from multiple perspectives, writing ongoing analyses, and testing the viability of my inferences by checking their status with participants. The bottom line of interpretive research, it had seemed to me, was to ensure that one's portrayal of "the other" was highly credible. Without this epistemological warrant, interpretive research would be untrustworthy (perhaps "untruthful" is the muted accusation?) (Erickson, 1986; Guba & Lincoln, 1989). And so it seemed to most science educators publishing their interpretive research in the 1980s and early 1990s!

Fortunately, my established interpretive research paradigm was about to be radicalized. But because I had been moving along a particular research trajectory of late, the process of radicalization was welcome and empowering. It was my reading of John van Maanen's[4] (1988) text, *Tales of the Field: On Writing Ethnography*, and a timely visit to the Salvador Dali Art Museum,[5] which gave me at that critical juncture an immediate sense of elation, for I had found a new way of conceiving the epistemology of my inquiry and a methodological way forward: research as the writing of "impressionistic tales of the field." I was able to bring together a number of disparate strands of my own life experience, including a growing interest in the role of the self in research, my fascination with the paintings of the Impressionists, and my early love of expressive writing and poetry, which had been "nipped in the bud" as I moved into the emotionally arid mathematics-science stream at high school and university. I was to find in the ensuing years that this trail would lead me ineluctably to the expressivism of narrative, story, and autobiography, and then to the quicksilver of existentialism and phenomenology.

My writing in this chapter focuses on that period of just a few years ago when a new door was opening for me, a door that enabled me to interpret my commission from Southern American University as not to record an accurate "topographical map" of the cultural terrain, but to generate (pedagogical) impressions of striking social activity among the "natives." As befitted the context of my inquiry, I chose to represent in the form of *impressionistic tales* the experiences of my brief forays into specific cultural sites—the lecture theaters, laboratories, and offices of professors of college science and mathematics.

But here I am caught on the horns of a dilemma: to present firstly the tales for your reading and reflection or to avail you firstly with a rationale for their legitimacy as science education research. Perhaps I can offer a choice.

A Hyperlink Metaphor

Before the advent of hyperspace as a communication medium, written communication (letters, books, reports, scripts) was limited to a two-dimensional textual space. The author wrote with the assumption that the best way to communicate with potential readers was to construct an essentially linear "story," comprising an introduction, a middle, and an end (in that order). Of course, linearity has been used to great effect by novelists, playwrights, scientists, and poets; albeit in different directions depending on whether it is English, Hebrew, or Mandarin. On the other hand, linearity can be a repressive structure that helps to propagate powerful cultural myths that limit our freedom to use text in unique and imaginative ways.

Now, however, hypermedia offers multidimensional textual spaces and nonlinear genres, thereby opening up new possibilities for research, especially for the way that we communicate our experiences. What I am referring to here is, for example, the multilayered nature of a hypermedium such as a website on the World Wide Web. Many websites comprise a number of electronic "pages" (using that old print metaphor) that are "hyperlinked." Although on each page short passages of text continue to demand a linearity of reading (such as the text of the first pages of this chapter), the order of reading the pages is no longer necessarily linear.

In relation to this chapter, which is largely constrained by the linear characteristic of the printed text medium, I am endeavoring to create an opportunity for you to evoke a hypermedium metaphor to shape your subsequent reading experience. If you turn first to the next section, *Two Tales of the Field*, then you can read the tales according to your own literary and pedagogical standards "uncontaminated" by my subsequent considerations of the status of the tales as research or by my interpretive commentary on their pedagogical significance. Alternatively, if you turn first to any of the subsequent sections *A Feeling-Thinking Fieldworker*, *Impressions of Lived Experience*, or *Tales as (Phenomenological) Research*, then you can read first my account of the rationale for regarding the tales as science education research.

Tales of the Field

A Stern Tale about Learning

There, I heard it again, an increasingly familiar declaration!
"That's where the student will be doing most of the learning—at home with his textbook," declared Dr. Stern.
I had just witnessed Dr. Stern lecturing to his Health Science class, a small class of about fifteen students. Now we were seated in his office and were well into a conversation about his teaching.
"In college, the learning takes place at home, and the function of the classroom is to explain difficult concepts that can be learned at home. If the student doesn't learn at home the whole thing falls!" continued Dr. Stern, as if to leave me in no doubt about his view.
Hmm, I wondered silently, the whole thing falls! What an interesting metaphor. What does that signify—learning as the building of an edifice, perhaps? Quite a contrast to the New Physics view of the universe as being in constant evolutionary transition and of learning as personal evolution or becoming, as Skolimowski reminds us.
The theme, that learning should occur outside, rather than inside, class, was beginning to emerge from my inquiries. I had detected it in conversations with other college professors.
Today, it was emphasized boldly and unashamedly by Dr. Theodore Stern, a professor of health science and a man of imposing physical stature whose deep bass voice and declarative opinions signaled an aura of exceptional confidence in the authority of his own expertise, both in his teaching and in the conversation we were having right now.
"So how," I contemplated, "does Dr. Stern construe the purpose of his lecture class? And how does it relate to his philosophy of a college education? And what does he mean by learning?"
I wanted to ask these questions all at once, but it was clear that Dr. Stern preferred to answer one question at a time. He seemed to have a

high regard for correct answers (it is very apparent in his teaching) and wanted to ensure that he could deliver to me an answer that was in need of no revision.

I recalled his response to my earlier assurance of confidentiality in relation to the audio recording of our conversation. He declared confidently that he was certain he would not need to retract anything that he might say to me.

Clearly, Dr. Stern was not in the habit of making mistakes!

That afternoon, I had observed Dr. Stern explaining to his Health Science class, with unwavering certainty, the intricate details of cell membrane transportation. With frequent reference to the textbook's diagrams, Dr. Stern decorated his blackboard with sketches and notes about what happens when cells are placed in solutions of varying saline concentrations (a hypothetical empirical account), what official labels should be attached to the different concentration states (the official language of science), how the cells would appear physically as a result (a further hypothetical empirical account), and the nature of the biochemical mechanism that accounts for changes in cell size (the theoretical account).

It was as though this scientific account, whose unquestioned authority reflected on Dr. Stern, was an accurate description of the real world, rather than a constructed account. The context of the account was historical, independent of preceding theory, and apparently, empirically verifiable.

Ah, the old Empiricism of the seventeenth century lives on! Only sharply refined senses are needed when studying Nature. But didn't Kant's critique of pure reason identify the active mind as central in imposing structure on perceptions of the world? I thought from the side of the room. Therefore, observations are governed by theoretical suppositions. So, it is our theories on which we need to reflect critically when making sense of our observations. And, if that is the case, isn't it important for the teacher to understand students' extant theories in order to understand what sense they are making of what they "see" in class and observe in the lab? And, therefore, shouldn't the teacher spend some "lecture" time questioning students and letting the replies moderate his explanations?

At no stage did Dr. Stern relate the significance of his scientific account to the world outside of college. Nor did he engage the students in a process of inquiry that might have enabled them to construct scientific understandings by first pondering on their understandings of the phenomenon to be explained.

Why not give them a few raisins and grapes to stimulate an inquiry into why one is shriveled and the other is smooth and juicy? I wondered at the time.

It was not until students commenced their next laboratory class that they got to test the permeability of membranous material and consider the implications for medical processes such as dialysis. However, the "experiment" consisted of following routine instructions, recording predetermined observations, performing standard numerical calculations, and inferring a conclusion identical to those of the rest of the class. Where was the inquiry in this "cookbook" approach to laboratory learning?

Later, Dr. Stern talked to me about the complex relationships that existed among the multitude of concepts and principles of his field. Yet in his teaching, he did little to enable students to construct their own relational knowledge. He reduced his own knowledge into bite-size chunks and spent much time explaining it to the class. How students were to see the important relationships he valued was somewhat of a mystery, even to him. But one thing was certain: It wouldn't occur in class.

"I would be surprised if the student really learned [in class].... I think they learn this little bit, and how this little bit connects to this little bit, and the big picture develops on the sidewalk after they've left class or when they are preparing for their exam or something sometime, I hope."

As he was talking, several thoughts flashed through my mind. Is Dr. Stern laboring under a legacy of behaviorism, which evokes an image of good teaching as the ability to present component pieces of detached factual knowledge and to measure students' reproductive performance? Although he talks about concepts and their inter-relationships, his concern seems to be not so much with students linking new ideas to their existing ideas and understandings, as with transmitting his concepts to students and checking whether they have received them. How they go about receiving and relating them seems to be a non-teaching issue.

Dr. Stern bemoaned the lack of self-motivation of many of his students, blaming their immaturity and blaming their high school teachers. "It's absurd to think that freshmen are capable of inquiry!"

But where was his sense of moral responsibility for stimulating students' inquiring minds and, thus, their attitudes toward learning science?

How can conventional science teachers stimulate students to reflect on their own knowledge when the explicit purpose of their teaching is to "deliver information" or "present the material" that students can take away with them in order to "truly learn" outside of class, presumably in the library or at home? I wondered.

I found Dr. Stern to be rather intimidating, especially, I noted, if I allowed myself to be drawn in to assenting to his construction of others as "in deficit." I could feel it in my bones. I wondered how his students felt when confronted by the frequent questions that punctuated his commanding classroom delivery. I had watched through the viewfinder of my video camera as most of the class continued to reject Dr. Stern's implicit invitation to expose their "soft underbellies" to his unremitting authority. If you got it wrong, you were left in no doubt that you were deficient. It was safer to pretend to be reading your textbook rather than to make eye contact with the professor at moments like these.

I couldn't help thinking, His questions seem designed to reveal students' ignorance and to justify him pouring his expert knowledge into their empty minds, rather than building on what they do seem to know and drawing them forward in self-reflective inquiry into the quality of their own understandings.

During our conversation, I was careful to "dance" with Dr. Stern in a way that neither antagonized him nor allowed me to be subjugated by the controlling power that he seemed to direct at me. I resisted the temptation to contest his opinions, especially the arrogance of his attitude toward the silent majority of students whose failure to engage with him in class was labeled haughtily as undue passivity instilled by their high school teachers.

This was not like other conversations I had had with college professors, conversations where dialectics had a place, where I might expose and explore conflicting interests and values. Nor was it a conversation of co-participation in which I might offer my own professional opinion in the interest of mutual learning.

I felt that Dr. Stern was trying to control our conversation by willing it to be an exposition of his uncontestable expert opinion. It was an experience that I had had before with medical practitioners who insisted on diagnosing my health on the basis of theoretical understandings of how, in their opinion, I should feel. I reminded myself not to let my discomfort escalate to irritation.

It seemed to me that any time a group of students met with a teacher for an hour it was inevitable that learning would occur. Clearly, Dr. Stern and I had different views on the relationship between teaching and learning, and on learning itself!

"We are getting [students] prepared to learn...our efforts are spent in getting the students the things they will need in order to truly learn after they leave the classroom," I was told again.

This proclamation continues to echo in my mind.

Reflecting on Dr. Stern's preferred way for students to come to know, I am reminded of a metaphor of mining. Students attend class and are given apparently immaculately conceived semiprecious gemstones, which they must take away and add value to by cleaning, polishing, and mounting in prescribed settings. The final exam evaluates their mechanical skills at each of these steps.

The problem with this metaphor of learning is that the students don't get to know how to be miners (i.e., how to construct concepts from within a matrix of inquiry and need), how to design the settings (i.e., how to construct the relationships between the concepts), or how to appreciate the beauty of the jewelry (i.e., standards of parsimony, elegance, and explanatory power).

The term lecture seems to be truly apposite as a label for the classroom actions of science teachers such as Dr. Stern, whose right to "profess" their knowledge is inscribed in their professional title by the culture of higher education.

Familiarity Breeds Intimacy

It was time for lunch, as Mary and I walked toward the student cafeteria, chatting amiably.

Mary was intent on restoring her energy level after a particularly busy morning and did not yet feel ready for the intensity of the interview that she anticipated having with this person who had come to learn about her teaching methods.

"What will he want to know? How shall I explain myself clearly enough? Will he really understand? He did seem to be open-minded when we talked yesterday. What did he think of my teaching? I think he prefers fancy problem-solving methods, but this class cannot be like that.

He was looking at his watch. Did he expect me to start the class on time, even though most had not arrived? I hope he didn't mind my joke about the class running on Hispanic time! How much of the Spanish could he understand? Could he tell they were a little shy today?"

These and other questions darted through her mind, but she forced them aside and continued to engage Peter in idle conversation.

That morning, Dr. Mary Buenos had taught two 95-minute mathematics classes without a break, one of which, of course, was her favorite—her bilingual class. This class prepared students for a state-controlled scholastic aptitude examination that served as a filtering mechanism for entry into professional career courses in the state's universities.

Mary was especially proud of the class and was so pleased yesterday when Peter had shown interest in visiting it.

As usual, after each class she had stayed behind to advise students, and the morning had rushed by. Arriving back in her office, she had been drawn by colleagues into the usual end-of-semester crises. Thankfully, Peter had had the grace to leave her alone for half an hour and she had managed to either resolve the problems, or at least defer them until later.

Sometimes she wondered whether the entire department would self-destruct if she were not there to serve as everyone's consultant and mentor, a role that had been recognized by the college when she had been awarded a "personal chair." But then she reminded herself of how much she loved her work.

Ah, but teaching is not just a job; it is my way of life! And the young people, especially, they are the joy of my life! she mused, and an intense feeling of fulfillment and compassion displaced her concern about Peter and the interview.

Mary wanted to help young people as best she could. So many needed her care, particularly the Hispanic students, strangers in a strange land.

Her doctoral study, completed five years earlier, had confirmed what her common sense had told her at the time, that Hispanic students were grossly underrepresented in university courses. Reflecting on her own experiences, she had surmised that they needed special assistance to succeed in their studies, and her research had confirmed that teaching bilingually was the key.

Since then, she had offered one bilingual class in mathematics each semester, a class in which she and the students moved freely between speaking Spanish and English. It was a class that never failed to attract fewer than thirty students.

After the first week, she knew them well enough to tell who had been born in the United States and who had arrived as an immigrant. After the second week, she knew how many hours of paid employment each worked to support themselves and their families as they grasped the life-enhancing opportunity of a college education. After the third week, she was drawn into their lives, learning about their dreams for the future and their struggles to survive in neighborhoods where crime, racism, poverty, and violence are rampant.

As they entered the cavernous student cafeteria, Mary was deaf to the rock music pulsating from speakers high in the open roof space. She inhaled the atmosphere of the room, savoring the luxury of having lunch away from her office for the first time in months.

In the space of a moment, she reflected on the path she had taken since arriving in the United States at the age of fifteen, speaking twenty words of English. Silently, she thanked God for having been selected as one of the privileged few who could do His work.

She became lost in her reverie as she guided Peter instinctively toward the buffet area.

I was intent on giving her space. I walked quietly beside her and recalled how consuming of my own energy teaching had been. Sometimes there had not been enough left for my body to combat the marauding viruses that continually pursued the children and that laid me low for a week at a time. That was twenty years ago! Today the challenges were of a different kind.

Right now, I was hoping that Mary would choose to eat lunch outside. I preferred fresh air and was concerned about the impact of the resonating music on my planned recording of our lunchtime conversation. Like most small tape recorders, the one that I was carrying in the palm of my left hand preferred recording background noise rather than the person being interviewed.

I had just spent the last hour relaxing in one of the shaded gardens of the college, taking the opportunity to reflect on what I had seen of Mary's teaching and to prepare myself for the interview.

The unbearable sticky heat of a few weeks ago had been replaced by the freshness of cooling air and the warming sunshine of a Floridian fall; it was no longer essential to seek refuge in air-conditioned buildings.

The mid-morning outdoor college environment, with the rich aroma of freshly brewed coffee, a comfortable chair to relax into, and the greenness of exotic trees and shrubs, was exceedingly pleasant; too pleasant, in fact, to dwell on negativities.

Although I was critical of the image of mathematics implicit in Mary's teaching—rule-oriented and largely decontextualized—I had sensed something vibrant and exciting about the class, something that usually was absent from conventional mathematics classes.

There was a spirit of "joie de vivre" in Mary's mathematics class. As students arrived, Mary appeared pleased to see them and greeted each one personally. The formality that distances many teachers and students was absent here.

Rather than sitting in isolated silence or conversing in hushed tones waiting for the class to begin, students were voluble, relaxed, and smiling. They conversed in Spanish with Mary and among themselves, sharing their life stories, at least those parts that they wanted to make public.

Today was an important class. It was to serve as a review for next week's final exam. Mary knew that she must help the students to tie together the seemingly infinite number of algebraic rules they had been learning during the preceding fourteen weeks.

She also felt a need to pass on the accumulated wisdom of her twenty-two years of experience, especially about how to prepare for the multiple-choice questions of the examination, questions that required an accurate recall of algebraic rules and their simple application to single-answer problems.

She didn't much like this type of formal algebraic mathematics. It was so difficult to relate to students' everyday lives. Nevertheless, this was her assigned role, and she must do her very best to help them jump over an obstacle that blocked their path to university courses and professional careers.

As she conducted the class from the front of the room, the students listened attentively to Mary's mathematical explanations, memorizing her tips and problem-solving advice, and feeling bolstered by her entreaties and strategies for dealing with the forthcoming examination. The condescension toward students that can occur in conventional college classes was replaced here by a genuine concern and a caring attitude.

The tone of Mary's voice, her gestures and facial expression, and her emotional intensity communicated to her students that she wanted them, indeed was willing them, to succeed. Whereas many teachers badger students about the need to develop a positive attitude toward their studies, Mary chose to speak respectfully about the demands of study.

For the most part, students responded spontaneously to Mary's questions, questions that sought to help them recall mathematical theorems and procedural rules. For some reason, there was little sense of shame in getting a wrong answer. Many students were prepared to take a risk in this class.

In turn, students were not reluctant to ask questions of Mary, questions about the very work they had been struggling with last night. Mary either replied directly with an answer or referred the question to the class. Students' questions were welcome here. They helped Mary to judge the quality of their learning. She used this insight to shape her teaching.

Sometimes, students sought help from one another in quiet tones as they tried to avoid disturbing the class. This was legitimate activity. In Mary's class, being a successful student was the rationale for whatever activities occurred.

Throughout the class, the discourse was vibrant, rich, and inclusive. It wasn't just a few bright or brave students who spoke up, as often is the case in conventional classes.

And, when students did speak up, often it was in their native Spanish tongue. The only person in the room for whom this was an unusual experience was me.

As I carried my tray of food through the cafeteria door and followed Mary into the garden beyond, I recalled a sign that I had seen that morning in Mary's office. It read:

FAMILIARITY BREEDS INTIMACY.
SINCE INTIMACY IS SO SCARY TO SO
MANY OF US, WE SEEM MORE
COMFORTABLE WITH A PROVERB THAT
KEEPS A DISTANCE.

Mary seated herself at the table and looked longingly at her carefully selected plate of tasty Cuban food.

How am I going to eat and talk at the same time? she wondered to herself. She had so much to tell Peter. She had planned to explain that her teaching was based on three valued principles that guided her own life: faith, hope, and charity.

It was obvious to her that a spiritual dimension was missing from the lives of so many of her students, impoverished lives, lives of despair, directionless lives. No wonder so many of her students were unable to cope with the traumas they experienced. The world had changed so much since she was a child. The world today was a difficult place in which to grow up. The simple truths and values of yesterday had all but disappeared for so many, and in its place was a vacuum.

Mary had committed herself to communicating these simple but life-enhancing principles through her teaching.

I placed the tape recorder on the table between us and asked my first question: "Mary, what do you value most about your teaching?"

A Feeling-Thinking Fieldworker

Here I present a brief account of my epistemological turn, from a rationalistic to thoughtful-emotional fieldworker. I make a case for valuing both reason and emotion for generating a critical reflexive understanding of the lived classroom experiences of teachers and students. My focus is on the *representational* aspect of interpretive research (Denzin & Lincoln, 1994), that is, the writing of the research in which I, as the fieldworker, attempt to portray, through the dialectical interplay of my own experiences and the life worlds of others, something significant about the ineffable quality (or *essence*, as the phenomenologists call it) of the relationship between good teaching and learning. But in this reconnaissance, I get only as close to my elusive quarry as a consideration of various "opportunities to learn" associated with contrasting pedagogies. Nevertheless, the (partial) view that I gain is one that owes as much to my emotional sensitivity as to my intellectual sensibility.

The Rationalistic Fieldworker

And for all that concerns ornaments of speech, similitude's, treasury of eloquence, and such like emptiness, let it be utterly dismissed.
—Francis Bacon, *Paraceve*

Research in science education underwent a revolution during the 1980s when it intersected with a number of seemingly alien fields of scholarship. From cognitive science came a unique interest in understanding what goes on in the minds of teachers and students, an interest that took us behind the surface features of behaviorist-oriented process-product research. Cultural anthropology supplied powerful ethnographic research methodologies for examining the dynamic relationship between teachers' and students' patterns of thought and their social classroom roles. And sociology made us aware (through the lens of social constructivism) that the social reality of the classroom is pluralistic, dynamic, and contingent, rather than singular, static, and immutable. Thus, science educators adopted a qualitative-interpretive researcher persona and began to represent in their research accounts the multiple meaning-perspectives (i.e., beliefs, values, perceptions) of the teachers and students with whom they participated in school classrooms (Gallagher, 1991).

In time, some interpretive researchers took a "reflective turn" and realized the centrality of their own meaning-perspectives in portraying the lived experiences of others. The inextricable co-participation of self and other in the "hermeneutic circle" (Gadamer, 1976) means that these portrayals are inescapably partial, intersubjective, and context-dependent. In subsequent attempts to account openly for their interpretive activity, researchers have sought to make visible their own (orienting and constraining) frames of reference. In science education, it is not uncommon for interpretive research reports to contain a theoretical/interpretive framework section that precedes the results or data analysis section. Depending on the duration of the study, interpretive-researcher-as-learner also might reflect critically on their initial frame of reference and weave into the research report an account of their own personal maturation (e.g., see Taylor, 1993a, 1993b; Dawson & Taylor, 1998). Thus, interpretive research reports seemingly can tell as much about their authors as they do about those who were the initial focus of the research.

In science education, however, interpretive research accounts remain, by and large, unashamedly rationalistic. The rich descriptions, vivid vignettes, and multiple voices of many interpretive research reports are assembled with a dispassionate rigor that signals the hegemonic influence of modernist science with its sacred (but disguised) standards of value-neutrality and unfeeling objectivity (Milne & Taylor, 1998).

Although personal experience methods and narrative modes of inquiry (Bruner, 1990; Clandinin & Connelly, 1994; Geertz, 1989) have become commonplace in educational research, in science education they remain highly susceptible to the colonizing influence of the sociocultural myth of "cold reason" which strips away the swirling currents of feeling that flow between people (Taylor, 1996). A "cold" view of cognition (as prescribed by Francis Bacon in the epigram above) eschews the emotionality of human experience while privileging the logical reasoning associated with a mind-as-machine metaphor. When cold reason prevails, interpretive research reports are characterized by a *logico-scientific* reporting style (Bruner, 1986), which creates an emotionally neutral textual "space" for portraying (distorting?) the complexity of human relations.

The reader of such reports is not able to judge the impact on the research process of the emotional interplay between the participants, most notably between the researcher and the researched. Thus, we are led to believe (often by default rather than design) that the "good" interpretive researcher does not let his or her emotions influence communicative relationships in the field. What, then, of the interpretive researcher who wishes to establish a trusting relationship with a teacher whose classroom he or she wishes to examine? Regardless of whether the relationship is premised on a nonparticipatory observational role or a fully collaborative (action) research relationship, is there not a need for an ongoing flow of empathy between the researcher and teacher? Logic, alone, seems a woefully inadequate means of sustenance.

What, also, of the interpretive researcher who is interested in studying the quality of the communicative relationship between a teacher and her class, one that exemplifies a relational ethic of mutual care and reciprocal understanding? How is the researcher to "understand" this interaction? Should he or she rely on reason alone or attempt to include a "feeling" response to the emotional climate of the classroom? And, what also of the emotions that can "flare up" when a teacher contests a researcher's interpretation (perceived as negative judgment) of the efficacy of his or her pedagogy? Should the researcher retreat to a rationality of transcendence or, perhaps, wonder whether an attitude of cool reason has not, in fact, precipitated the crisis of confidence in their relationship?

In each of these (not-so-hypothetical) cases, there are implications for the reporting of interpretive research. If we acknowledge the importance to fieldwork of emotionality, then we need a literary genre that provides an appropriate language of emotional expression.

The Thoughtful-Emotional Fieldworker

> *But as for the beauty of it, the Microscope manifests it to be all over adorn'd with a curiously polish'd suit of sable. Armour, neatly jointed, and beset with multitudes of sharp pinns, shap'd almost like a Porcupine's Quills or bright conical Steel-bodkins; his head is on either side beautify'd with a quick and round black eye.*
>
> —Robert Hooke, *Micrographia*

That emotions can and should have a legitimate role in research writing is evidenced by the growing interest being shown of late among ethnographers.

In a recent review of *representation* in ethnography, John van Maanen (1995) describes the emergence of nonrealist styles of ethnography in which the fieldworker makes the subject of research the fieldwork itself. In *auto-ethnographies*, the writer's own culture is textualized and his or her *confessional tale* offers "a passionate, emotional voice of a positioned and explicitly judgmental fieldworker," thus obliterating the distinction between the researcher and the researched (van Maanen, 1995, p. 9).

Drawing on symbolic interactionism, Sherryl Kleinman and Martha Copp (1993) recognize the distracting "emotion work" of sociology fieldworkers whose rationalism forces them, in their writing, to disguise and suppress negative feelings toward participants. Kleinman and Copp urge fieldworkers to eschew this scientific detachment, to avoid privileging the cognitive and behavioral and, instead, become more aware of their feelings and tell how they actually helped to understand the field setting.

But as we all know, the emotions, especially in excess, can be troublesome. Kleinman and Copp point out that a major challenge for ethnographic fieldworkers is how to include the emotions while avoiding the construction of a *self-absorbed self* who loses sight of the *culturally different other*. On the other hand, Ellis (1997) argues (passionately?) for self-absorption, which can serve as part of the emotional introspective homework necessary for getting in touch with our own contradictory and ambiguous thoughts. So, the question arises: Have I overly indulged myself in creating the perhaps idiosyncratic textual characters of Dr. Stern (my whipping boy?) and Dr. Buenos (my guardian angel?)?

And Williams (in this volume) adopts a Jungian perspective and warns of the closely related problem of the fieldworker *projecting* unconsciously onto others his own emotionally intense *inner energies*. For example, in the case of Dr. Stern, was I guilty of a negative projection of my own inner archetypal *shadow* (or underdeveloped aspects of my ego) onto another, resulting in an overemphasis on his negative traits? Or did I project my inner *soul-woman anima* onto Mary, resulting in a romantic (blind?) attraction to her positive traits?

In a realist epistemology, self-absorption and projection might be serious threats to the validity of (truthful) portrayals of participants. In the writing of impressionistic characters (discussed below), however, the evocation of cultural archetypes from one's own (collective) unconscious can serve a useful purpose if they work as intended for the reader. The construction of a credible textual character who can engage a reader in *pedagogical thoughtfulness* is the aim of my impressionistic writing. I would argue that the act of projection could be harnessed to create a compelling character who conveys to readers an authentic sense of their own (usually invisible) cultural-situatedness. To the extent that this is achieved, then, the problem of the author's self-absorption fades quietly away.

So, now, I turn to the metaphor of Impressionism and explain its connection with John van Maanen's (1995) *second moment* of ethnography, when the fieldworker is writing rationally and emotionally about his fieldwork experience.

Impressions of Lived Experience

I wanted to develop a document, a text that came as close as possible to painting with words.
                    —Sara Lawrence-Lightfoot, *A View of the Whole*

Van Maanen (1988) provides a powerful and colorful means of portraying the complex interplay of reason and emotion that creates the mood or ambience of a social setting. Van Maanen draws on the Impressionist art genre, that ill-defined revolutionary art school (of Degas, Monet, Renoir, Seurat, van Gogh, et al.) which rejected the realist landscapes and portraiture of the Classical school in favor of a figurative approach with a focus on the *sociology of emotions*. Earthy unposed scenes *in situ* are preferred over formal studio portraits; tangled wheatfields over roses and vases. The familiar is made strange by valuing the unclassified, shuffling conceptual categories, producing incongruities. The artist tries to evoke in the viewer an open and participatory sense, at times, to startle the complacent viewer. The associative link to fieldwork writing lies in the artists' self-conscious use of their materials, particularly through striking use of color and close attention to light.

An appreciation of its raison d'être can be gained from examining paintings of nineteenth-century French Impressionists (Thompson & Howard, 1988). In Monet's lake-scape, *La Grenouillere*, there is a recognizable public space inhabited by people whose social habits are identifiable but of minor concern to the artist. Monet's main concern is to express the mood of this particularly sunny day. The choppy brushmarks of pure color result in a mesh of light and atmosphere that is the true subject of the painting.

More dramatically, Van Gogh's land/sky-scape, *Starry Night*, is a partly imaginary or *composite* image that amalgamates elements from two of his earlier paintings as well as a fictional church spire drawn from memory. These devices contribute to a synthesis of the artist's present and past experiences. But the painting is not so much about landscape forms as it is about the theme of a wonderful restless life force that sweeps upward through the trees and swirls across the sky, a theme that is conveyed through self-conscious use of vibrant color and heavy brushstrokes. These paintings engage the viewer as a participant in the rich lived experiences of the artists.

For ethnography, van Maanen's impressionistic mode of writing—*tales of the field*—is an attempt to bring the knower and known together in representational form as a means of "cracking open the culture and the fieldworker's way of knowing it so that both can be jointly examined...[keeping] both subject and object in constant view" (van Maanen, 1988, p. 102). The fieldworker draws on his or her experiences to write stories about remarkable and memorable (rather than recurring) events, which are made striking by skillful use of words, imagery, phrasings, and metaphor. The writer aims to draw readers into the story, to have them relive the experience from beginning to end, to work out its puzzles and problems. To intensify the relived experience, the writer may exaggerate, be entertaining, be uncharacteristically kind (or unkind), or use crude figures of speech typically forbidden.

Literary Standards

Impressionistic tales draw on literary (poetic, rhetorical) standards of narrative rationality and, so, should be *plausible* or *believable* rather than accurate (in the sense of scientific rationality). They should *resonate* with our sense of lived experience. Thus, the criterion of *verisimilitude* replaces that of accuracy of representation (or trustworthiness or truthfulness). An impressionistic tale should be judged, therefore, in terms of its *interest* ("does it attract?"), *coherence* ("does it hang together?"), and *fidelity* ("does it seem true?"). Van Maanen suggests the use of the following literary devices for writing impressionistic tales.

Textual Identity. The writing should engage readers in mimicry of the fieldworker's lived experience, as though they were peering over the shoulder of the fieldworker and doing the seeing, hearing, and feeling of/in the event. Interpretive commentary should be kept to a minimum within the tale.

Fragmented Knowledge. The learning process of the fieldworker can be exemplified by a novelistic style that lets events unfold in a way that makes the reader uncertain of meaning or destination.

Characterization. The fieldworker-writer must take a stance (rather than feign disinterest) by expressing, for example, befuddlement, sensitivity, compassion, anger, or skepticism; and characters must be given names, faces, motives, and lines to speak.

Dramatic Control. The fieldworker-writer should recall events in the present tense, avoid giving away the ending, and build a degree of tension; provide condensed but contextual descriptions; use artistic nerve, unusual phrasings, fresh allusions, rich language, cognitive and emotional stimulation, puns, and quick jolts to the imagination.

Art Metaphors for My Tales

How did I make use of the metaphor of impressionism in writing the tales of Dr. Stern and Dr. Buenos? Reflecting on Impressionist paintings and on van Maanen's (1988) literary guidelines for impressionistic writing, I was guided by the following artistic metaphors to help me to breathe life into the writing of these tales.

Illuminating the Mood. Taking as a metaphor the artist's concern with the relationship between his subject and the light that illuminates her, I made visible the role of my own feelings and values in illuminating chosen features of teachers' pedagogies. In *Dr. Stern*, my own stern (archetypal?) voice speaks out in protest at the figurative character of a college science professor whose cold, hard pedagogical grasp of his students reaches out to assault my own sensitivities. I allow myself to rebel on behalf of all repressed students (especially my historically repressed self). I reveal my disquiet about a professional educator who seems to thrive on a pedagogical spirit of competitive individualism, and I portray my experience of it during an interview in his office.

In contrast, I speak with warmth and praise of the care and compassion of Dr. Buenos, whose pedagogical humanity was palpable in her bilingual classroom. My celebration of Mary is due in no small way to my recollection of the few moments of true connectedness that I recall experiencing in the brief company of one or two compassionate teachers during the impressionable years of my youth. I talk also about my impact on Mary's thoughts and feelings as she finds a space for me somewhere between her busy daily schedule and her altruistic dreams. That this reciprocation is not represented in Dr. Stern is indicative of the distancing I experienced in my brief relationships with a number of male college professors.

Composing the Subject. Taking as a second metaphor the composite character of van Gogh's landscape painting, *Starry Night*, I constructed from my fieldwork experiences a semi-fictionalized and composite character. Dr. Stern is an amalgam of the disparate features of a number of (mostly male) college teachers who were united in the common belief that the place for "real learning" is not inside their classrooms. Dr. Stern speaks for teachers who have little respect for students as people, who are captives of a technical rationality that underpins their unremitting didacticism, and whose scientific attitudes privilege the smug authority of their own knowledge. I speak with passion about the factual-fictive Dr. Stern because he reminds me of the way some of my own teachers treated me (perhaps unwittingly?) with contempt by ignoring my learning needs and my personhood at critical times in my life.

In a subtle sense, I also composed Dr. Buenos, especially in the way I chose to illuminate certain selected aspects of her real-life character. And in *Getting Ready*, I composed my own character, choosing to emphasize certain of my experiences while masking others, and semi-fictionalizing my inner discourse (for who ever thinks in syntactically correct sentences?).

Showing the Brushstrokes. Taking as a metaphor the visibility and texture of the brushstrokes in Impressionist paintings, I wove together into the same tale elements of the methodology and findings of the study. In *Getting Ready*, the tale is largely methodological in its emphasis, but in the other tales I mention the ways in which I interacted with the real-life people and their workplace settings. I paint the physical and social context of the college environs as I wait for Dr. Buenos and as I accompany her to the cafeteria for a lunchtime conversation about her pedagogy. The problem of the tape recorder was paramount at this time.

Of course these metaphors do not stand alone, independent of one another. They work together, sometimes overlapping in an attempt to portray evocatively a sense of mood or ambience and a thoughtful understanding of the key pedagogical issue of the provision for students of opportunities to learn within the classroom.

Although based on real-life people, and my experiences of them, the characters of the tales are not meant to represent "mirror images" (as might be intended by a naive realist). Rather, I was in pursuit of the faintest glimmer of the essence of good pedagogy (often suggested by its absence) while trying neither to predetermine its nature nor bracket out entirely my own predispositions (see the following section entitled Tales as Research). For it was my own pedagogical sensibilities and sensitivities that served as the main "sensor" in my inquiry. I find that Sara Lawrence-Lightfoot captures this issue in a poetic description of her experience of gazing upon several portraits of herself at different stages of her life:

> These portraits did not capture me as I saw myself…they were not like looking in the mirror at my reflection. Instead they seemed to capture my very essence—qualities of character and history some of which I was unaware of, some of which I resisted mightily, some of which felt deeply familiar. But the translation of image was anything but literal. It was probing, layered, and interpretive…the piece expressed the perspective of the artist and was shaped by the evolving relationship between the artist and me…in searching for the essence, in moving beyond the surface image, the artist was both generous and tough, both skeptical and

receptive…the portraits expressed a haunting paradox, of a moment in time and of timelessness. (Lawrence-Lightfoot, 1997, p. 4)

Tales as (Phenomenological) Research

Phenomenological human science sponsors a certain concept of progress…the progress of humanizing human life and humanizing human institutions to help human beings to become increasingly thoughtful and thus better prepared to act tactfully in situations.
—Max van Manen, *Researching Lived Experience:*
Human Science for an Action Sensitive Pedagogy

To what extent can I claim that my tales have arisen from legitimate research activity that they differ significantly from literature or journalism? For it must be quite apparent that I have transgressed (some of) the canons of recognized ethnographic-type research, especially those which stipulate that the fieldworker should spend considerable time in the field (in a particular setting) and that the credibility of his interpretations depends on checking them with those whose perspectives he purports to portray! And although I have argued that impressions differ from interpretive inferences (or assertions), I have not presented an alternative set of epistemic standards for judging their legitimacy or quality as fruits of research. In this section, I shall argue that the tales of *Getting Ready, Dr. Stern,* and *Familiarity* are narratives of *phenomeno-logical research,* and are closely aligned with Max van Manen's (1990) pedagogically sensitive *hermeneutic phenomenology.*

The type of phenomenological[6] research described by van Manen brings together in an intimate relationship the central concerns of *phenomenology*—sensitive description of lived experience, particularly the everyday experiential realities of teachers and learners—and *hermeneutics*—understanding how others (teachers, learners) make sense of their everyday (pedagogic) experiences. The goal of phenomeno-logical research is to discover the *essence* of a phenomenon, the nature of an experience, in language that "reawakens or shows us the lived quality and significance of the experience in a fuller or deeper manner" (p. 10). Thus, to do phenomenological research, we must be engaged with the phenomenon; and our engagement must be the subject of critical reflective inquiry.

And because the phenomenon concerns us as *educational* researchers, we need to be sensitive to and question the way we orient ourselves pedagogically to *being* in the world. Thus the fieldworker should enter empathically (with due care and concern) and existentially (with a heightened awareness of being and becoming) into a shared lifeworld experience. After the French existentialist Sartre, phenomenological understanding involves understanding (through sensing and thinking) one's situatedness in a shared social reality:

> I see myself because somebody sees me. I experience myself as an object for the other. (Sartre, 1956, pp. 252–302; cited in van Manen, 1990)

This *pre-reflective* experience generates in the fieldworker a rich and insightful *descriptive* (or phenomenological) understanding, but one that is subvocal and embodied. Subsequently, in representing the meaning of this experience—the thoughtful *bringing to speech* of something—the fieldworker generates via some text (i.e., spoken or written language) an *interpretive* (or hermeneutic) understanding which, because it presents a particular reading, remains necessarily partial and contingent. The writing process is doubly interpretive inasmuch as it entails consideration of the quality of understanding to be generated by the reader in meshing his or her lifeworld with that of the author. Van Manen's is a holistic approach to fieldwork, one that involves the researcher's whole sensing-thinking-being self.

In the fieldwork that precipitated my tales, I was phenomeno- logically concerned with understanding some essential aspect of college teachers' pedagogues. I knew that it had something to do with *the quality of student learning*, but I resisted defining in advance what that essence might be. Instead, I sought to discover it in the dialectical relationship between my fieldwork and my writing of fieldtexts and vignettes, activities that were interspersed across a seven-week period as I traveled to colleges, returned to base, and set out once more. Somewhere in this consuming process, the key emergent issue was distilled: Where did college teachers assume that learning primarily took place—inside or outside of their classrooms? This sharpened the focus of my subsequent inquiry to: *the quality of opportunities for students to learn within the classroom*. I sensitized myself to this pedagogical issue by reflecting on my own previous lived experiences as a college student. While gazing at the teaching and learning in various classrooms and while in (felt) discussions with college teachers, I imagined what it would be like to be a learner situated in an educative relationship with these persons.

I resonated to the openness, warmth, and vitality of the real-life (female) teachers from whom I constructed the textual character of Dr. Buenos, both in their classes and in our private discussions, and I felt their pedagogical sensitivity toward and liberating aspirations for their students. This afforded me an enhanced opportunity to make sense of their pedagogical beliefs and values. With the real-life (mostly male) teachers from whom I constructed Dr. Stern, I sensed a much cooler (sometimes cold) distancing quality in their communicative relationship with me, and I witnessed its manifestation in their classrooms, where they construed learners more as group members than as persons. In conversation with these teachers, I found myself being drawn into experiential realities where regret, frustration, animosity and visionless pragmatism were the defining features of their professional worldviews. I felt compelled to endow Dr. Buenos with a pedagogical essence and to mark it by its absence in Dr. Stern. Thus, the two tales are united in their complementarity (yin and yang); neither can stand alone. But for reasons that I shall now explain, I resisted the temptation to represent (cripple) that pedagogical essence with a defining propositional statement in an accompanying interpretive commentary.

Dialogical Text and Standards

Dialogically constructed texts allow us to recognize our lives in the mimicry of stories and conversational anecdotes...they allow for a certain space, a voice, which teaches by its textuality what the sheer content of the text only manages to make problematic...The reader gains something more important than a definition...the experience of being oriented (turned around)...in a way that is profoundly conclusive.
—Max van Manen, *Researching Lived Experience: Human Science for an Action Sensitive Pedagogy*

The "final" writing of the tales (which continues to this day as I shape them for successive audiences) involved me in a (Hegelian) dialectic as I constructed and reconstructed my experiential reality and manifested it in the textual characters of Dr. Stern and Dr. Buenos. At times, I engaged in hermeneutic inquiry of the fieldnotes, vignettes, and video/audio recordings. What could they tell me about the meaning-perspectives of the teachers I had met? Other times, I indulged in a phenomenological inquiry of my own lived experience: as a radicalized risk-taking fieldworker, a (still-developing) teacher of twenty years, a lifelong learner, a social critic and reformer (for as long as I recall), and an erstwhile student. What was it like to be a student? What would it be like to be a student in these teachers' classrooms? What quality of meaning would I have been able to construct from day to day? The more I wrote the more my inquiry blossomed; the emergent text presented me with increasing opportunities for critical reflection and creative inspiration. And, as I wrote, I considered the desired impact on my readers of my emergent tales. How did I wish to engage them? In what type of experiences?

For van Manen (1990), to do phenomenological research involves an inseparable relationship between writing and research; after Sartre, "writing is the method" (p. 126). Thus, the inquiry reaches a zenith in the writing of the research. But we write not to map some aspect of the world, in an analytical-propositional sense, not to speak abstractly *of* the world. We should write using a poeticizing language that "authentically speaks the world" (p. 13) with a moral force, that moves the reader to find buried memories and to rediscover his or her own pedagogical *being* in the world. Phenomenological research writing is concerned not with reporting *findings*, but with developing a "critical pedagogical competence: knowing how to act tactfully in pedagogic situations on the basis of a carefully edified thoughtfulness" (p. 8).

And so we are interested in our text establishing a particular *dialogic*, or educative, relation with the reader, one that engages, involves, and requires a thoughtful response. Dialogical writing should avoid defining (in a propositional sense) issues such as pedagogy because of its ineffable quality, which is not capturable by or in text, but which text can at best allude to. Instead, the *textuality* of the writing should open up, in an indirectly teachable way, questions of pedagogy; to engage readers in vicarious and mimetic experiences that aim to teach them something profound about their pedagogical selves. Thus, a change of orientational metaphor for research writing signals a major shift toward an emphasis on *research as reading*, to transform by engaging readers in critical self-reflective thought and action (i.e., praxis).

Now, to return to the question that opens this section, phenomenological research/writing should be judged in terms of its "power and convincing validity" (van Manen, 1990, p. 151). Van Manen presents four evaluative criteria for the dialogical textuality of phenomenological inquiry.

Oriented Text. The text should be oriented to answering the question of how the researcher as educator stands in relation to life: What are the valued beliefs that shape the educator's lifeworld?

Strong Text. The text should be committed to a strong pedagogical perspective that addresses the question of how we should be and act with children.

Rich Text. The text should provide rich descriptions (anecdotes, stories, and narratives) that explore experiential phenomena that cause the reader to be engaged, involved, and thoughtfully responsive.

Deep Text. The text should enable the reader to explore reflectively the depthful character of his or her pedagogical nature beyond what is immediately experienced, to appreciate the inherent complexity, ambiguity, and mystery of life.

These are standards that, for me, fit well with John van Maanen's literary metaphors for writing/reading impressionistic tales of the field—illuminating the mood, composing the subject, showing the brushstrokes—but that extend them by adding a unique pedagogical dimension for engaging the reader in a critical reflective experience.

Last-Minute Confessions

> *Ethnography is no longer pictured as a relatively simple look, listen and learn procedure but, rather, as something akin to an intense epistemological trial by fire.*
> —John van Maanen, *Representation in Ethnography*

Although phenomenological writing might be designed to speak for itself, I could not end the chapter having made that point. For I believe that no particular discourse can claim a privileged status, and I feel that, in closing, I should like to say a little more about my intentions in writing the tales and providing justification for this relatively new research approach. So I shall end in the spirit of John van Maanen's confessional tales.

Life in college classrooms is much more complex and chaotic than many science education researchers are able to account for in their interpretive activity. If we are to understand richly and deeply what is going on in college science classrooms, then I believe that we need to include in our research an emotional dimension that sensitizes us to the mood or ambience of an educational setting. As well as a thinking being, the fieldworker is an emotional being, and it is legitimate to listen to his feelings. On the central question of how to legitimate such a seemingly unholy marriage (and thus reunite our masculine and feminine selves), I have explored impressionistic tales and sought to legitimate them as research within the pedagogically rich framework of phenomenological inquiry. In doing this I learned two new metaphors—*research as writing, research as reading*—metaphors that orient the researcher toward a phenomenological understanding of lived experience and a desire to engage dialogically the reader's imaginative self.

Having conducted a relatively brief phenomenological inquiry into college science teaching, I wrote a pair of complementary impressionistic tales that allude (I hope) to the essence of pedagogies that provide contrasting opportunities for students to engage in quality learning within the classroom. One pedagogy is embodied by Dr. Stern, a composite character constructed from my experiences of visiting a number of college science teachers at work. This construction of a composite or fictional character is not unprecedented in science education research (Tippins, Tobin, & Nichols, 1995). In this volume, Craig Bowen drew on his experiences to construct the fictional character of Diane, and Sue Mattson *crystallized* the events of many research team meetings to create *condensed description.*

In the introduction to this chapter, I outlined some concerns about the impact on prospective teachers of conventional college teaching. By this, I mean strongly didactic teaching shaped by an implicit commitment to an objectivist epistemology and a behaviorist psychology of mind. While not wishing to suggest that all conventional teaching equates with Dr. Stern's teaching, I do want to point out that his teaching is composed of nothing less than elements of conventional teaching that I witnessed during my reconnaissance. With the use of Dr. Stern, I have, to some degree overplayed my hand. I have chosen to represent a *worst case scenario* of a college teacher; a teacher with apparently few redeeming features; and a teacher who provides students with questionable opportunities to learn meaningfully or deeply within his classroom. But I have found that Dr. Stern is a believable character. This has been demonstrated on many occasions when I have introduced his tale to teachers in my classes and workshops. There are those who recall, with knowing and regretful looks, that he is very reminiscent of (some of) their own past teachers. Bad learning experiences are recounted in animated ways.

Dr. Buenos closely resembles a particular and very memorable college teacher. Even so, as I constructed (composed, illuminated) her textual character, a character that differs necessarily from the real person, I was mindful of other college teachers with similar characteristics. What struck me about these teachers was their enthusiastic commitment to ensuring that students had opportunities for rich learning experiences *inside* their classrooms. They seemed to provide this opportunity by engaging the class in an *open discourse* (Taylor & Williams, 1993), that is, a discourse that encourages and rewards students for disclosing their ideas, no matter how tentative. Such a discourse entails communicative classroom relationships based on mutual respect. For the teacher, this involves active pedagogical attention to the life experiences of students in ways that connect meaningfully with scientific or mathematical activity. Whereas Dr. Stern constructed learning as a largely cognitive activity (involving "cold reason" and "hard control"), and used language as a vehicle to transmit his knowledge, communicative teachers add an important social dimension to their images of students as learners and engage in classroom discourses intended to communicate positive human values as well as scientific and mathematical ideas.

Mary[7] is a particularly interesting case, not only because of the bilingual nature of her classroom discourse but because she bases her teaching on a strong set of explicit moral virtues that compel her to adopt an enduring *duty of care* (Noddings, 1983) toward her students. This is not a cloying type of care that breeds emotional dependence, but a care that aims to foster respect for self and for other. The resulting learning environment is rich in possibilities for enhancing, in a reciprocal way, the personal evolution of both students and teacher. But I must hasten to add that Mary is not intended as the epitome of exemplary mathematics teaching. Indeed, I recoil from the idea that grand narratives of "best practice" are desirable or even possible, preferring to believe instead that the ineffable quality of good pedagogy is context-dependent.

Nonetheless, I continue to harbor a perspective that favors a role for *critical discourse* aimed at deconstructing repressive sociocultural myths (Taylor, 1998). And Mary's teaching is notable for the absence of a discourse that engages students in a critical awareness of the historiocultural or sociopolitical webs of significance within which teacher and students are suspended (unknowingly). Absent from her classroom discourse is any concern with considering the contingent nature of mathematical knowledge or of the ways that (state-mandated) technical rationalist curriculum imperatives (summative examinations as filters to higher education) distort the image of both mathematics (rendering it as decontextualized algorithms) and of what it means, therefore, to *be* a successful participant (involved in non-inquiry instrumental learning). On the other hand, Mary's open discourse is, I believe, a necessary precursor to critical discourse; without a climate of trust and security epistemological (self) critique can be very intimidating.

And so, my intention in offering the textual characters of Dr. Stern and Dr. Buenos is to engage the reader in an act of pedagogical thoughtfulness about the future of college science-related education. With the realization that newly certified teachers of our children are most likely to imitate the teachers of their own lived experiences, I feel that we are compelled to ask the following of college degree programs:

What (lived) educational experiences should we value (most) for (our) children's prospective science teachers?

Peter C. Taylor
Science and Mathematics Education Centre
Curtin University of Technology, Perth, Western Australia, Australia

EDITORS

METALOGUE

Evoking Deeply Felt Reflections by Telling Tales

KT: Peter's chapter is an exemplar of a new way of thinking about research on teaching and learning. In some respects we gave him the impossible assignment: to visit colleges around a big state like Florida to get *tales of the field* that would inform us about the teaching and learning of science and mathematics at the college level. We were grounded in interpretive research and it was clear that a two- to three- days site visit was insufficient for a credible interpretive study. The decision to use an approach based on phenomenology and impressionistic tales was risky but most informative. How else could we have dealt with a situation of teachers whose beliefs in a transmission model of practice made it just common sense to teach as Dr. Stern? We have all had teachers like Dr. Stern, and speaking for myself at least, I have taught just like Dr. Stern—but for all of the altruistic reasons. People teach that way because it makes sense for them to do it that way and it does not make sense to do otherwise.

PT: As I wrote the tales, I had in mind the need to cater for multiple stakeholders, not only the funding agency but also the many professionals—college science teachers, science teacher educators, graduate research students—in need of highly readable texts that might evoke critical reflective thinking. My intention was not to be informative about college science teaching, in a normative sense; rather, it was to write a *dialogical* text that might elicit long-suppressed memories of bad (and good) learning experiences and provoke a debate about the need to transform the long-standing dominant epistemology of practice of college science teaching. Other authors in this book (Craig Bowen, Sue Mattson, Wolff-Michael Roth and Ken Tobin, Ben Cunningham, Mark Williams, Penny Gilmer) have made use of *alternative genres* with a similar aim in mind (see Chapters 2, 6, 10, 14, 16, 17).

PG: During Peter's sabbatical in 1995, several times he visited Ken's graduate class on evaluation, in which I was a student. On one occasion, he spoke with us about the impressionist tales he was constructing of the context of teaching mathematics and science in community college classrooms in the state of Florida. Peter was traveling around the state, visiting community colleges, while interviewing various faculty and attending their classes. Peter drew on the work of John van Maanen (1988) on writing ethnography, quoting him, saying "striking stories materials are words, metaphors, phrasings, imagery, and most critically, the expansive recall of fieldwork experience" (p. 102). I was struck by reading his, at that time, incomplete story of Dr. Stern. Peter's text included all these elements, and it evoked strong feelings in me. I felt and saw an image of Dr. Stern in his health science classroom. Dr. Stern did remind me of some of my own teachers in the past and to a certain degree of myself when I was in a more objectivist frame of reference.

I could also sense how Peter felt when he interviewed Dr. Buenos and attended her bilingual mathematics classroom at the community college. Peter quoted again from van Maanen (1988) that he was attempting "a representational means of cracking open the culture and the fieldworker's way of knowing it so that both can be jointly examined...Impressionist writing tried to keep both subject and object in constant view. The epistemological aim is then to braid the knower with the known" (p. 102). Also, from van Maanen, "the well told tale will always go behind the bare bones and embellish, elaborate, and fill in little details as the mood and moment strike" (p. 17). Peter did paint us rich impressionist tales of both Dr. Buenos and Dr. Stern.

KT: What we know extends far beyond what can be written or spoken of. One of the things that appeals to me about the impressionistic tales you have written is that they are holistic expressions of what *you* know, and they manage to provide a context *for me* to reconstruct my knowledge of teaching in ways that conjure up a significant emotional content. Because of the ease with which I can connect to narrative, I have always found narrative genres appealing. In the case of impressionistic tales the focus on salient issues is a not-so-subtle way of engaging more than the knowledge we are conscious of and can reflect on.

PG: Peter's powerful style of drawing on impressionist art opened up for me new ways of representation, rejecting formalist and realist approaches. I found this intellectually very stimulating and exciting. On the evening after Peter, Ken, and I had a farewell dinner before Peter returned to Australia, we drew up our plans for constructing this very book. Later that evening I had a dream that reminded me of a full-color impressionist Van Gogh painting with three individuals, obviously me with Peter and Ken, having a picnic in the park, with checkered tablecloths, parasols, and a picnic basket containing food and a bottle of wine. This was a sign to me that this collaboration would not only be productive but also the learning would be fun. It was a good sign. (See Chapters 16 and 17 for more on the role of dreams in research.)

The Partialness of Evocation

KT: Referring back to my claim that "What we know extends far beyond what can be written or spoken of," I wonder whether the most important parts of what you learnt were incorporated in the two tales or just what was easier to tell in the form of tales? I wonder that about Monet's water lilies too. As I look out from the bridge and onto the river that may have inspired many of his paintings, I wonder if he was painting what was of greatest significance about his lived circumstances or just those aspects that are easiest to capture through his art. What are your thoughts?

PT: Well, my tales tell only about those aspects of college science teaching that are tellable within the framework of impressionistic research. Other stories remain to be told within other frameworks (both qualitative and quantitative). There are likely to be a host of other teaching-learning issues in college science teaching that remain to be identified, some of which may yield to impressionistic research, some to orthodox interpretive inquiry. And we should recognize the dependence (and significance) of those issues on the subjectivities/standpoints of the researchers who pursue them and the genres they employ to communicate their research.

This is very much in evidence in a number of chapters in this book, which bear witness to a variety of researchers' standpoints and epistemologies of practice, both orthodox and radical. And it is heartening to see that some of these researchers have adopted a *critical reflexivity* that acknowledges the way in which their research has been shaped by their particular standpoints. Chief among them is Ben

Cunningham, who acknowledges the unavoidable *ironic validity* of his self-representations (Chapter 14).

Auto-Ethnography?

KT: Let me continue to press this issue. How much of the tales reflect your own experiences in all those colleges you visited and how much reflect your lived experiences in science and mathematics classes from back when you were a student in grade school through to your present activities as a researcher?

PT: The tales are as much about me as they are about the many professors and their students whom I encountered during the fieldwork. But I did try to ensure that the tales were not accounts of my own self-absorption, as Mark Williams warns (Chapter 16). In the actual process of writing the tales, I interrogated artifacts that I had brought back from the field (recordings, notes, maps, receipts, souvenirs) to evoke my memories/images/feelings whilst, in a recursive manner, I constructed both my authorial self and the textual characters of the college professors, Dr. Stern and Dr. Buenos (after Richardson, 1994). So, the tales are about my living in the world *in relation to* the lived experiences of others. It is significant that my greater sense of connectedness with the professor who gave rise to the character of Dr. Buenos afforded a richer portrayal of her own lived experiences.

As impressionistic accounts, the tales are as related to everyday life as are the recognizable forms of Monet's ethereal haystacks and bridges or Van Gogh's spidery church spire and cedar trees. But these forms do not constitute the greater part of the significance of the paintings in which they appear. Undue concern with the representation of form can distract and distance the viewer from the mood of the painting wherein the ineffable plays teasingly on the sensibilities and sensitivities of the viewer (as it did for the early skeptics of Impressionist painting). Likewise, the form of the characters of my tales are not, for me, in and of themselves, the central issue; rather the significance of the tales concerns the professors' pedagogies; especially their contrasting actions in relation to the implicit question of whether learning should occur inside or outside their classrooms.

Thus, it matters not that the tales are *trustworthy* (in the orthodox sense prescribed by Guba and Lincoln, 1989). But it does matter that they are believable, because it is in their *believability* (or *verisimilitude*) that their power lies to evoke in readers a resonance with their own lived experiences and then (hopefully), a critical reflection on their own (and perhaps others') epistemologies of teaching practice (Denzin & Lincoln, 1994).

I must say also that the writing process afforded me somewhat of a cathartic experience, as it must have done for Harry Wolcott (1990) when he wrote that very challenging ethnographic account of a painful chapter in his own life, "The Sneaky Kid." I relived some of my own less-than-delightful experiences as a learner of college science, and came to understand better my own early biography as a teacher of high school physics.

Evocative Teaching

PG: I could strongly sense the ethic of caring that Dr. Buenos showed to her students. She cared about them as people, and this motivated the students.

KT: Dr. Buenos was not a native speaker of English, and she knew firsthand how difficult it could be for students such as those she taught to make sense of mathematics. To what extent can her success be attributed to the fact that she had lived circumstances similar to those of her students?

PT: What I hope the tale of Dr. Buenos indicates is that she is strongly empathic with students as people who are struggling (against the odds) to realize their full human potential. Her propensity to include this genuine concern in her teaching can be attributed, I believe, to her own lived experience. For Dr. Buenos, teaching mathematics is a means to a greater end. And in giving expression to her concern, she seems to evoke in her students a willingness to succeed academically and in life, an outcome that is manifested in the classroom as a collaborative and respectful ethos of co-participation.

NOTES

1. The role of a title is to provide a theme that flags the central issue of a tale. But sometimes the title can distract the reader from engaging with the text in a relatively unfettered manner. I chose to provide a title for this tale retrospectively because I want readers to encounter the text largely from their own perspective, to be, perhaps, somewhat puzzled about the phenomenon of the tale, to be drawn into the tale with their predisposition toward tales (stories, narrative) largely intact. It is only later that I wish the reader to become analytical about the nature of tales.

2. This term is used as a generic descriptor of both colleges and universities.

3. The common term of *professor* is used in the United States, but I have chosen to use the generic term *teacher* because I wish to emphasize the practice of teaching rather than professing.

4. As a point of clarification, I refer in this chapter to the work of both John van Maanen and Max van Manen, whose surnames bear an uncanny resemblance.

5. The museum in St. Petersburg, Florida, contains an excellent representation of Dali's life's work, including his provocatively metaphysical Surrealist period and his powerfully evocative Classical period, which, for me, captures key essences of Western philosophy (*The Hallucinogenic Toreador* sits above me as I type these words). As I left the gallery I was moved to exclaim (to myself), "If only I could write like that!"

6. Although van Manen describes his research approach as "hermeneutic phenomenology," he prefers to use the term "phenomenological research" in order to distinguish it from other qualitative research. Van Manen describes phenomenology as the study of the lifeworld—the world as we immediately experience it pre-reflectively rather than as we conceptualize, categorize, or reflect on it.

7. A literary strategy that distinguishes Dr. Buenos from Dr. Stern is that the former has a first name—Mary—which makes her a much more accessible character.

BIBLIOGRAPHY

Bacon, F. (1620/1968). Paraceve. In J. Spedding, R. L. Ellis, & D. D. Heath (Eds.), *The works of Francis Bacon: Book 4*. New York: Garrett Press.

Bruner, J. (1986). *Actual minds: Possible worlds*. Cambridge: Harvard University Press.

———. (1990). *Acts of meaning*. Cambridge: Harvard University Press.

Clandinin, D. J., & Connelly, F. M. (1994). Personal experience methods. In N. Denzin & Y. Lincoln (Eds.), *Handbook of qualitative research* (pp. 413–427). Thousand Oaks, CA: Sage Publications.

Dawson, V., & Taylor, P.C. (1998). Establishing open and critical discourses in the science classroom: Reflecting on initial difficulties, *Research in Science 28*(3), 317–336.

Denzin, N. K., & Lincoln, Y. S. (Eds.). (1994). *Handbook of qualitative research.* Thousand Oaks, CA: Sage Publications.

Ellis, C. (1997). Evocative autoethnography: Writing emotionally about our lives. In W. G. Tierney & Y. S. Lincoln (Eds.), *Representation and the text* (pp. 115–139). Albany: State University of New York Press.

Erickson, F. (1986). Qualitative methods in research on teaching. In M. C. Wittrock (Ed.), *Handbook of research on teaching* (pp. 199–159). New York: Macmillan.

Gadamer, H. G. (1976). *Philosophical hermeneutics.* Berkeley: University of California Press.

Gallagher, J. J. (Ed.). (1991). Interpretive research in science education. *Monographs of the National Association for Research in Science Teaching 4*, 3–17.

Geertz, C. (1989). *Local knowledge: Further essays in interpretive anthropology.* New York: Basic Books.

Guba, E., & Lincoln, Y. (1989). *Fourth generation evaluation.* Newbury Park, CA: Sage Publications.

Hooke, R. (1665/1961). *Micrographia.* New York: Dover.

Kleinman, S., & Copp, M. A. (1993). *Emotions and fieldwork: Qualitative research methods series.* Newbury Park, CA: Sage Publications.

Lawrence-Lightfoot, S. (1997). A view of the whole. In S. Lawrence-Lightfoot & J. H. Davis (Eds.), *The art and science of portraiture* (pp. 3–16). San Francisco: Jossey-Bass Publishers.

Milne, C., & Taylor, P. C. (1998). Between a myth and a hard place: The application of critical theory to science education. In W. W. Cobern (Ed.), *Socio-cultural perspectives on science education: An international dialogue* (pp. 25–48). Dordrecht, The Netherlands: Kluwer Academic Publishers.

Noddings, N. (1983). *Caring: A feminine approach to ethics and moral education.* Berkeley: University of California Press.

Pinar, W. & Reynolds, W. (1992). *Understanding curriculum as phenomenological and deconstructed text.* New York: Teachers College Press.

Richardson, L. (1994). Writing: A Method of Inquiry. In N. K. Denzin & Y. S. Lincoln (Eds.), *Handbook of qualitative research* (pp. 516–529). Thousand Oaks, CA: Sage Publications.

Taylor, P. C. (1993a). *Teacher education and interpretive research: Overcoming the myths that blind us.* Invited keynote address at the International Conference on Interpretive Research in Science Education. Science Education Center, National Taiwan Normal University, Taipei.

———. (1993b). Collaborating to reconstruct teaching: The influence of researcher beliefs. In K. Tobin (Ed.), *The practice of constructivism in science education* (pp. 267–297). Hillsdale, NJ: Lawrence Erlbaum.

———. (1996). Mythmaking and mythbreaking in the school mathematics classroom, *Educational Studies in Mathematics 31*(1, 2), 151–173.

———. (1998). Constructivism: Value added. In K. Tobin & B. J. Fraser (Eds.), *The international handbook of science education* (pp. 1111–1123). Dordrecht, The Netherlands: Kluwer Academic Publishers.

Taylor, P. C., & Dawson, V. (1998). Critical reflections on a problematic student-supervisor relationship. In J. A. Malone, B. Atweh, & J. R. Northfield (Eds.), *Research and supervision in mathematics and science education* (pp. 105–127). Mahwah, NJ: Lawrence Erlbaum.

Taylor, P. C. & Williams, M. Campbell. (1993, March). *Critical constructivism: Towards a balanced rationality in the high school mathematics classroom.* Paper presented at the annual meeting of the American Educational Research Association, Atlanta, GA.

Thompson, B., & Howard, M. (1988). *Impressionism.* London: Bison Books.

Tippins, D. J., Tobin, K., & Nichols, S. E. (1995). Constructivism as a referent for elementary science teaching and learning. *Research in Science Education 25*(2), 135–149.

Tobin, K. & McRobbie, C. J. (1996). Cultural myths as constraints to the enacted science curriculum. *Science Education 80*(2), 223–241.

Van Maanen, J. (1988). *Tales of the field: On writing ethnography.* Chicago: University of Chicago Press.

———. (1995). An end to innocence: The ethnography of ethnography. In J. van Maanen (Ed.), *Representation in ethnography* (pp. 1–35). Thousand Oaks, CA: Sage Publications.

van Manen, M. (1990). *Researching lived experience: Human science for an action sensitive pedagogy.* London, Ontario: SUNY Press.

Wolcott, H. F. (1990). On seeking—and rejecting—validity in qualitative research. In E. W. Eisner & A. Peshkin (Eds.), *Qualitative inquiry in education: The continuing debate* (pp. 121–152). New York: Teachers College Press.

CHAPTER 2

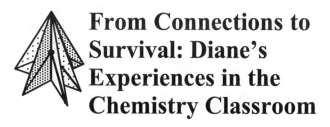

From Connections to Survival: Diane's Experiences in the Chemistry Classroom

Craig Bowen

My perception before I got in the course was sort of a chemistry or science appreciation. When I got in it was stated to us that it was science appreciation. At this point there is no appreciation for it. It's just trying to figure out the teacher's methodology to try to get through the course. It's quite a disappointment because I was looking for some sort of overview—a way to connect to science.
—A comment made by Diane about the class.

Diane was one of several students who had agreed to participate in an interview about the non-majors chemistry class that they were taking. Little did I know that we would collaborate on discussing issues about the course that culminated in the writing of this chapter (Bowen, 1993).

During that first interview, Diane explained many of her frustrations about the class on issues ranging from teaching methods to evaluation. She also explained how she utilized various resources to try and survive the class. But although other students were using tutors, she explained that she could not afford one. I offered my services for free on the grounds that, by helping her study and work through some of her problems in the class, I would be getting a better sense of her experiences. So, through the remainder of the semester, we met about two hours a week to talk about some of the ideas that the instructor presented in class to try some sample problems, and to predict what sorts of questions might be on upcoming exams.

Besides being a home economics major, Diane is also very involved in dance. She choreographed a piece in the "Evenings of

Dance" that was performed on campus. She suggested that I might like to attend one of the performances. The performance was interesting to see, because at the same time, I was thinking about the chemistry class and how it related to dance. Diane had mentioned during class one day that something in chemistry (the motion of molecules) was similar to some of the things that take place in dance. From that experience, Diane and I decided to conduct an interview about her experiences in the chemistry class and how they related to her world of dance.

During the interview, we discussed several things. First, we talked about dance, choreography, and the chemistry class. We touched on numerous topics, including teaching/choreography methods, performing/learning methods, and the need for connections among participants in both the realms of dance and the chemistry class. We also carried out a task intended to enable Diane to categorize her experiences in the chemistry class.

Based on interviews with other participants in the class, and other observations in the classroom, I formulated a list of about fifty words that participants used when talking about the class (Chi, Feltovich, & Glaser, 1981; Lakoff, 1987). Diane was asked to sort the cards, develop a category label, and pick two salient examples of the categories she formed, to try to represent how she organized her experiences in the chemistry class. Diane formulated her categories in a temporal fashion. The category labels and the salient examples included:

- GENESIS   1) Quality of learning 2) Interested
- HMMM???   1) Confused   2) Chemical equations
- OOPS — PANIC! 1) Chaotic   2) Bored

Finally, the third area of discussion that we had during the interview was the writing of this case study. I explained to Diane that a case study would fit nicely next to the overview of the chemistry class, and I asked if she was willing to participate. She agreed because she thought the class was not well done and that students were ignored when it came to such courses. She wanted to try to help me write something for chemistry faculty who teach such courses to assist them to understand how a student might be thinking and feeling.

What follows is Diane's story about her time in the chemistry class, and an epilogue to those experiences. The story is organized along the temporal lines that Diane formulated with themes that, for her, ran throughout the course.

Genesis

Dear Dale,

I've started classes this week and I'm very excited. I am particularly interested in the chemistry course, it seems to be designed for people like me who have little to no experiences in the sciences. It has been described as a general overview, or rather a science appreciation course. I assume I will see some significant history and basic principles that I can apply to my everyday life. I'm not afraid of this challenge. My excitement really lies in the idea that I will begin to be in touch with another discipline in my life—the "World of Science." I've always been fascinated with it, the questions, observations, and the answers. It reminds me of art history in that I will get to see the most profound contributions and mistakes and gain insight as to why they are important to my life and the way I see the world. I'm really hoping to be inspired by this course, to maybe realize I have another aptitude. There are incentives today, you know, to be a science teacher—who knows?? It's just a thought, a small flame.

I'll update you soon.
Diane

When I first started back to school I met with my adviser and found out that I would have to take a chemistry course. On one hand I was apprehensive about it because I had not taken any science in a long time, and I had never had chemistry. On the other hand I thought that taking chemistry would help me to make connections to other disciplines. All these years being in dance and the arts, and in the restaurant, that's all I knew. Like in anything you do in life, that's all you know for a period of time. You tend not to branch out, and not to cross over and that sort of thing. Not to integrate. Coming back to school to go into home economics education would let me try some new things.

When I walk into something, I try to walk in as fresh as possible and empty as possible. There's always going to be a preconceived notion about something, but I go with the attitude that I'm going to listen to every word, and I'm going to get something out of everything, at least make some notations for myself about the most important things, and then from that follow some direction. That's what I look for when I'm taking a class. If I knew the information and if I knew the directions I wouldn't take the class.

In this situation, being in college, there are certain directions I don't want to go in, but if it's guided enough or if it's outlined enough I would do it for the purpose of completing what the task is. For example, in my textiles class, some of it was interesting and some of the assignments were really annoying, but I do trust professors. As in my first experiences in college, if they were telling me to do something, whether I thought it was worthless or not, I'd do it. I do it by the book, and once I've completed it I can usually find that there was something to be had. There was something, even if it wasn't directly related to the class. So I go in with that idea today.

But getting back to integrating ideas. Patterns in the world are important to me. Craig and I talked about integration and how it related to my life and experiences in chemistry. I have this idea that everything is connected, and processes within everything are very connected also. The more I learn, the more interesting life becomes, and the more I feel I can accomplish. I'm happiest when I am learning new things and applying them to my life. It's almost like rather than saying a mystical thing, there is this bigger scientific picture of how the world operates emotionally, psychologically, technically, whatever. I just need to figure patterns to everything, and being able to learn those patterns in another discipline and integrate those ideas. This integration can be both between disciplines (like chemistry and food science) and within a discipline (like chemistry).

I had not realized that so much of the home economics education program was related to chemistry. In retrospect I realize that it's all chemistry. But one of the first things that popped into my mind was when we hear the word "chemistry" or "chemical" it always has a negative connotation to it. We don't think of our bodies as chemistry. We don't think of our food as chemistry. And it is all chemistry. And there is a lot of good stuff in there. We look at food additives as something negative. But they are there for a very good reason. And there are instances there when you can compare a food that has been processed with one that has not been processed and it may come out better. It may come out on top. Certainly we love natural things, but even in nature it's chemistry. At least I did get out of Chemistry 101 that this is nature. Certainly there are things going on in labs that are what we think of as unnatural but they made major contributions to where we are now—things that do affect everyday life. Maybe I don't feel as negative about the class anymore.

Connections and integration of patterns in the world are important to me. How do I discover them? That's where the idea of connections between people—students and teachers, for example—becomes very important. There are several aspects involved in making connections.

First, you and the others you are trying to connect with (teachers included) need to be committed to coming out of their own worlds. It comes down to a real personal side of a person—a desire for expanding your knowledge beyond your expertise. Because if I am there and I have to come out of my dance world, my home economics world, to broaden my horizons in this discipline, you as the person teaching that discipline have to get out there too and do the same thing.

Second, in a sense, making connections with others involves learning each other's language and past experiences. A good example of this happened with my food science instructor. Very quickly he caught on when we were misunderstanding his language. Sometimes he used his big dictionary words, or whatever—he likes to expound and show his knowledge—and it was great. But there was a young woman in the class who was very raw. She could say, "Now what do you mean by that?" She'd repeat his words, and say, "What does that mean?" And so he caught on very quickly, and rather than changing his language he added to his language. He would talk the way he wanted to talk, and then he'd say, "Another way of saying that is," and then he'd come back around and say things a couple of different ways, but let you know that he was saying them a couple of different ways.

Finally, by coming out of your own world, and trying to learn how to communicate with other people, you are starting to be in the position for making connections and building a relationship. Hopefully, the outcome of such connections is the creation of something together. It's like gathering threads and sewing something. It can be beautiful. For me, making such connections helps me to discover patterns in the world.

I titled this section Genesis because it represented a beginning. As I pointed out in the letter to Dale, I was hoping to come in contact with another discipline. The course description in the bulletin made it sound like I would be exposed to many wonderful ideas in science. On the first day of class, the instructor said that the class would be a science appreciation course.

I enjoyed the first class because I understood the professor when he talked in generalizations. Perhaps he did this to draw us in, to be interested in the fact that here is a discipline that affects our everyday lives. He talked about how a lot of conclusions are developed. He talked about experimentation, and sort of enlightened us that this didn't just pop up—a lot of work went into developing the chemistry. And then, just the family tree of the sciences, and things like that. I liked that it was general. It was general and for what I understood, what I was coming in with, and it could give me some bases for exploring further. I got some sense of what I could get involved in. Unfortunately, my appreciation stopped after the first day of class. I guess when I started the course I really thought it was the type of course that was going to be this exchange of information, or at least my own personal inquiries would be satisfied. But they weren't.

Hmmm???

Dear Dale,

I hope you are well, I miss you and just wanted to update you on "life in college at 30." Classes are going well for the most part. I've got some great professors who are very well prepared and enthusiastic about their expertise. Although there is one exception, chemistry! My excitement has changed to anticipation. I'm waiting for knowledge to kick in. I'm just not getting it. And I don't really know if the professor is delivering it! I read and review constantly but I can't even understand my own notes. There are lots of symbols and no explanations or vice versa. I suppose I need to figure out a new approach, because I'm not putting this all together. As for the professor, I suppose it's like any relationship where you have to look at someone's style and patterns in order to achieve connection. I just don't know if I'll do that in one semester's time.

So that's "life in college at 30." Until a new insight develops—stay well.

Diane

The classes were like conversations rather than lectures—of course it was a one-sided conversation. I think a lecture is a presentation of information with a beginning, a middle, an end, afterthoughts, and future thoughts. A good lecture is very satisfying to sit there and take in. But the class was not like that. Instead it was almost like there was so much information in the professor's head, and he just let it spill out, but he did not give the conversation a chance to develop with the other side—it takes two. He never listened to us so we could hold a two-way conversation.

The first day of class the instructor talked about general aspects of science. But then, all of a sudden, it became very technical without any introductory level. And I still think that a lot was assumed on his part, like that we had seen most of this before. And I hadn't seen any of it, but I felt like a good portion of the class had.

We needed the kind of how-to's and terminology first. Instead he would go to the concept and 10,000 examples, and then he came back after giving tons of information. Suddenly you would realize, "Oh yeah, we are talking about electron transfer. OK, now I know. OK, I got to start connecting that." I had to really listen to hear how he presented his concepts. Take a breath, listen to him ramble, and then pick up when he came back. So I mean I feel like rather than learning chemistry, I was learning how to deal with this person's presentation style.

One of the approaches he used was to give examples. One time he did that when talking about "Lewis Dot Structures." He put up a lot of Lewis structures on the overhead but without enough reinforcement about where the examples really came from. So I had pages and pages full of other notes, and then Lewis dot structures, but conceptually a lot of the stuff was not written down. So, even though I tried to write down all the examples, I missed what I needed to understand it. There were tons of examples on the board. I mean, I got tired of writing that after a while. I stopped writing because I would get home, look at my notes, and I would have all these examples, but I had nothing on what the examples were really about. And trying to listen in between—it was really hard to do both. I was not getting the highlights.

He also did experiments or demonstrations, but that's the least that I remember from that class. What I remember most about them was that they took a long time and they did not really work. There was just this pure silence while waiting for these things to take place. Sometimes he was really laborious about it. The most effective one I think is when he did the balloons, although I wasn't quite listening enough. But somebody explained to me, and I got a clear understanding of mixtures. And the idea of oxygen being on the outside of a balloon. Other than that, I didn't get a whole lot out of it. Sometimes it just took too much time. One time we had a thing where he was dissolving Styrofoam in acetone. Well, it was a large thing of Styrofoam with a little cup of acetone. And he took a good ten minutes to make sure that he showed us that it could dissolve the whole thing. And it was just frustrating—a big time waster.

Once in a while, people would ask him a question. It just seemed like questions were not addressed. Or, there was one question where somebody had asked it and he started answering it and giving illustrations. And then, somebody asked another question. And then another question was asked. And then he became frustrated. There was obviously no sort of formative check to see that the information was learned originally. And he was not hearing the differences in the questions that were asked about the same thing. I heard a lot of that going on in the classroom. And so rather than pushing it off as, "Well, I just answered that," why not jump into where the confusion is, or discuss the confusion that you see, "You all are asking the same questions, but there is something about not putting it all together?" He did not try to figure out what was happening. He was answering individual questions instead of trying to figure out the problem.

One time I went to his office, early in the semester when we were learning about entropy. I had a question I wanted to ask him about a controversy in evolution. I wasn't looking for brownie points, or looking for him to incorporate our conversation into the class. It was about religious people getting into the second law of thermodynamics, and I thought, "Oh, goodness. You know, this is science appreciation, let's get into it." And even in his office he just cut me off. He kept going back to, "Well, what I was trying to say." And I was like, "Let's sit and chat, let's talk about it." I mean, I felt like I had an intelligent observation in support of what he was saying, and I don't even think he was keying into that. He just thwarted it. He was totally blowing me off. Every time I tried to get the conversation going, all he kept saying was, "Well, it's a real controversial point." The point I was trying to make is that he was getting back into the science and I was feeling, "Hey, I'm connected to what you are saying and I want to have a scientific discussion about it," but he was just not open at all.

Finally, unlike my food science instructor, the professor in the chemistry class did not change his language. Maybe he just didn't have the expertise to do that. He had his set of expertise and that's what it was. If people did ask questions he would essentially say the same thing he had said previously. Eventually students stopped asking questions until they got pissed off when they saw their grades, then they came to the help sessions. I could see people getting really angry when the questions were being repeated and they were like—I remember a few instances where people pushed him to answer a question, but he continued to answer it in the same way, as though he was not hearing the question at all.

Craig showed me a few models of teaching and learning that we talked about. Maybe the instructor tried to bring students and the material closer together, but for the most part I want to say that the students have not been taken into consideration at all. And to compound the problem, I really feel it was a deficiency on the part of the teacher not knowing how to improve his delivery, and how to deal with such a large population. I think a lot was assumed, certainly on the teacher's part, that the students brought with them some knowledge from high school. That's pretty evident. I mean there was constant reference to "You've seen all these things before." A lot of them said they survived the course because of previous chemistry teachers, that they were going backwards, and thinking about their teachers and what their teachers taught them, but that they were not getting the ideas from the instructor. They would get the concept and then go home and deal with it. Now I don't know if that makes sense, but some of us don't have that prerequisite.

A related idea that I talked about with Craig was students teaching themselves. On the one hand, there's nothing wrong with that. But taking into consideration the population, the majority of the people were not going to be able or are not going to want to do that. I mean, we are all responsible. I was as responsible as the professor, but this is where I think taking the population into consideration is important, especially for an intro-level course. This is designed for liberal arts majors. Why do we need to do that if we are all struggling with our own majors? For whatever reason, the system sets up that we need to take a lab science, which is understandable. But I can't be expected to go into a library and study just chemistry.

Those were some of the things that went on in the class. One of the things that Craig and I talked about was some of my experiences in dance and how I used them to understand some of the things in the chemistry class. During my time in the chemistry class that semester, I was also involved in choreographing a piece in the "Nights of Dance." It was an improvisation piece involving two people who were meeting and getting to know each other for the first time. I talked with Craig about different approaches that are used in choreography and the way I worked with my dancers.

One thing we do as choreographers is we break things down. We don't assume anything. We have auditions—you have the dancers present something and you look and see what they know. If they don't know something then you have to either reorganize or get a new group of people. There is only so much forcing that you can get out of a body. We take real elementary ideas and make them do it, like walking. I make them walk. And they all start looking like pretty little dancers, and I say, "No, you are dancing. I want you to walk." It freaks them out, but they begin to do it, and then you build from there. Sometimes even the most wonderful dancers do not know how to walk when they are asked to. Knowing the elements of dance is important.

I tried to do a similar thing for myself in the chemistry class. I attempted to simplify. That's basically how I do most things—I simplify. And I now wonder if that is not possible in chemistry. Or, maybe simplify by getting some fundamentals to be able to understand something in a broader sense. Because I thought there were simplified ways of learning how-to, and once I got that under my belt I could absorb more of the lecture. That's why I went out to get the additional book. I think maybe that what was being attempted in the classroom was not real broad but the overall semester was broad. But when he got into things, we had to learn specific concepts that I tried to simplify, and I don't think it was possible. It's like to sum up an idea in chemistry and be tested on it. You might take an idea, but so many more specifics were being asked that I couldn't get that deep. I think I was being too simplistic about it. Maybe I had more of a high school approach to that because I was taking down notes that sort of summed up what was being said, but not quite all the intricacies about what I had to understand about the big thing that was being said. Sometimes I managed the details but sometimes it was so quick that I couldn't complete it. It was like nothing was ever complete for me where I could walk away and say, "OK, I can deal with that. I can answer a question about that." I just felt like my notes were always incomplete because I couldn't keep up.

There are at least two ways for choreographing and teaching dance. Somebody either teaches class or they give class. Giving class is standing up, showing the dancers the movement, and almost in effect turning your back and saying, "OK, you do it." In giving a class the teacher is sort of disconnected. If a choreographer is not really a participant, for dancers it's kind of a mystical thing. They either get it or they don't—for me that was always very disorienting.

Teaching class is breaking the routine down into components, walking around seeing that people are picking up the components. If they're not, you use all your different methods, like touch. Dance, although very variable, very expansive, can have basic building blocks. I feel it's preferable to teach the basic units. Define units so that there's going to be groundwork, and later on you can break those units or break the rules, the laws of some of those units. A teacher who is teaching class or a choreographer who is very integrated with the students is walking around the class and doing more than just demonstrating movement. They are breaking it down and showing you the relationship between things. They participate very directly with individuals. If there is a student who is not able to execute something, then they'll use touch to do that. They call upon pedestrian things. They do anything they can to get the dancers to be able to execute the movements.

For example, when I was working with Holly and Scott, I had to try different things. Because of the time constraints I had with the choreography process I had to quickly deal with what they were not responding to, and find something that would make them respond. Sometimes, I just had to stop and think about it for a moment. Like they were having a really hard time with the playful side. So I asked Scott to think about a fraternity song—you know, how they might have some secret little song that is really comical. I said, "I want you to think of a song." Immediately he burst out laughing. I said, "OK, I want you to sing the song while you are moving or at least think about it." Then, another approach was my calling out real childish things while they were doing the movements, like, "You are at the beach splashing water, throwing sand, and stuff like that." I needed to do the same thing for being sensual. There was a particular movement they were doing, just as a movement, and I wanted it to be a real sensual thing, and I just flat out had to say, "You're in bed right now. How does that feel? How do you respond to that?"

In the chemistry class, I feel that he gave class. He was not involved with the students. Not involved with individuals on any level unless they approached him, and even then it was a limited amount of interaction. I do keep dwelling on the negative side. I've tried to think of what he did that is more toward the teaching approach and I still haven't come up with a whole lot. I don't know if I intentionally just block it out because I really did not like his method at all.

It's clear to me that he was not really listening to anybody who was talking, in the few times that anybody spoke. And it was evident that he really didn't care about individual interests. Maybe a few times, when we talked about car batteries or whatever, I guess he thought that was his effort at making connections, but it surely didn't make much sense to me. Being older—which is why I wrote that letter about life in college at thirty—I was looking at this person, and sitting there thinking, "He doesn't give a damn about any of us. He's not interested. He probably looks at the whole population as eighteen- or nineteen-year-olds, and that if there was some one person enlightening in that class, he doesn't have his eyes open enough to see it." He wasn't interested in expanding his life in any way through his teaching. He was not trying to make connections.

When students did ask him questions to try and break through his wall, he got angry. He took it almost as a form of a personal attack. That was a problem for him. And it was a real problem for the students too. And, okay, students have this responsibility for understanding what learning is, so what about the teacher too? What is his responsibility? Maybe about really looking—maybe jump out of chemistry for a minute and jump into the classroom to see what is going on. And that I felt did not happen at all, ever. Not that I could see.

Besides taking questions as attacks, the instructor did not let us know what was happening with his life. I was thinking about when he went to deliver his paper in Austria or wherever, I don't recall him coming and back and telling us anything about what he had done. I thought, "Well, he left us, and went and did his thing, didn't even come back and share it." He did not reserve any of his enthusiasm or experiences about his trip for us. He was in the class, gone for two weeks, and back in the class without anything for us.

I used a number of different approaches when it came to learning in the class. It was a real tough, quick learning process for me—how to deal with the course. I went out and got an extra book. I talked to whomever I could, which was not a lot of people that are into this kind of thing. The extra book was sort of a how-to, but that did not work for every concept. I went to the professor twice in his office, stayed after to ask questions, and I asked questions before the class. And, basically, shared a lot of frustrations with other people, and through that maybe I got one bit of information that helped to get me to answer a question. I would have pursued a tutor, but I could not afford one.

I did a lot on my own. Unless it's somebody who really has some expertise, I'm too frustrated to study with somebody, struggling as much as I am. So I tried to do what I could on my own. But for at least the first half of the semester, Sue and I sat next to each other in the class—up front where she would tape record the instructor so she could get everything down. She and I would share our frustrations about the class. I remember one time he started out with what we were going to talk about, and I listened and then he said, "And first off we'll first look..." He never stated the topic. I turned to Sue and I wrote down on a piece of paper, "He's going to do a song and dance, to get to 'valence shell electrons,'" or something to that effect, "and he's going to come back to this idea." So, I told her this is what he is going to try to explain. Really, she was clueless as to what he was talking about. Well, sure enough, about five minutes later he got to the concept of valence shell electrons that he said we would talk about.

Because I was having a hard time dealing with the instructor, I tried a number of ways of taking notes. For a while, I thought maybe I just ought to write down every word I could grasp, and then take it home and digest it. That wasn't working. So then I decided to try and sit back a little bit, absorb just the conversation and try to highlight in my head, and then write down quickly what it is I think the overall concept is. And honestly I know that neither method was really getting me through for what the professor demanded on exams. I wrote notes while I was reading, and highlighted. I looked at how I was sitting in class, and I had to think about how he was delivering and how I was receiving. I realized I was trying to receive too much. So I tried sitting back and looking at him going in a circle and waiting till I could make a connection. Other times, I tried to get every word in. I went out and looked to see if I could buy a tape recorder, but could not afford it at the time. I talked with other students, but everybody else was as clueless as I was.

A second approach involved trying to integrate the book and the class. I was trying to write from the book what I thought major points were. But then, when I was getting to class, I realized I was overcompensating and getting things from the book that were really not being addressed. That was what I tried probably in the first two or three weeks, it was just becoming a waste of my time. Reading the book, highlighting the book, and then going back and writing. So, I guess I tried adding to my notes. I think I was rewriting notes from the class at some point, with additional notes from the book. I was trying to take information from the book as well as the lecture to try and understand it. I still was not able to clarify the ideas.

Finally, I went out and bought a second book. The class text did not help because there were no "how-to's" in it. And that was what I was desperately searching for. There were how-to things in the second book. Not for every unit, not for every concept, but it got me through a little bit. There were practice examples and answer keys. And in the answer keys, if I got it wrong, like if it was a multiple-choice type thing, or whatever, there were explanations for everything. That was really helpful for me. Getting the how-to's from the book allowed me to understand some of what he was saying in class.

To study, I set out my class notes in front of me, and the two books that I had. Initially, I just liked to read and highlight for this kind of course. Then, I would go back and take a concept, and grab whatever was in the notes and whatever was in the book, and try to combine that into some form of understanding onto an index card. And that's basically what I did. I tried, initially, to answer the questions that the class text offered, but when it's not multiple choice or matching, they did not provide you an answer key. So, I had no clue if I was right or wrong.

During the tests, I was really nervous—especially the first one. This was totally nerve-wracking for me. I didn't know how to study for this particular unit dealing with chemical reactions. So, I tried to memorize things, like I had for his generalizations about how science operates. But somehow, at the last moment, I was to understand that we weren't going to memorize things. And I think that total nerves just threw me off at that. So when I got to the test I froze. I could not explain what I had memorized. Nor could I give him sort of an idea of what it was about. There were particular units that I had to skip over completely in the book, because I could just not grasp. And that had to do particularly with this unit—"Alpha, Beta, Gamma Particles." Afterwards, we had gone into that a little bit more, and I have some sense of nuclear reactions. But, at the time, it was still very confusing.

Some of the things on the exams we did not see until after the tests. For example, a few of us insisted that we had not seen elements with atomic numbers and masses written in a particular way, nor had we seen the positive and negative charges. But there was a question on the test about it. Interestingly enough, in the lecture after this test, we did that—in the lecture and the chapter we had to read. I brought that to his attention and he said it was introduced, but that he would go back to his lecture notes to double-check that. Of course, I never heard anything on it. So I didn't like that because I really hadn't seen it.

Of course, my problems with the tests were also tied up with the language that he had. I had no idea where one question came from about kinds of nuclear reactions. We had spent a lot of time talking about reactions, and the properties of reactions, and all this stuff. I really spent a lot of time trying to understand it, and when I read the test question I didn't connect that this was saying nuclear reactions. We had not had a lot of discussions about nuclear reactions, so I felt it was totally unfair for him to assume that we would know the difference between chemical and nuclear reactions.

Oops—Panic!

Dear Dale,

Just say "No!" to Chemistry. Talk to you soon.

Disgusted,
Diane

The transition to this phase had built rapidly after the first two exams. What led to this change? One of the most frustrating things about the class was that I didn't know where I was in terms of my grade. He was the only teacher I had so far that did not adequately outline how he graded, and how those grades fit into an overall scheme as to what we could expect at the end. And I really thought I was failing the whole time. But I ended up with a C. But that's where the panic was because I thought, "Goodness, I'm actually going to fail a class."

It culminated in the third exam with the "Tampa" question. He had told us that he was going to put a question on the exam about the power consumption in Tampa because the book had made a mistake. That was the day he also got mad at the class for making noise. Anyway, during the test I just broke down laughing because I had no idea what the question was.

Eventually, I finally said, "Enough!" I kind of let go of the chemistry class because my other classes were suffering. I so much enjoyed them and wanted to get something out of them. I started shutting down. And I kept saying, "Well, if I come out of this at least with a D I'll be happy." I was totally discouraged at that point. I know I have responsibilities. And I know that I have to put in the time, but I also have learned when to say, "No, I'm not doing it." Because when you force something, nothing happens. That's just a lesson I've learned in life.

While I have responsibilities for the class, I'm beginning to wonder about the purpose of such classes. For one thing, the population of people who take such classes is very diverse. Because of this, one of the things the instructor for these courses should do is consider that diversity. Consider as an educator, consider in your mind, what is the level of the course? Go and read the course outline and description from the catalog. I am not saying make it easy. You can't make something easy for somebody in the sense to just get them through it. But really think about why they are there. If it's a non-major course—not even talking about a teacher prep course—we know it's a requirement from the state university system that students have to be there to take a science with a lab. Just don't be too intellectual. Give students a chance to explore. If it's such a varied population, then there's got to be varied interests. And what I may find interesting sitting here, you may not know you are interested in sitting over there. The instructor should play mediator of knowledge and get involved in the interests of the students—or, at least, let the student go with it if it's something the instructor doesn't know much about.

Maybe ask them to hand in something at the beginning of the semester to start the class thinking about something in chemistry that would be interested them. The instructor should read those papers. Take them in—I don't care if it is 200 students in the class—go through them. The instructor should write one sentence about it so you know where he or she is coming from. From those papers, be open to maybe redesigning the course and teaching approaches.

Back to some of my feelings during this phase. My relation with Sue changed during this phase. She still sat up by the overhead, but I sat further over toward the side and one or two rows back. She was not happy with me, and there were two ways that I looked at that. One was that I was resentful that she had the opportunity to go and get some of her questions answered by reading, or however it was she studied. I really felt like I couldn't do any more at the expense of my other classes, and having to work and all that. On the other hand, I thought she was taking it too seriously. And this is where I have to kick in—another part of me has to kick in and say, "I know in another year, you won't be worrying about this class." I don't know if that was just insecurity, or whatever, but I really sat there sometimes and was hysterical. And that's when she got really upset with me.

The tutoring that Craig gave me also helped. That was like a godsend because I knew I wasn't going to be able to afford a tutor. We could go back and forth and talk about it. And I could say, "Is this what that is?" Or he could say, "OK, go ahead and try it," without the pressure of being graded. Just to be able to understand and keep up.

An Epilogue

Dear Dale,

Oh, by the way, about all those heads of lettuce you grew and froze for the winter but ended up losing, well, this is what happened:

You see, H_2O (water) has strong covalent bonds within each molecule (shared electrons—harder to break). Between H_2O molecules exist weaker H bonds (hydrogen), which are really more of an attraction. This weaker bonding causes expansion so when you freeze a food like lettuce, which is more than 90% H_2O, the results are a loose expanding structure that will break through the cell walls of the food. Upon thawing, the broken cell walls cause excessive dripping of H_2O, taking with it texture, flavor, and nutrients, thereby decreasing the quality of the food. So, in essence, by freezing the lettuce, you killed it! I'm enclosing a list of fruits and veggies that are high in H_2O content. If you have any other questions, drop me a line.

Yours, covalently bonded,
Diane

Craig and I thought that an epilogue to my experiences in the class might help to end this story on a more positive note and leave the reader with a sense that the course was useful (but only after it was over). A couple of things come to mind about the positive aspects of the course—things I've learned about chemistry and things I know about teaching.

One of the main things I got out of the class is some of the language of chemistry and some of the how-to's. This was particularly important in my food science and nutrition classes that I took the following semester. For example, from the biochemistry part of the chemistry class, when we talked about different kinds of sugars and polymers I learned how to look for details in structures of molecules. This was really useful when we started talking about preservatives and vitamins, and other sorts of chemicals. I understand that looking at the details of molecules is important.

Another place the chemistry class was useful was in understanding terms like boil, pressure, heat, temperature, and energy. A few weeks into the semester following the chemistry class Craig and I sat in the Union and talked about some of these ideas that had come up in my food science class. We talked about what happens when something boils—molecules gain enough energy to leave the liquid phase and go into the gaseous phase. This helped me to understand how a pressure cooker works. As more and more molecules of water fill the space above the liquid water, it will take more and more energy to boil off additional molecules because the pressure will be increasing—the gas phase pushes onto the liquid. So, as more energy is added to the cooker, the temperature increases until the water begins to boil at a higher temperature. So, boiling water inside a pressure cooker becomes hotter than boiling in an open pot, so food can cook faster.

Although I did not necessarily understand all these ideas in the chemistry class, they started making more sense when I could relate them to things I was already familiar with (like cooking). This helped me to realize that the instructor's role is very important in at least two places: providing students with experiences, and clarifying things by breaking them down into elemental parts.

Experience is probably the thing I want to focus on the most. I am finding it to be a weakness in the way I approach things. For example, there was a problem when I approached my class at State High. I mean, I was sort of delivering. I was either delivering, or I was expecting the children to connect with what I was saying with their own personal experience, all by themselves. I was expecting something that they have no practice in doing. By the end of the week, though, I had critiques done by the teacher and professor, and so I reorganized. And at the end of the week it was—what they were doing was real functional. They were having hands-on experience. I really do believe that that is the key: the experience. Everything I have ever learned in my life has been a total experience.

Craig and I talked about the role of experience in learning. I explained that experience involves doing. For example, having students develop a list of resources pertinent to an interest in their life. Suppose a little girl wants to become a police officer. Where does she go to seek out how she is going to do that? You facilitate by handling and classifying the resources for the students and showing them differences. And then, guiding them or prompting them to some pertinent questions that might need to be considered. And then, maybe, leave them alone for a while to ponder some questions. Their experience is choosing one resource, going to that resource, finding some information, like an address or phone number, whatever, and calling the person and asking them some questions.

What about the experiences in the chemistry class? Craig asked me about experiences in there. There have to be experiences. Except when he asked me, I just did not know what those experiences were. I think that's why I'm having a real hard time answering that. Because I have to think creatively of how I'm going to get my students to do something. You know, what are the activities that I'm going to involve them in? When it comes to chemistry, I don't know. I don't know what the tools are to do that. Without a lab, I don't know. I can only think of it in terms of reading. And that's not considered really an experience, is it? There are other ways of having useful learning experiences. Like panel discussions, and buzz groups sharing ideas.

Getting other opinions. Actively participating in a group where they have to think and form their own opinions. Reading, I guess, is an experience, but not one of the higher-level experiences, I guess that's what it is. But I do know that useful experiences facilitate self-empowerment.

To summarize some of my thoughts about the class, I started the class with a somewhat eager anticipation for making connections to another discipline, but it changed to one in which I just had to let go because I couldn't connect with the instructor. Although I did not enjoy what happened, I learned some useful terms and ideas that became more meaningful in some of my later classes. The class also gave Craig and me a forum to talk about teaching and learning, not only in the context of the chemistry class, but also the classes I taught and will be teaching in the future.

Craig Bowen
Department of Academic Affairs
Towson University, Towson, Maryland, USA

EDITORS

METALOGUE

Strategies for Ethical Research

KT: Reading this chapter was a very emotional experience for me. Craig manages to bring Diane to life in his poignant reconstructions in likely scenarios. The methodology was a real breakthrough when Craig did the study. This was fiction, but based on actual data from intensive research. He worked closely with Diane and also shared the chapter with the teacher in the study. At the defense of Craig's doctoral dissertation, of which this chapter is a part, a committee member said: "I have one question. Who is Diane?" Diane, who was present at the defense, raised her hand in what was an emotional moment for all of those who had read the dissertation. Yet the use of this approach, while very powerful, can raise significant issues about doing research with others. Craig used pseudonyms in his study, and he worked as well with the ethical issues as he could at the time.

PT: The ethics of research on teaching is a serious matter for consideration in any study. Certainly the broader educational community can benefit from examples of both good and bad teaching. But the cost of reporting on bad teaching might be too high if it impacts negatively on those whose teaching practice is being documented in the greater interest. We would not want such research to result in retribution. In this book we can read about innovative research strategies for protecting teachers, including the writing of fictionalized stories and the creation of composite characters that mask the identities of participants (see Metalogue of Chapter 6). Another way around this problem is via *action research* that combines research and professional development. The restorative power of action research involves the teacher as a reflective learner who investigates his own practice and acts on his own self-critical judgments. A social constructivist perspective can provide the teacher with a set of criteria for improving his teaching practice *and* for validating his claims about the impact on student learning of his teaching innovations. Action research works best as a collaborative activity in which teachers serve as critical friends for one another.

Decentering from the Logic of the Discipline

PG: I too found this an emotional chapter to read, partly because it is about teaching chemistry, which I dearly love. I'm afraid that there are many students who feel as Diane did about Chemistry. Being a chemist, I find when I introduce myself at social functions that there is only about one person in ten who has anything positive to say about chemistry. However, it is hard for the chemistry teacher to see it from the student's point of view. He feels that he has the knowledge, and the students need to listen and learn. The problem is that for students to learn, they need to be active participants as they construct (or reconstruct) knowledge from their prior learning.

KT: I still cannot read "Just say 'No!' to Chemistry" without crying! That letter fills me with utter despair. Here was a student who came to class with a strong desire to learn and to make every effort to make sense of chemistry, in and out of class. Even when Diane went to the instructor's office to seek help, the teacher was defensive. Is this because of the culture of science, in which argument over data interpretation is quite common? Is there an initial tendency for scientists to deal with questions aggressively and defensively? This chapter shows as clearly as I have seen it shown that a teacher's way of dealing with students can make a very positive or very negative impact.

PT: Put simply, the logic of the discipline (of chemistry, physics, biology, etc.) is not sufficient as a logic of teaching. Science professors who wish to become successful at both research and teaching need to develop multiple logics and to learn how to employ them selectively and skillfully in their teaching. Social constructivism is a powerful referent for a logic of teaching. It highlights the need for professors to decenter from their own frames of reference and take account of the frames of reference of their students; building communicative relationships with students (individually and as a group) is central to such a pedagogy. This theme, which is central to the book, is developed further in subsequent metalogues.

Enriching Scientific Discourse Via Multiple Texts

KT: I want to take up on Diane's strategy of supporting herself by using Craig, her peers, and several textbooks, to make sense of science. Too often these days I see prospective teachers (in particular) ignoring the value of surrounding a learner with textbooks to get the perspectives of different writers on a given topic. There are even more opportunities to get diverse texts related to given concepts and phenomena with the growth of the Internet and the explosion of the numbers of available books and magazines. As a learning strategy I would rate Diane's approach very highly. It is the essence of social constructivism that learners access multiple texts and endeavor to co-participate in interaction in a process of building a science-like discourse.

PG: I agree with your point, Ken, about the importance of co-participation. Since taking two courses from you in which co-participation and developing a discourse community were critical lynchpins to my learning, I have incorporated more and more co-participation and student discourse into my own classes. It is exciting that Diane knew to co-participate and work with Craig and utilized co-participation to increase her learning of chemistry.

BIBLIOGRAPHY

Bowen, C. W. (1993). But I came here to learn: Students' interpretations of their experiences in a college chemistry class for non-science majors. Doctoral dissertation, Florida State University, *Dissertation Abstracts International 54*, 474A.

Chi, M. T. H., Feltovich, P. J., & Glaser, R. (1981). *Categorization and representation of physics problems by experts and novices.* Learning Research and Development Center, University of Pittsburgh, Pennsylvania, Technical Report No. 4.

Lakoff, G. (1987). *Women, fire, and dangerous things: What categories reveal about the mind.* Chicago: University of Chicago Press.

CHAPTER 3

 **What Does that Word Mean? The Importance of Language in Learning Biology**

Noelle Griffiths

Communication is always the creation of community.
—Lemke, *Talking Science: Language, Learning, and Values*

Language is the knot that ties us together. It is the common bond that allows people to form communities, to work toward the common good of humankind. Indeed, when one thinks of a social system, one very important aspect of that system is the language used to express ideas. One might also say that without the ability to share a common language there would be no civilized social communities.

Habermas (1978) speaks of the evolution of the human species, noting the importance of language and arguing that it is the act of speech that separates human beings from our evolutionary forebears. The very nature of the word *social* implies language and communication.

So what is language? Lemke (1990) refers to language as "a system of resources for making meanings" (p. 14). People use vocabulary and grammatical structures to produce relationships that create meanings for ideas and concepts. The semantics of a language refers to the different relationships that allow concepts to have meaning. Indeed, semantics can be considered as the study of meaning as it is expressed through language (Lemke, 1990).

Teaching relies on effective dialogue between all participating communities in the classroom. Traditionally, the exchange is between teacher and student; however, learning need not be restricted to such a two-dimensional domain. Indeed, it is simplistic to assume that all learning or content is "transmitted" from the teacher to the student. Such a condition might exist in a classroom with only one student and one teacher. However, our nation's classrooms typically have more than twenty students per teacher. In such a highly social setting, understanding the use of language and how it facilitates learning becomes critical.

In science, there are specific vocabulary words and specific phrases that have the desired meaning only when put into a scientific context. The task of learning science then becomes tightly linked to learning the language of science. Just as any language cannot be learned by simply knowing the terms and phrases, understanding science involves more than a surface-level familiarity with science vocabulary. However, this task of learning to speak the language of science is quite formidable, considering the emphasis often placed on vocabulary in science classes. Yager (1991) argues that there are more new vocabulary words introduced in a science class than in an introductory-level foreign language class. In such an environment, does learning science mean that students are able to apply the concepts from class to their daily lives and think critically about scientific phenomena? Or, does learning science merely become a drill in terminology memorization? A close analysis of how teachers and students talk in classrooms is useful in coming to understand how scientific terminology is used, and how students and teachers make sense of the language of science.

First, it is helpful to determine the nature of the dynamics of a science classroom. Roth (1995) discusses different communities that exist in social systems and in science classrooms. The flow of information is a critical factor in understanding how these communities are able to interact.

Science classrooms can be viewed as being constitutive of a variety of communities. One community is populated by students, another by teachers, another by scientists, and so on. Each community has distinct goals and beliefs about science and learning, and these goals and beliefs shape what happens in the classroom. When talking about different communities, it is not only useful to speak of those who populate them, but it also becomes convenient to speak of borders to the communities. Most systems, regardless of what distinguishes them, have some parameters that act as borders. These borders, according to Giroux (1992), are not impenetrable, but (to use a scientific term) are selectively permeable. It is possible for members of communities to step outside their "community of practice" and participate in other communities so long as the possibility for shared values and beliefs exists. These are called "border crossings," as ideas and beliefs are negotiated across gaps between different communities. The extent to which these crossings are successful depends on whether or not a shared language can be found and used. Lemke (1990) contends that when people talk science, they "create, or re-create, a community of people who share certain beliefs and values" (p.). For some groups, crossing borders is sometimes too risky or too difficult; and this can set up a barrier between two or more communities. What happens when there is a barrier between two communities in the same classroom?

What does it mean to share a language? I mentioned that border crossings are associated with sharing a language or sharing common goals. Each community has its own language, a "native register," so to speak. Senge (1990) describes a register as an outline of the specific rules and nuances of a language. To share a language means that people can communicate and build understandings of phenomena, even though they might be members of different communities. According to Senge, it is important for people to keep their native language register, as it allows for diversity within a community (and a collection of communities) and allows the mood to be open to new ideas.

The purpose of this chapter is to examine the discourse in a general biology class offered to prospective elementary school teachers, and to explore how the social interactions between the constitutive communities of teacher and students influenced learning.

Reforming Elementary Science Teacher Education

Science education in the United States has a history marked with reform efforts and is a field saturated with literature on what ought to be done, what ought not to be done, and why. The educational system is often the target of much criticism when national reports show that our nation is lagging behind other countries in the areas of industry, commerce, and science. Reports such as *A Nation at Risk*, published in 1983 by the National Commission on Excellence in Education, strongly advocated reform in education at all levels, particularly in the areas of science and technology. The National Research Council's Board of Biology published a report in 1990 that discussed numerous issues concerning science education and the teaching of biology at the K-12 level. One of the claims made by the council was that teachers of science in elementary schools need to become conceptually stronger in the sciences and to dispel their "anxieties about science." Two important contributing factors for this claim are that "most elementary school science teachers have hidden behind textbook-centered lessons that stress vocabulary and memorization of facts," and that in college only a minimal amount of science instruction is taken by elementary education majors (National Research Council [U.S.], 1990, p. 16).

Elementary school science education has been a critical area of reform efforts for several reasons. Bennett (1986) calls for a "revolution in elementary school science," suggesting that students at this level are naturally curious about the world around them, yet this curiosity is not utilized. This fact is reflected in the students at middle and high school levels who are bored with science, take fewer science electives, and have little interest in how science affects their lives. Also, elementary school teachers take very few science courses during their college career, thus opportunities are limited for them to develop a strong background in the sciences.

In recent years, universities and colleges, with support from the National Science Foundation (NSF), have designed new science courses for prospective elementary school teachers. This is often done with collaboration between faculty from education and from the sciences.

A Study of Language Use

In this chapter, I draw on the results of a study that I conducted recently of a biological science course offered at Grand State University, located in Grand City.

During the time of this study I was extremely interested in language acquisition and investigating the uses of language in classrooms, specifically science classrooms. A graduate student in the Science Education Program, I became involved with an ongoing study of a series of courses designed for elementary education majors. There were other research projects being conducted on various aspects of an earth science course and a physical science course, and there was the potential for additional projects to be done involving the biology course.

The course was taught by several university professors, and this study examines the teaching practice of one of them—JoEllen Watkins (a pseudonym). The class size was kept small (less than twenty-five students) in order to promote interactions among participants. All the students listed elementary education as their major or intended major. The driving force behind the planning, implementation, and evaluation of this innovative course developed in direct response to calls for reform in science education. The course was developed to immerse prospective elementary school teachers in science, in effect to strengthen both their content knowledge of biology and their confidence to teach science to elementary school students. In light of this goal, the purpose of my study was to examine the teaching practices of one of the professors, focusing on the patterns of language and science terminology that she used in class, and to develop an understanding of the dynamics of the two classroom communities (teacher, student) as they relate to science language and student learning.

I used a variety of data sources in an effort to optimize the validity of my interpretive assertions. The primary data sources consisted of videotape recordings of each class and each field trip, including five tapes of JoEllen's lecture sessions. I transcribed each of the tapes. I also obtained the lecture handouts that JoEllen had given to the class. They were quite helpful in establishing what materials the students had access to, apart from their class notes. In addition, I conducted and transcribed an audiotape interview with JoEllen. Before the interview was held, copies of video transcripts were sent to her so that she might refresh her memory of the course. On completion of a draft of the final report on the study, I conducted an interview with JoEllen in order to obtain her response to my assertions. Subsequently, I incorporated into the final report many of her concerns and comments. All of the transcripts were analyzed and coded according to the patterns that emerged. Insights came in discussions with other researchers doing research on similar courses.

The transcriptions of the videotapes were essential to this study, because a close analysis of the text revealed several language patterns. A prevalent pattern was JoEllen's heavy use of scientific terminology. This was intriguing; and to further study the pattern, I selected all the science terminology from the lectures. For example, for each transcript, a list of scientific vocabulary was generated, accompanied by the context in which the term was imbedded. This exercise provided a great deal of insight into how terms were defined and presented to the students, and it was a good indicator of the number of scientific terms that were used during the lectures. A formidable list was compiled. The list was especially useful to this research, because there was no textbook issued to the students, so any definitions for the terms had to come from their lecture notes. Textbooks were brought to class each day and taken out after each class, but the students had little class time to review the texts.

A Constructivist Perspective

Constructivism, a way of thinking about teaching and learning, was used as a referent for interpreting the data. Tobin (1993) explores the tenets of constructivist theory. The basis of this theory is that learning is something that people construct for themselves, using their own previous experiences as building blocks. Knowledge can exist only in the minds of the knower (Tobin & Tippins, 1993). Learners have unique experiences, which shape their thinking and knowledge. Learners build knowledge in a process that is influenced by experience, previous knowledge, values, beliefs, and sociocultural factors associated with life in the communities that individuals populate. Although knowledge cannot be separated from the knower, there are some commonly held values and beliefs, and these common ideals help to define communities of practice. So, in analyzing the data generated for this study, it was helpful to use a constructivist framework when thinking about the language that was used and the learning that occurred.

Throughout the research process, I used a *fourth-generation evaluation* interpretive approach (Guba & Lincoln, 1989). Fourth-generation evaluation is a methodology that allows the researcher to negotiate meanings from the data. Assertions and claims that are made by the researcher are evaluated utilizing a constructivist "lens." This type of interpretive research is characterized by member checks, peer debriefing, and attention to all stakeholders. Member checks occur when a researcher shares her data and assertions with the stakeholders for the purpose of determining their credibility. The test of credibility does not focus on finding the real "truth," but instead focuses on "establishing the match between the constructed realities of the respondents (or stakeholders) and those realities as represented by the evaluator and attributed to various stakeholders" (Guba & Lincoln, 1989, p. 237). In the study, there were several stakeholders, including JoEllen, the students, future elementary education majors, teachers, science faculty, and other science education researchers also involved in their own research about this general biology course and the other two general science courses (physical and earth science). Peer debriefing is a practical method of validation whereby assertions are discussed with colleagues and developed further. As a result of peer debriefing and member checks, there was an emergent design to the study in which assertions and data were continuously reconstructed and redeveloped to create an accurate account of what occurred.

Keeping Discourse Communities Apart

JoEllen gave five lectures during the course, including an overview of evolution and plant and animal life cycles. She focused the bulk of the material on plants and plant life cycles. The main topic of her lesson plans was "alternation of generations," and how different plants and animals incorporate alternation of generations to some degree into their life cycles. With respect to plant evolution, there is a distinct trend in the reliance on alternation of generations, and JoEllen felt that this phenomenon was important for the students to understand. The discourse that JoEllen used, however, was not easy for students to understand. She went through the life cycle of sea lettuce, mosses, ferns, pines, and the flowering plants, all of which incorporate an alternation of generations during the life cycle.

There was a prolific use of terminology, terms that students had not studied before, which created tension and confusion for them. Their confusion was evident in the questions they asked and in their responses to JoEllen's questions. In identifying science terms, I made use of my own background and experiences in biology, of the way JoEllen verbally emphasized the terms while she spoke, and the terms in course handouts that were printed in bold typeface.

JoEllen's lectures were rich with vocabulary, but the motivation behind the use of the terminology was frequently not entirely clear. Often, there were times when I was unable to follow how the terminology fitted into the lecture. Some of the terms students were hearing for the first time. As with any novel item, understanding comes slowly and with practice. JoEllen presented some of the terms with the assumption that the students had heard them before (either in previous lectures from different professors in the course, or from high school classes). For most of the students in the class, it had been several years since they had taken biology or any type of science class. These assertions are illustrated in the following discourse analysis of vignettes of JoEllen's lectures, which highlights students' difficulties in understanding the concept of "alternation of generations."

A Rich Source of Confusion

In the following excerpt of a video transcription, JoEllen has just begun the section on plant life cycles. To accompany this lecture she is using a large poster with the sea lettuce life cycle illustrated in color.

> But what happens with this organism is, you see the key components, fertilization and meiosis here, okay? Let's begin with fertilization. A zygote is produced. Just like with us, it grows into a multicellular diploid organism. But the difference is that when meiosis occurs, these little spores are produced, and there are two different kinds of spores, each of which grows into a full organism but is haploid. So it's as if your egg and sperm grew into a whole individual, and then produced the sex cells that united to form the next generation. They grow up into a whole individual, and that individual is haploid, that individual produces the gametes. So it's life you produce, the spores that germinate and grow up like your sperm, like a man's sperm, grows up into an individual that produce the sperm that actually produce the zygote, and this is called alternation of generations. Because we go from a haploid to a diploid organism to a haploid organism. These organisms produce the gametes, which in this case are plus and minus, you can think of them as egg and sperm, they're identical, you can't tell them apart.

In this segment, JoEllen explained the life cycle of the sea lettuce and, judging by the looks on their faces, many of the students were confused. However, it is not unusual for students to be confused the first time they encounter novel concepts. This was the first time that the majority of the students had heard about alternation of generations and, immediately after this explanation, JoEllen asked the students to join with a partner and explain the life cycle of sea lettuce and how alternation of generations applied to the sea lettuce life cycle. As the term suggests, a cycle consists of stages that occur in a certain order and repeat over and over again. Alternation of generations refers to the existence of two distinct generations (one haploid, the other diploid) that alternate back and forth, each one comprising a different stage of the life cycle. However, the preliminary explanation from JoEllen was confusing in that it did not resemble a cycle, nor did it show that the process keeps repeating. Also, she did not give a clear definition of the concept of alternation of generations. Before she explained it further, the students were asked to explain it to each other and then to the entire class. Their confusion was evident in the way they described what alternation of generations means to them:

JoEllen: Who would like to tell me now? How about you [gesturing to a student]? Tell me what's the difference between the life cycle of a human being and this seaweed.

Student: Okay, the seaweed goes through the spore stage then... [pause]...produces another organism, and then goes through the breeding stage and that produces ...[interrupted by JoEllen].

First, rather than ask for a basic description of alternation of generations, JoEllen asked instead for a comparison between sea lettuce and human life cycles. This added another level to what the students were expecting that they had to explain. The difference between the two life cycles that JoEllen expected to hear was that the sea lettuce life cycle has evident alternation of generations, while a human life cycle does not. The student used her own words, "spore stage" and "breeding stage," to explain her understanding, and did not use the words that JoEllen had used in the lecture. The student showed that she was willing to struggle through an explanation with JoEllen's feedback; however, this feedback did not really come. Without exploring or building on what the student meant by "spore stage" or "breeding stage," the student's discourse was cut short, and another student was asked to describe alternation of generations:

JoEllen: Seems strange, doesn't it? It really seems funny. Does anyone else, how about you?

Student: Uh...meiosis first takes place in the zygote and produces the diploid cell, which then the meiosis comes and there's two spores, which makes two individual organisms with half the number of chromosomes. Then they make the gametes, which make another organism.

Here, the student tried to appropriate the words that JoEllen had used. The student was trying to cross the border from the students' community to JoEllen's community, but something got lost in the translation. First she said that meiosis takes place in the zygote, when in fact meiosis occurs after the zygote has grown into a full individual (after the individual has grown into a mature adult). Further, meiosis is a form of cell division that results in a reduction of chromosomes by half the original number. Thus meiosis results in the formation of haploid cells from diploid cells. Here, the student states that meiosis produces a diploid cell, which is not correct. Understanding how meiosis reduces chromosome number is critical to making sense of life cycles because, essentially, it is the chromosome number that is going through a cycle. The chromosomes, in turn, direct the growth and expression of the actual physical individual. Also, an understanding of meiosis is crucial to understanding reproduction and alternation of generations. The rest of what this student states about cells with half the number of chromosomes producing the gametes that form another organism is accurate.

Again, without exploring the student's response and giving feedback, JoEllen follows up with a set of questions addressed to the class. The questions she asks are sequenced in such a way as to lead the students to an anticipated correct answer, a scientific term, which happens to be mitosis:

JoEllen: So how do they get from the zygote to this? How do you
              go from the zygote to the multicellular organism? What
              process produces that?
Students: Cell division.
JoEllen: Cell division...which is...[searching for another key
              term]...which is...[not getting it from the
              students]...mitosis, right?

Rather than build on the students' previous response about meiosis and the life cycle of sea lettuce, JoEllen asks a series of questions about a process that is closely linked with meiosis, but is quite different. Mitosis is another form of cell division that occurs not in the sex cells, where meiosis takes place, but in all other body cells. It is evident from this dialogue that JoEllen was hoping to hear mitosis. When the class responded with "cell division," she then transformed this into mitosis. A student taking notes might have written "cell division = mitosis." This is not entirely false, as mitosis is a type of cell division. However, meiosis is also a type of cell division, but the two are different processes and occur in different cell types. Though the terms are typically included in the same textbook chapter or the same unit on reproduction, they are not synonyms. Considering that the students have just begun going over meiosis and life cycles, to have mitosis added suddenly to the vocabulary list could have been confusing.

Absence of Student Voice

JoEllen continued to lecture on plant life cycles, going through each life cycle several times and explaining how alternation of generations fits within the various cycles. She explained the sea lettuce life cycle three times, the moss life cycles twice on one day and three times the following day, and the fern life cycle three times one day and once more on the following day. This totals twelve times that a plant life cycle was explained, and consequently twelve times that alternation of generations was explained in detail. After the second explanation of the moss life cycle (which was the tenth explanation of alternation of generations overall), during the second day that JoEllen lectured, the following whole-class dialogue occurred:

JoEllen: Now, any questions?

Student: Can you explain alternation of generations?

JoEllen: Okay, alternation of generations means you have—and we do not have alternation of generations. That's where we started, remember? We talked about most animals do not have alternation of generations. We have one cell that is haploid, and that is the egg cell and the sperm cell. Let's back up one step. We've got to reduce the number of chromosomes in order to have a combination for the next generation, is that clear? We have to do that, why do we have to do that? Yes?

Student: Because…[indicating that she cannot answer JoEllen's question]

JoEllen: You don't know why? What would happen if we did not do it? How many chromosomes would we have in the cell the next generation?

In this exchange, several interesting things are going on. First let us examine the first question that is asked by the student, "Can you explain alternation of generations?" This student, and others who nodded their heads after the question was posed, was unsure of what the phrase meant. This was after almost a dozen detailed explanations. When JoEllen began to answer the question, she started with a half sentence of what would probably have been yet another detailed description, but then she broke off and began to explain the human being life cycle, which does not even have a pronounced alternation of generations component. There was little connection between the question that was asked and the response that was given. Further, the students were still confused and could not answer JoEllen's next question of why there needs to be a reduction of the chromosome number.

Not only did she not respond to the original question, she jumped into the human life cycle, when all along she had focused on plant life cycles. JoEllen continued and proceeded into a lecture explaining why sex evolved, answering her own question of why a reduction in chromosome number is so important. At the end of this mini-lecture, one student asked if all plants have an alternation of generations, to which JoEllen replied, "Plants have it—and I am going to show you how it changes, it becomes less conspicuous." She then went through the moss life cycle one more time, at the request of the students. This explanation continued for some time and, following this short lecture, another student asked, "So not all plants go through that exact cycle?" This is essentially the same question that started off the lecture, when a student asked if all plants have an alternation of generations. This is clear evidence that students did not have a firm grasp on what alternation of generations meant, or how it fitted into plant life cycles. Again, JoEllen commenced an answer with a half sentence, beginning with what might have been an explanation, then jumped onto another related idea, namely the moss life cycle. The postures and gestures of the students showed that each time JoEllen began another long detailed response it was as if their minds turned off. They wanted a brief, concise response. Looking closely at many of the questions, the students seem to have wanted a simple yes or no response.

There was much confusion regarding the meaning of plant life cycles, alternation of generations, fertilization, and meiosis. There are several factors that facilitated this discrepancy between the students' understanding and JoEllen's perception of the students' understanding. Of great importance was the language that JoEllen used to explain the concepts. Technical terminology characterized JoEllen's speech, and the ease with which she used the terminology helped to portray her as an expert in the fields of biology and botany. In JoEllen's discourse community, she is able to speak with colleagues and even upper-level biology majors without encountering any errors in translation. In any community, the members are so comfortable with the language and discourse pattern that sentences and phrases don't necessarily need to be finished for other members to understand. When speaking with other members of a community the discourse can often contain slight mistakes. These do not need to be corrected because it is extremely likely that any others participating in the discourse will mentally correct the error, or know what the speaker means regardless. However, this type of discourse is much more difficult to maintain across borders, or between two different communities. In this class, JoEllen often began sentences without finishing them, or stopped halfway through phrases in order to pursue other thoughts. In a science community, this is acceptable because there is no loss of understanding. However, there was a loss of understanding in the community of elementary education majors. They did not have the background knowledge to complete JoEllen's sentences by themselves, or to follow her from thought to thought. This made JoEllen's role as a teacher very difficult, because the students had limited access to the language that would have told JoEllen that there was confusion, and that the students were having trouble understanding. So, not only was JoEllen unable to mediate the material in a language that the students could assimilate, but also the students were unable to communicate to JoEllen their confusion.

JoEllen used a lot of specific terminology in the lectures, such as diploid, haploid, meiosis, mitosis, gametophyte, sporophyte, haploid generation, diploid generation, gametophyte generation, sporophyte generation, gamete, zygote, and spore, just to name a few. As I mentioned before, her manner of lecturing included half sentences and unfinished phrases, and for a student trying to follow along, it could easily have become incomprehensible. Especially when words with very distinct meanings, like diploid and haploid, or meiosis and mitosis, were used so often in the lectures and in the half sentences, it was difficult to pick out which half of the sentence referred to which term. The end result, related by a student, was that the lectures were confusing, and it was difficult to make sense of the terminology:

Student: I remember being frustrated and confused and dazed, and Dr. Watkins was just going on and on, and I just felt

myself falling behind. If there was a question that I
could ask to make it all better, even if I stopped and said,
"I don't understand," she would say, "Well, what don't
you understand?" That's just her method. She's just so
intense. There's nothing I could do; I was lost. I was
drowning. That's how I felt, and once I drown, I don't
resurface very easily and so off goes the brain and I say,
"I don't want to think anymore. I'll just do the
assignment, it will be put together shabbily, and I really
won't care."

Interviewer: But it's not learning.

Student: No, I can't recall anything that I learned during that unit.
I don't know any more about plant reproduction than I
did five years ago.

Why this student did not learn anything more about plant life
cycles could have been brought about by many different factors. The
student implies that she did not know any question that she could have
asked to clarify the concepts. In her words, "There's nothing I could do, I
was lost. I was drowning." This student felt powerless during JoEllen's
unit, because she did not understand, and she could not find a way to
show JoEllen that she was lost. JoEllen continued the pace of the class,
which for this student was very "intense." Apart from this student's
belief that she had no power or control during JoEllen's unit, another
factor influencing her lack of understanding plant life cycles could be the
manner in which JoEllen used language, specifically the language of
science.

A Distancing Epistemology

A constructivist theoretical framework can interpret many aspects
of teaching and learning depending on what is the focus. To say that the
students and JoEllen shared the same beliefs about teaching and learning
would be presumptuous and an invalid claim. The next issue is whether
JoEllen shared constructivist beliefs about teaching and learning, or if
her teacher methods resembled a transmission model of information from
her to the students. This topic came up during our first interview
together, when JoEllen explained her beliefs about science and learning

science. She uses a metaphor to contrast ideas of constructivism and transmission of fact in relation to learning:

> So I guess it's the whole approach of science as standing on the shoulders of those who went before. You have to understand what came before, before you can add your little block to the building that's being built.

This is consistent with constructivist thought, in which a student takes what has happened previously, and adds that to his or her knowledge base. She comments that learning is a building process that is dependent on what went on before. In contrast, she goes on to show how her belief, consistent with constructivist thought, differs from the current paradigm in science education:

> I see what you're approaching in science education as a different thing. You're attempting to go back and just look at the real basics and say, "Let's just throw out the whole thing and start over. And let's not expect to have to build, that this brick doesn't have to fit here, they can all be lined up on the ground instead of building on top of each other. You can put them in any direction you want, they are not going to fall over if they are only one brick high, you see what I mean?" But if you are trying to build a wall or a tower, what's at the base has got to be solid before you can add more to it. I think of it as the difference between an annual plant and an oak tree. The oak tree gets bigger and bigger, and adds twigs all out here on the end, whereas an annual flower grows up, blooms once, and that's it. Next year it grows up, and never gets any bigger than this, let's say.

Here, JoEllen explains how her theory of learning is like that of an oak tree, as students continually add to the twigs and "stand on the shoulders of those who went before." Yet in reviewing the transcripts of her lectures, and the limited types of interactions that she had with the students, it seems that her lectures were more transmission of fact and knowledge to the students, and there were few checks to see if the students were actually grasping the material. Were these students adding leaves and twigs to their "oak tree of knowledge?" or were the students just trying to absorb ideas, facts, and concepts "transmitted" from JoEllen? In a class where the dominant technique was lecture, it would seem that the latter is the case. The student in the interview stated that she did not learn anything during JoEllen's unit on plants (i.e., relatively little was built). Also, after several explanations of alternation of generations, the students still felt the need to ask JoEllen, "What does that word mean?"

The technical language that JoEllen used created a barrier that prevented members of the student community from crossing the border into JoEllen's world of science. The students were unable to access a language register that could have demonstrated to JoEllen that there was confusion. This inhibited meaningful discourse between the two communities and, as a result, there was little or no shared language.

JoEllen had a discrete set of goals that she wanted to accomplish in the class, and more specifically in the unit that she taught. The main goal for the class was to immerse elementary education majors in doing real science, and to increase their content knowledge and confidence in science (particularly biological science). This goal involved tailoring the class to include a lot of laboratory work, field trips, and alternative assessment strategies, enveloped in a team-taught atmosphere in which the students could feel comfortable in approaching any of the teachers. One of JoEllen's personal goals was to involve the students in scientific discourse regarding plant life cycles and evolution, and to involve the students in her community. JoEllen was very comfortable in her community, but she also wanted very much to cross the border into the community populated by the elementary education students. In fact, she would have had no trouble sharing a language to bridge that gap, but unfortunately the students were unable to vocalize the right signal to JoEllen. These students also had their own discourse; however, it was not characterized by the same traits as JoEllen's discourse. They too wanted to cross the border into JoEllen's community of science.

In order for JoEllen to cross the border, she had to realize that her discourse could not be readily understood by the students. In order for the students to cross the border, they had to show JoEllen what type of discourse they could use and understand. The problem that surfaced during JoEllen's unit on plant life cycles was not because the students did not want, or were unable, to learn the material, nor was it because JoEllen chose material that was too difficult. The problem surfaced because the language that JoEllen and the students used was not sufficiently similar to facilitate a shared discourse.

In Need of a Shared Vision

From this brief glimpse into one teacher's involvement in a general biology course for elementary education majors, what can be said about discourse communities, terminology, power, and learning in science? The pattern that emerged from the data suggests that there are communities that exist within classes, and that those communities, by virtue of each having a boundary, have gaps between them. These gaps can be very wide or close enough that the borders are blurred and discourse is not hindered by language barriers.

JoEllen is an expert in botany, and by virtue of that expertise she had a great deal of power and control. JoEllen therefore had a strong voice in the class, and she had the majority of control over the discourse and material. The students, however, lacked that expertise, and thus were left without a voice. This is not to say that every student in the class had a negative learning experience. There were students who enjoyed the extensive laboratory work, field trips, and material. However, enjoying the class and learning can be two different things. In reference to learning, the evidence suggests that the students did not understand the big topics even after repeated efforts by JoEllen to explain the information.

Although her espoused beliefs about teaching and learning resembled a constructivist viewpoint, her practice in the classroom more closely resembled an effort to transmit knowledge to students. The language used for that transmission was too technical for the students to assimilate. The emphasis on terminology in JoEllen's lectures widened the gap and strengthened the borders between the two communities. This also created an imbalance of power and control in the classroom. It takes a lot of patience and practice to develop a working understanding of something as complex as the language of science. Likewise, it takes just as much patience to deconstruct language so that border crossings are possible.

In a follow-up discussion with JoEllen concerning these findings, she mentioned that she was surprised that the students had so much difficulty. A pretest had been administered to all students prior to beginning the units of study, which had revealed that a majority had taken high school biology. JoEllen and the other teachers assumed that the students had had previous experiences with these terms. She also recognized the importance of increasing prospective elementary school teachers' content knowledge in science, and agreed that this should be a major goal of elementary teacher education.

Indeed, it is difficult to decide how to prepare teachers. Should the focus be on content or on methods? How can methods and content be effectively combined to provide teachers with the best possible preparation? Such questions are and will no doubt continue to be investigated at universities around the world. This particular course has been evolving and improving since its beginning, taking into account feedback from students and researchers and discussions between members of the education and science communities.

To quote Lemke (1990) again, "Communication is always the creation of community" (p. 21). It seems appropriate to mention that, whereas communication can create a community, specific patterns of communication can also distance and isolate communities. It is quite a challenge to find the right balance between retaining a native language register and sharing a common goal between two or more communities, thus enabling discourse to occur. Senge (1990) adds that sharing a language is not enough to achieve goals. There must also be a shared vision, for "you cannot have a learning organization without a shared vision. A shared vision provides the rudder to keep the learning process on course when stress develops" (p. 209). It takes a lot of negotiating and willingness to step outside the cozy boundaries of a community of practice so that a shared vision and language can exist. However, for all the work that goes into border crossings, the rewards are grand.

Noelle Griffiths
College of Education
Florida State University, Tallahassee, Florida, USA

EDITORS

METALOGUE

Teaching as Co-Constructing Linguistic Bridges

KT: Noelle's chapter shows that a well-intentioned teacher making sense of her roles in terms of objectivism can emphasize the transmission of knowledge in the form of facts about science in ways that make it difficult to learn. The issue in this chapter is not so much the intentions of the teacher but the characteristics of the worldview that is brought to the planning and enactment of curricula. I am sometimes appalled that teachers can have such a reductionist perspective on science (as facts to be learned) and of learning (absorption of these facts). This perspective is often associated with a tendency to disconnect what is to be learned from the learner. In such cases science facts are stripped of contextualizations and science becomes barren, to my way of thinking.

PG: Noelle included some quotes from Lemke (1990) that "[c]ommunication is always the creation of community" (p. 21), and language is "a system of resources for making meanings" (p. 14). The problem was that JoEllen, the science teacher in Noelle's research, did not understand that the language she used did not foster communication and meaning making. The students only heard it and were not given the opportunity to use it and make sense of the knowledge. The barriers were too high for students to cross the borders into science. The specific pattern of communication distanced the students from science.

KT: What about a teacher making mistakes in her content delivery? If a teacher is but one source of learning and if there is a spirit of inquiry in which nobody has to know everything, then I can be very forgiving of errors in content. But what is the situation when a teacher such as JoEllen makes errors in her content delivery? Is she to be held to a higher standard than those of us who make sense from a social constructivist perspective? Let me answer that myself. Obviously not. But is there a culture of asking about evidence in this class? When knowledge is passed along as truths to be learned as such, the spirit may be different than when science is seen as a discourse in which perspectives are hotly contested, as is evident in Penny's chapter (Chapter 17).

PG: When a teacher lets the students be part of the discourse of making sense of science, it is hard for the teacher because you have let go of the power that you have all the knowledge. You must allow your students to see you as a learner. It is easier to be the giver of "truths" than to be a co-learner with your students. Of course, your students know you know more than they do, as you have thought about the topic and often experienced it directly in scientific research. When you have a social constructivist approach to learning, your students see you learning and reconstructing on the basis of new information or a question from one of your students. You model that for your students. This is what Mark does in Abdullah Abbas's chapter (Chapter 7).

KT: What is the status of incorrect understandings? From a constructivist perspective something that is not known can be the springboard for learning. My colleague Judy McGonigal has taught me that we should celebrate what we don't know as the beginning of a journey into deeper learning. From many other perspectives incorrect knowledge is something that counts against a student, to be pushed into the background and preferably to be extinguished. Hence from a social constructivist perspective one of the key roles of teachers and students is to ascertain the extent to which knowledge is viable. The role of what is known by students is so important that every lesson needs to be considered from the perspective of making connections to what students know and can do. JoEllen was more concerned with covering content and having students know right facts rather than connecting to what the students already knew. This is very consistent with a deficit model of education rather than a model in which what is known is seen as a form of capital on which future learning is built.

PT: Noelle's chapter links well with Hedy's (Chapter 4) inasmuch as they both highlight the problematic issue of the relationship between power and knowledge. Where teachers are not inclined to share with students control of the learning environment, then the teacher is not allowing herself the opportunity to listen carefully to the (learn) students' native language register. Without this understanding it is virtually impossible for the teacher to build a linguistic bridge between her own scientific discourse community and that of the students. And so the students are left to construct their own linguistic bridges, a hit-or-miss affair that depends on the (often impoverished) linguistic resources that students bring with them to class. Often when teachers are in the early stages of developing a social constructivist perspective on teaching and learning, they concentrate their teaching efforts at the conceptual interface between the ideas and understandings students bring to the classroom and the canonical ways of knowing that serve as their primary teaching goals. But Noelle's chapter argues forcefully for teachers to take account also of the linguistic interface between the two communities. Indeed, as Sfard (1998) argues, a rich social constructivist perspective makes use of two metaphors: *learning as knowledge acquisition* and *learning as participation in a discourse community.*

BIBLIOGRAPHY

Bennett, W. J. (1986). *First lessons: A report on elementary education in America.* Report given by the U.S. Secretary of Education, September 1986. Washington, DC: U.S. Government Printing Office.

Giroux, H. (1992). *Border crossings: Cultural workers and the politics of education.* London: Routledge, Chapman and Hall.

Guba, E. G., & Lincoln, Y. S. (1989). *Fourth generation evaluation.* Newbury Park, CA: Sage Publications.

Habermas, J. (1978). *Knowledge and human interests* (2nd ed.) J. J. Shapiro (Translator). London: Heinemann.

Lemke, J. L. (1990). *Talking science: Language, learning, and values.* Norwood, NJ: Ablex Publishing Corporation.

National Commission on Excellence in Education (1983). *A nation at risk: The imperative for educational reform.* Washington, DC: U.S. Government Printing Office.

National Research Council (U.S.), Committee on High-School Biology Education (1990). *Fulfilling the promise: Biology education in the nation's schools.* Washington, DC: National Academy Press.

Roth, W.-M. (1995). *Authentic school science: Knowing and learning in open-inquiry science laboratories.* Dordrecht, The Netherlands: Kluwer Academic Publishers.

Senge, P. M. (1990). *The fifth discipline: The art and practice of the learning organization.* New York: Doubleday.

Sfard, A. (1998). On two metaphors for learning and the dangers of choosing just one. *Educational Researcher 27*(2), 4–13.

Tobin, K. (1993). Constructivist perspectives on teacher learning. In K. Tobin (Ed.), *The practice of constructivism in science education*. Hillsdale, NJ: Lawrence Erlbaum Associates, Inc., pp. 215–226.

Tobin, K., & Tippins, D. (1993). Constructivism as a referent for teaching and learning. In K. Tobin (Ed.), *The practice of constructivism in science education (pp. 215–226)*. Hillsdale, NJ: Lawrence Erlbaum Associates, Inc.

Yager, R. E. (1991). The constructivist learning model. *Science Teacher 58* (6), 52–57.

CHAPTER 4

 Dynamics of Power in Teaching University Biology: Influence on Student Learning

Hedy Moscovici

I enjoy learning from and with someone who cares, respects, and appreciates my knowledge. If I knew science and I was science literate, I would not have come here in the first place! I came to the university to be involved in my education. ...I do not believe in "getting an education" because everything evolves through negotiation and participation in learning. There is no getting or giving. Education is a process of sharing and learning from and with others. Learning is a partnership.

This was the way Jackie, a prospective elementary school teacher enrolled in an experimental university biology course, described the process of learning, or education. According to her, the process of learning is a partnership. It means that all the participants earn something, in this case learning, in the process. It also means that the partnership involves people who are ready to negotiate during this process based on caring, respecting, and appreciating each other's attributes.

Kari, another prospective elementary school teacher enrolled in the same course described learning in a way similar to Jackie's statement. Kari thought of learning as a journey that involves partnership and participation. During the journey, partners discuss and negotiate their different experiences while recognizing and appreciating each other's areas of expertise.

The purpose of this study was to examine how two students experienced two professors teaching a course designed for prospective elementary school teachers. This study introduces Dr. Mendelson (male) and Dr. Parker (female), the main instructors for the experimental biology course. Dr. Mendelson was selected to teach the course because of his credentials as a biology researcher and teacher. Dr. Parker was chosen because of her experience with teaching biology courses at the non-science major level.

I was a graduate student at the time who studied the process of how faculty implement a new science course designed for prospective elementary school teachers and how students respond to the instructors and the new course. My primary role during this study was that of participant observer (Patton, 1983). While videotaping each class meeting, I wrote field notes, listened and carefully observed various interactions, and followed observations with formal and informal interviews. I developed friendly relationships with participants during the course that contributed to communication and the quality of the study.

Most university courses are imposed partnerships. In this paper, I follow the journey of self (i.e., of the two students and myself) through a biology course as it involved different types of partnerships. I focus on the students' feeling of autonomy to express and follow ideas that make the subject matter more relevant and the learning more personal.

The Biology Experimental Course

This study follows two female target students (Jackie and Kari) enrolled in a biology course for prospective elementary school teachers that was developed and taught as part of a National Science Foundation initiative at a Research II university in the northwestern part of the United States. Maximum enrollment was twenty-four students. Teaching was divided mainly between two instructors (Dr. Mendelson and Dr. Parker), according to their areas of expertise, interest, and availability. Classes were taught in a science laboratory that was adequate for presentations, laboratory activities, and group work. All names are pseudonyms.

Design and Procedures

The class met for a semester, for two hours, three times a week. I videotaped and later transcribed and/or coded more than ninety hours of class time. Formal and informal interviews with participants as well as analysis of artifacts (e.g., class syllabus, overhead notes, students' portfolios, and tests) contributed to the process of data construction.

The study utilized an interpretive research design (Eisner & Peshkin, 1990; Erickson, 1986; Gallagher, 1991; Guba & Lincoln, 1989).

I used the *hermeneutic cycle*, as described by Guba and Lincoln (1989), in order to verify "discoveries" with the different *stakeholders*, or "persons or groups that are put at some risk by the evaluation, that is, persons or groups that hold a stake" (p. 201). Comments were discussed and appropriate changes were made in order to accommodate different interpretations. I also conducted follow-up interviews and interpretive sessions with participants (e.g., reviewing a videoclip, reading part of a video transcript) in order to validate or refute assertions. This procedure contributed to the viability of the study.

Jackie and Kari, the two target students, were selected using SPSS cluster analysis on students' responses to a slightly modified version of the survey (Moscovici, 1994) described by Tobin (1993). The survey explored the dimensions of the learning environment that are instrumental in the constructivist classroom: students' involvement, autonomy, commitment to learn, and relevance of the subject matter. According to the results from the cluster analysis, Jackie and Kari experienced their learning environment in "extremely" different ways.

Historical Dimension of Self

Self must be treated as a construction that, so to speak, proceeds from the outside in as well as from the inside out, from culture to mind as well as from mind to culture.

—Bruner, *Acts of Meaning*

Interactions and reflections are the foundations of *self*. The mind, or the *brain at work* (McCrone, 1991, p. 11), uses the *inner voice* to make sense of experiences while interacting within the context (in physical and personal symbols, or *objects* as labeled by Piaget, 1967). Accordingly, our self changes in response to interactions between the historical self and the context.

The students do not enter the classroom empty, or in a neutral state. In the biology course, the prospective elementary school teacher's self is based on his or her image of science, of being a teacher, and of the undergraduate science courses for non-science majors at the university level. Other factors, such as family and friends, influence the concept of historical self. I concentrate only on the images that are essential components of self in a study of learning to teach elementary science.

At this point I will introduce my initial values about power relations in the classroom based on my experiences as a scientist, a teacher, and teacher educator. Born in Socialist Romania, I was encouraged to follow subject matter that I loved: science and mathematics. There were no gender separations, nor was socioeconomic status an issue (education, including books and materials, was free, and entrance to high school depended on the results of entrance examinations). I grew up in a society in which individuals chose their area of study according to preferences and aptitudes and were encouraged to reach their goals.

During my studies (in Romania and Israel) I found scientists to be just people. Joining the community of scientists while pursuing my bachelor degree in biology and master's in microbiology allowed me to understand the scientist in the research laboratory environment—curious, questioning, experimenting, uncertain, and planning. I had a wonderful time in the research laboratory. My work was described by my oldest son as "my mom was playing with rats," which in many ways was true. I learned because I could follow my own ideas and had professors to advise and guide my research in a collegial and friendly manner. The professors knew more about the subject in general, while I knew more about the specifics of my experiments. Learning was a process of sharing and gaining knowledge.

Historical Self and Science

Science is often presented as a subject that is distant from everyday life (National Research Council, U.S., 1990; Rutherford & Ahlgren, 1990; Tobias & Tomizuka, 1992) and described as a series of unquestionable facts, exposing an objective reality. The gatekeepers of science, the scientists, become a special, elite minority of "intelligent people [who] can really understand it" (Lemke, 1990, p. 139). Using the third person in reporting of the results in scientific papers magnifies the presumed objectivity of science, distancing findings from the initiators' subjectivity such as emotions and other human "nonrational" attributes. The elite group of scientists does not typically include teachers, who are regarded merely as presenters of science to the public (Lemke, 1990). This image of science is not surprising, as the historical dimension of science lies with "witchcraft" and the supernatural (Aicken, 1991), where there was no need or interest to allow "simple" people to understand the logic of science. Their lack of understanding was a tool to further scientists' elitist positions.

Jackie and Kari were no exceptions as they learned during their formative schooling the "science that knows" (Latour, 1987, p. 7). Jackie's memories of her elementary school science did not amplify her understanding of science. Watching the movie *The Red Balloon* caused Jackie to question the properties of gases. She had an inquiring mind. She did not stop asking questions of herself, her colleagues, and her teachers. This affected her experience with science in the high school where she was a teaching assistant in anatomy and physiology. In her new role, she had the opportunity to respond to her students' questions and answers on tests. These experiences enriched her knowledge of and confidence in science. In an experimental university physics course Jackie experienced physics in an environment that strengthened her knowledge and ability to think scientifically. These successes contributed to Jackie's high expectations about the biology course.

Kari's experiences in informal science (i.e., observing and trying to understand what happened to her beans after getting overwatered, or the difference in properties of sand and oatmeal when mixed with water) provided the foundation for developing an inquiring mind. Learning science during her formative years emphasized memorization of facts—such as the length of the bloodstream without understanding the advantages of having such a long bloodstream. Knowing the length of the bloodstream frightened her, and the mysteries of science expanded with the years. Kari did not consider her "informal science" at home as science. What she called science was the subject matter taught in school. When time allowed, even in school, she would follow her curiosity, as when she cut the stems of blue and red colored celery stems in order to follow her questions. Upon entering the science courses at the university, Kari was "apprehensive about the quantity of facts that [she] was expected to know before contemplating a 'scientific problem.'"

Historical Dimension of Teaching

Both Kari and Jackie taught during their formative years. In Kari's case, she taught her sister and the other children in her neighborhood. She not only enjoyed it, but had recognizable success in her sister's case:

> When I "played school" with my friends, I always assumed the role of the teacher. I used my own teachers as role models. I was experimenting with the different teaching styles, without understanding then that I was trying to figure out the kind of teacher I wanted to be in the future. I played the

> role of the teacher with my sister. I taught her how to read and do simple arithmetic. When she entered kindergarten, she told the teacher that she already knew what the teacher planned for kindergarten. Members of the faculty were amazed to find that my sister was not lying; she really knew the kindergarten subject matter very well....This early success as an "uncertified teacher" introduced me to the joys of teaching and the enormous rewards that my chosen profession offered.

Jackie's experience of being a teacher's aide in high school not only enhanced her understanding of science but also led to rich discussions with the cooperating teacher. Discussions included negotiations and understanding meanings and goals beyond class observations. She looked at the in-class activities from the perspective of what Eckensberger and Meacham (1984) called *action theory*, or looking for explanations and meanings beyond behaviorist observations. She was also interested in goals and meanings of actions from the actors' points of view. Both Jackie and Kari entered the biology classroom with a strong conception of self in relation to teaching.

Historical Dimension of Science Courses for Non-Science Majors

It is no secret that undergraduate science courses for non-science majors do not generally produce scientifically literate individuals. This fact is substantiated by the report of Caprio, McIntosh, and Koritz (1989) on the results of a questionnaire distributed by the Society for College Science Teaching in 1988. Eight problems were identified: past negative experiences, motivation, what to teach, abstractions, how to introduce relevance, staying power, inquiry material, and defining objectives. The committee found that it was easier to define problems than find solutions. Solutions took the form of "reasonable avenues toward solving the problem," and the committee prepared a public report with "the larger goal of encouraging and supporting a national dialogue on this topic, to create a route for sharing solutions" (p. 425). The report concludes with two examples of activities to be used in biology courses that may contribute to the solution. One is having the students construct a dichotomous key that allows them to identify specific organisms. The other example involves a field experience with the students studying near-shore tropical marine environments. Both examples focus learning activities away from the lecture hall and actively involve students in their own learning.

The task of restructuring science courses can be even more complex at research universities, where the evaluation and promotion of faculty is related primarily to their research productivity. A research university often overlooks promoting and tenuring faculty on the basis of their teaching. In fact, such faculty often refer to their "teaching load," in large part because teaching requires time away from their research, which does count for promotion and tenure (Merriam, 1988).

Jackie enrolled during her freshman year in an experimental physics course, specially developed and taught for prospective elementary school teachers, that was a part of the same grant as the biology course. In the physics course, Jackie felt appreciated by the instructor, the teaching assistant, and her peers. The subject matter was relevant to her because she was encouraged to ask questions and participate. She defined her role as "the person who asked the questions." Sometimes her questions would encourage the instructor to deviate from the topic being discussed. For example, she asked questions about the physics of music because she knew that the instructor liked music. This is an example of Jackie's power to divert the discussion in class or change the subject.

Kari was enrolled in the experimental biology course during the first semester of her freshman year. While Jackie had already experienced a different learning environment in her physics class, Kari came fresh from high school into the biology class. Her image of undergraduate science classes for non-science majors in a research university was of "large college lecture halls led by stiff uncaring instructors ready to deliver the material and then hurry off to their own research."

Jackie and Kari came into the biology course with different perceptions of self in relation to college science classes. Although both had a record of successful teaching experiences in an informal environment—Kari with her sister and neighborhood children, Jackie as a teaching assistant in high school—their perceptions of science courses at the college level were very different. Jackie had had a positive experience in the experimental physics course, and she knew what to expect by transferring her experience in physics to biology. Feeling comfortable with biology because of her science experience in high school provided Jackie with an even more confident self. Kari was feeling lost in a large university, with no friends and with negative science experiences during her formative schooling. Her strengths were her teaching success and her wish to become an elementary school teacher.

Interactions between Historical Self and Elements in the Biology Class

To better understand interactions in class in terms of historical self, I have clarified key terms and their meanings, as I understand them. Using Yukl's (1989) framework for power in organizations and Foucault's (1979, 1980) use of "power as knowledge dependent," and "knowledge as power dependent," I have analyzed different power relationships. I fully agree with Foucault (1979), who realized that learning takes place in the midst of power relations in all situations and societies:

> Perhaps too, we should abandon a whole tradition that allows us to imagine that knowledge can exist only where we situate the power relations, and that knowledge can develop only outside its injunctions, its demands, and its interests. Perhaps we should abandon the belief that power makes mad and that, by the same token, the renunciation of power is one of the conditions of knowledge. We should admit rather that power produces knowledge and that power and knowledge directly imply one another. (p. 27).

Noddings (1984) described a definite type of personal power, the power of care and the ways in which power influences interpersonal relations. Grundy (1989) and Freire (1990) explored situations in which power is challenged and where individuals fight for *emancipation* (Grundy, 1989) or *liberation* (Freire, 1990). Grundy used the three *human interests*, as examined by Habermas, and analyzed the curriculum in terms of *knowledge constitutive interests*. Freire, on the other hand, concentrated on the *oppressor/oppressed dialectic* in the work force, extending the principle to the educational system. Giroux (1989) and Giroux and Simon (1989) focused more on the role of the teacher in an environment with *emancipatory potential*, in which they differentiate between the role of the *teacher as intellectual* (when the critical questioning in class concentrates around the subject matter), and the *teacher as a transformative intellectual* (when the critical questioning in class involves not only the subject matter, but the social norms as well, with the goals to critique and to change).

Power Relationships

Relationships and interrelations imply positions of power that need to be negotiated between the different participants in a context. Yukl (1989) defined power as

> [A]n agent's potential influence over the attitudes and behavior of one or more designated target persons. The focus of the definition is on influence over people, but control over things will be treated as one source of power. The agent is usually an individual, but occasionally it will be an organizational subunit. (p. 14)

Lemke (1995) went beyond "potential influence" and explored the means by which ideology supports violence in a social relationship, namely potential to inflict pain: "Inflicting pain on others is the pervasive and fundamental mode of social control" (p. 14). In other words, the potential influence translated into the power to "inflict pain" on others.

During the biology course, different power relationships evolved between students and instructors. These relationships depended not only on the people in a specific context, but also on the historical dimension of self for each participant, as related to science, teaching, and university undergraduate science courses for non-science majors.

Yukl (1989) discussed power in organizations. He defined three categories—*position, personal,* and *political*—and related the different types to their sources. For example, position power includes power resulting from formal authority or "legitimate power," control over resources, rewards, punishments, information, and the physical environment (also called *ecological control*). Personal power is the result of expertise, friendship/loyalty, and charisma. Political power is the result of control over decision processes, sometimes involving the development of coalitions. Yukl explored the positive correlation between individuals' involvement in the decision-making processes and their commitment "to carry out the decision" (p. 27).

Different historical selves perceive the actual context differently. Kari was content with the amount of freedom and care she received. She felt empowered and felt that the students "have some degree of influence on the amount and the specific facets of the subject matter, also on the pedagogic methods of this class." Jackie, on the other hand, brought her expectations from the experimental physics course and expected more from the biology course. Here is how she expressed herself in an interview one year after the completion of the biology course:

> The students did not have any input in this course, as it was pretty much laid out for us. Each instructor had his/her agenda set with what and how to proceed during each session. When we were asking questions, often we felt that the instructor did not feel our question relevant, although we were asking because it was relevant to us. Some of the instructors looked at their watches all the time, implying that they are in a hurry, so we should not ask, or say anything...Dr. Parker said that our written comments would influence how the course will be implemented next time. Why did our comments not influence Professor Mendelson's unit that followed unit number two? I think that people [students] should have a say. I think if a class agrees that a teacher goes too fast, or that a subject is too difficult, students should speak up and be taken seriously. I am an education major. I feel like I should have a say. This is my profession. Of course I should say what I feel, just like I will expect my students to feel free to speak up, so I can better meet their goals, and mine: to have more meaningful and relevant learning.

Dynamics of Students' Selves in Lecture

The Lecture on Classification of Life Forms. Dr. Mendelson, the biology teacher, used a long lecture in an attempt to "shed some light" on the topic of classification of organisms. The presentation included an enormous number of scientific terms. Many students were in a passive position, since they were unable to break down the scientific language barrier that denied them access to the information. Although the concept of classification is important in everyday life, it was not presented as such. Many students had a difficult time copying scientific terms from the overhead projector. Dr. Mendelson continued to explain the philosophical issue of the reality of species versus the human mind's constructions for the other classification units. For many students, this lecture did not shed light on the topic of classification of organisms, and it did not enable students to find the important concepts using the scientific terminology. Their answers on the test stopped at the level of binomial system and did not extend into the typological aspect of it. For example, a question on the test was, "Explain the Linnaean system of classification." Jackie's response counted for only three of the ten allocated points:

> Linnaean system: Binomial names. Genus capitalized, while the name of the species is all in lower case, and it is a descriptor. The person who

described the organism used an initial at the end. Example: Flour beetle—
Tribolium confusem L.

The instructor put a red check on "Binomial names," which indicated the correct information in her answer. But her answer was incomplete because she received only partial credit. During the session in which Dr. Mendelson returned the graded examinations, he explained that the results on the tests were raw data, before he adjusted them according to a normal distribution "curve." He also stated that most of the students received a low grade on this question. Some students asked:

Yolanda: What were you looking for on Linnaean system?
Dr. Mendelson: A lot of you failed to remember that Linnaeus not only
                   created a binomial system of classification, but a
                   hierarchical and typological one. Those are the main
                   points that we were looking for. It's a nested hierarchical
                   classification system—boxes within boxes—and unlike
                   modern evolutionary thought, it is typological, which is
                   still reflected in the way that we do taxonomy, which is
                   subscribing name species.
Molly: What do you mean by typological?
Dr. Mendelson: What do I mean by typological?
Molly: Yes, do you mean like types?
                   Dr. Mendelson: Yes, each species is described on the
                   basis of a single individual of type, or prototype.
                   Remember?
Molly: [looking in her notes] Oh, yeah!

The discussion continued with "what were you looking for"– type questions, and Dr. Mendelson elaborated on the specifics for each question. The students provided much less information on the examination than expected. Many students felt "trapped" by the way the questions were phrased; they could not know the depth of the answer expected. As Jackie expressed herself in class: "I feel that we cannot win in this class. I feel like it doesn't matter how hard we study; we have to study harder." Furthermore, in a course without a textbook, although students had an open-book-type examination, the lectures were so diverse in the amount of detail that a single book did not suffice as a primary source. Some students even found contradictory information within different textbooks.

It seemed to me that Dr. Mendelson used his position of power to provide himself with formal authority and control over resources, rewards, punishments, and information. He was the one deciding the expected answers, how to evaluate them, and the grading procedures. Curving the grades also was entirely in his hands. Expertise in science served as undeniable personal power, and control over the decision-making process provided the political power dimension. Dr. Mendelson had power from all three sources discussed by Yukl (1989).

On the other hand, students did not feel they had any power from their position as students. Jackson (1966) states, "School is...a place in which the division between the weak and the powerful is clearly drawn" (p. 10). Jackson also emphasizes that a clear division of power influences perceptions of degree of responsibility, with many teachers feeling fully responsible for the learning environment. Tradition dictates that students have lower status relative to the instructor. In school, this differential begins with age difference, difference in expertise connected to the subject matter, and the salary given to teachers for their positions. There is no material reward for being a student. Usually, students are taking required courses without the freedom to choose the topic, the instructor, the materials, or the assessment strategies to be used to evaluate their learning.

Dr. Mendelson and his students did not develop friendly relations for two reasons (Yukl, 1989). The first relates to the relatively long time required to develop trust and respect for one another's attributes. Dr. Mendelson was only one of several instructors in the course. The other reason relates to the students who tried unsuccessfully to gain political power. Target students, like Kari and Jackie, tried to build political power through use of coalitions, such as helping the instructor develop the topic by supplying him with questions and answers. Dr. Mendelson made it clear, however, that effort did not count in college, only results counted. This comment seemingly ended target students' illusion of gaining political power.

During the lecture on classification, Dr. Mendelson initiated an attempt to use students' extant knowledge with a question on the definition of species. "What is a species?" he asked. Jackie tried to answer the question using everyday language: "It is an individual, it has similar characteristics, but [is] different from others. I don't know!"

Other students did not have the courage to add, or provide another definition. Although Dr. Mendelson said that there was something in her answer, he did not use Jackie's statement or aspects of it in the formal definition that he read to the class from his notes:

> There are, in fact, a number of definitions of species; not all of them agree. There are differences depending on the application to which the concept is put. The sort of classic definition of species was made by a man by the name of Myer. He said that a species consists of groups of potentially or actually interbreeding organisms. So, the capacity to interbreed and exchange genes is thought to be part of species characteristics. Now, that works well for animals, but it doesn't work well with plants and microorganisms. That is because in these groups, organisms that are clearly different species, or different families sometimes, or genera, can interbreed and the species has a different meaning. Now, another way of defining species is by descent. Groups of closely related organisms and that would be the family tree approach to what is a species.

As a result of actions such as this, it appears that Dr. Mendelson did not appreciate students' previous knowledge and did not use it as the foundation for his presentations. His statement above reveals the dilemma regarding the definition of species. The "classic" definition does not work with plants and microorganisms, while the second one does not define the meaning of "closely related." Students represented their misconceptions on the test. The lecture on classification did not involve the students because their previous knowledge was not appreciated, and it refuted the recognition of students' personal power. As already mentioned, Dr. Mendelson had power from all three sources described by Yukl (1989); the two target students did not feel they had any. Dr. Mendelson interpreted students' passivity as "not being interested and committed to learn. They only take it because it is required." Analysis of the surveys shows that most students enrolled in this class were committed to learn biology (3.9 with a standard deviation of 0.5 on a 1 to 5 Likert scale, with 1 being the low end of the scale; reliability factor 0.82). They could have chosen the regular, large biology class, memorized the facts, and played the game of "learning biology" by answering the multiple-choice questions that tested for memorization rather than learning. One of the twenty-four students enrolled in this experimental class decided to leave after only a few class meetings and to fulfill the biology requirement in a less demanding atmosphere in a very large class together with about 1,400 other students. The other students decided to stay.

Low grades challenged students' self-images and made them seek empowerment. Students spoke up during a class session in which Dr. Mendelson returned their examinations for one of his units. They challenged the content as related to Dr. Mendelson's teaching technique. Students knew that the continuation of the course depended on their development of biological knowledge, their feelings, and the extent to which they were meeting their goals. They also knew that what they would, or would not be doing in their classrooms centered on their personal feelings and understandings from this science experience. This session ended with a tacit promise by Dr. Mendelson to consider students' concerns and not let things go that far another time.

In the next session, however, Dr. Mendelson returned to the long lecture format, and the students returned to their passive state. Each blamed the other. Dr. Mendelson was frustrated that the students did not interrupt with questions or participate. The two target students were also frustrated since they had expressed their feelings, and Dr. Mendelson continued with the lectures that "went over their heads." They were writing in their journals, but Dr. Mendelson did not read them. He complained about the time required to read the journals that it did not allow him to do other things. Passivity returned. Dr. Mendelson continued to be involved in teaching "flour beetle" experiments, which were meant to illustrate population growth and natural selection. Students continued to be disturbed by the high level of statistical analysis required in the experiments. Some could not understand the statistical terms *mean, standard deviation,* and *normal curve.* Some were not familiar with letter representations (e.g., N representing total number of individuals). Many became confused by the different statistical tests used for analysis of results. On the test, students were asked to answer ten of the thirteen questions. None of them chose to answer the flour beetle question or the question related to difficulties that ecologists and evolutionary biologists might encounter in their work. When asked about this, Jackie answered that she did not want to answer Dr. Mendelson's questions and experience again the frustration of learning a lot and understanding but getting a low "uncurved" grade that would disappoint and discourage her.

Karp and Yoels (1976) presented students' reasons for passivity in college classrooms. Most of the reasons (except for not reading the assignment) fell into the lack of relevance category (i.e., "the course simply isn't meaningful to me"), the fear that their responses would diminish their image of self (i.e., "the chance that I would appear unintelligent in the eyes of the teacher and/or other students"), that it may affect their grade, or that ideas expressed would not be formulated well enough. The authors also report on the students' image of college instructors as "experts" ready to serve the "truth" through a prepared lecture. The quantity of material in such presentations is very large, and interactions or interruptions are not expected to take more than ten per cent from the lecture time (Macionis, 1993).

Genetics Lecture. Dr. Parker started her last teaching session by thanking the students for their interest in learning biology, their help in providing remarks and writings in their journals (which she always read and responded to), and their general enthusiasm and commitment. She was very appreciative of their suggestions. The students trusted her enough to talk with her about aspects of the course. Dr. Parker said,

> This morning, in fact, we heard several people say, "We're really coming a lot for this class. Thirty minutes early, two hours for class, three times a week." Those are the kind of things that we want to hear. We don't mean for this class to be five times more difficult than the usual offering of this course. So, you are going to help us with the course by putting comments about it in your journals.

Dr. Parker offered to help students prepare for the examination that they would take in the next session. Her room was very close to the laboratory where class usually met. Many students had already visited her, as the resource textbooks were in her office. Being involved in the regular biology course for many years, Dr. Parker was familiar with the students' age group. She also had experience with teaching biology at the community college level. Dr. Parker appreciated the students' previous knowledge, and her caring nature added to her personal power. The students volunteered information or answered questions by stepping into the discussion and taking a risk. After asking about examples of genotypes, Jackie answered, "Brown eyes." Jackie's answer provided the opportunity to develop the idea of phenotype versus genotype. Examples with heterozygotes and homozygotes were used in order to exercise the scientific words as expressed in both scientific and everyday language. Some students also confided in Dr. Parker that they had not taken genetics in high school, so her material was new for them. She encouraged them to ask questions if something was unclear, or come to her room so she could help them. Dr. Parker had position power, and she used it when describing the grading procedure, as a given fact:

> Each unit will be a portion of your grade. You have four units; you will have four grades. We are only in the second unit, and we will turn in a grade for each of you. You pretty much know what your grade is up to now because you've kept all the assignments and the test. Of course, the test counted for a lot more points, but you already know that. Some little assignments were five points, others were three, but the test was 100, and the next one may be over 100—no less. So, you pretty much know what your grade is, as we've already done one test.

Dr. Parker was willing to share responsibility with the students. Hers was the final decision, but she asked the students to put suggestions in their journals.

Dynamics of Students' Selves during Hands-On Experiments

Nails and Screws Cladogram. Dr. Mendelson chose to teach about classification by asking the students to find one example of each kind of nail that he laid on the overhead projector display. While Dr. Mendelson could see the different objects in three dimensions, students could see only their projected two-dimensional images as shadows. Dr. Mendelson did the classification, after spending time at home thinking about the best way to do the task. His position power and his "undeniable" expertise provided him with the "right" to do it. If trust had been established, students could have stated their concerns, but there was no intention on Dr. Mendelson's part to diminish his powers, nor was there any attempt to enhance students' power. He never shared mistakes, negative results, or puzzles. He never asked them to try and solve anything beyond pure technical tasks during class. Students were faced with an expert who knew the right answer. Their role was restricted to matching the expected outcome on the examinations. Lemke (1989) expanded the idea of the "right" to ask and assess students' answers according to one's expectations and the ethical aspects in this process:

> What sort of social power is it that lets some people get away with asking questions to which they already know the answers? Try it with a friend, or your supervisor, and see what happens. Who has the right, or power to test

someone else, to set criteria, or say what is the best answer? The note of
uncertainty in the student's answer can be heard as a small challenge to
the pretense of a game, as a hedging that signals the danger of a failure for
which one may be punished by someone with the power to do so (p. 222).

Lemke argued that this situation is partially the result of the
science classroom that has influenced the "wider social pattern" of
people dealing with "technical and scientific matters" (1989, p. 222).

Constructing the Bony Maroni. The instructors designed this unit out of
concern for the preparation of prospective elementary school teachers.
Dr. Parker had observed the exercise with a different group and brought
a video clip to class that showed the process of constructing a Bony
Maroni from scratch connecting all of the body systems. The name Bony
Maroni came from a popular song several decades ago. Each group was
in charge of a system, and they had to communicate with each other and
with other groups to build the human body. Jackie, Kari, and almost all
the students in class enjoyed the exercise because of its creative and
connective profile. Having the parts of the body represented by different
materials provided the students with opportunities to "guess" what part
each represented. A relaxed atmosphere encouraged participation and
sharing. Only one student expressed frustration with not knowing the
parts that the objects stood for. It may be that her group was too creative
for a more precise mind. Dr. Parker emphasized that there was not a
"right way" to create the Bony Maroni, that what counted was the
process, not the funny-looking product.
       Position power enabled the instructor to have the students perform
the task of creating the Bony Maroni, and power of friendship/loyalty
helped create the meaningfulness of the building process. Students and
instructor shared the power of expertise. Position power in terms of
control over rewards was not shared, and Dr. Parker decided to evaluate
students' participation in this exercise on a satisfactory/unsatisfactory
(S/U) basis. Since all students participated, all received a grade of S, so
that did not contribute to the grade. Some students expressed their
concern that such a valuable learning experience did not receive any
points (Moscovici & Gilmer, 1996). Students were in the position of
making decisions with respect to appropriate materials for the different
body parts. Feeling empowered, and not having to match instructors'
images, enhanced students' interest and participation. Their genuine
enthusiasm was a reward for Dr. Parker. As an experienced teacher, the

instructor learned with the students and facilitated their involvement in the task. The solutions to the students' problems came from them and stood for something that made sense to them. It is interesting to mention that five years later, Kari uses this activity with her third graders. It has been an undeniably successful experience for her students and for herself.

Summary

The subject of dynamics in power relationships between students and instructors is a fascinating field of study. Participants' historical selves enter the game of negotiation of new terms and images of self in the new context. This affects the way they learn and develop as individuals. Traditionally, scientists as the producers and knowers of science have distanced themselves from others by becoming members of a small elitist group. The separation between scientists and society affects society, resulting in fewer students choosing to enter the sciences, even when there is no lack of potential ability (Tobias, 1990).

Dr. Mendelson provided the image of the "expert" research scientist dispensing knowledge to the non-knowers, the students. His intention was to provide students with the most correct and actual results of scientific research (i.e., facts) through an efficient use of time (e.g., the building of the nails and screws cladogram). Using position power and expertise, he disempowered the students, reducing them to silent objects. He transmitted his knowledge in the form of projector notes to overloaded and thus unreceptive students by writing on the overhead and through voice. His scientific language and his content-driven presentations did not allow students to participate and critically challenge extant knowledge in terms of new concepts addressed. He did not appreciate students' previous knowledge. This denied students' development of personal power in terms of expertise. He thought that by lecturing students, they would automatically learn the material. But almost all the subjects that he "taught" remained at the level of "unsolved mysteries" or "puzzles," with students feeling somewhat comfortable at the technical level (e.g., students felt relatively comfortable recognizing the life cycle stages of the flower beetle), but very uncomfortable with the rationale of the "population growth" or "natural selection" experiments. What concerns me even more, as a science educator, is that the prospective teachers/students will avoid such subjects as biology in their own classrooms, as a direct result of their own lack of understanding and lack of comfort. The "passive classroom" is the result of not sharing classroom power with the students.

Dr. Mendelson knew the right answers, and he played a game in which he had nothing to lose. Position power allowed him to play the game, and expertise allowed him to juggle with the right answers. Friendship was not established because the students were in a disempowered position. Students' images of self were very low, and they decided to stop participating instead of being exposed to the risk of providing "wrong" answers and getting even more embarrassed in front of their peers and instructor (Karp & Yoels, 1976). Students' passivity interfered with their learning and their connecting of extant knowledge with new situations. Jackie noticed that the dynamics in Dr. Mendelson's class discouraged participation because of his use of time and his reaction to students' comments:

> I remember Professor Mendelson was constantly looking at his watch during his lecture to see if the pace was right to cover everything he had in his yellow notebook. This was a signal not to interrupt with questions. There was no time for discussion if he were to cover all of his material. We need to have an exchange of knowledge. Only three students had the courage to ask questions. I was one of them. Our remarks got corrected scientifically, or they were ignored.

Sharing power in the classroom provides the instructor with students' interest and participation. Students' sincere comments allow the instructor to be more competent and meet common goals. Power of care is returned as the students feel that they need to work harder and not disappoint the person who recognizes their potential and cares for its fulfillment. The sharing of power enhances students' responsibility for their learning and the sense of partnership among learners.

In the case of Dr. Parker's class, the power is dynamic, with opportunities for participation in the form of answers, questions, and clarifications. Students are encouraged to answer and clarify other students' questions and responses, even when they are only partly correct. Students in Dr. Parker's class feel more empowered, and their self-image is reaffirmed. Students have a place as individuals and are treated as such.

With respect to assessment, Dr. Parker believes in the absolute of grades as representations of what students know. The power to assess others, as part of the position power, is still unshared in the case of Dr. Parker.

It is interesting to point out that the distance between the students and the instructors can be measured by the names used in the interviews. While talking about Dr. Parker, the students often used her first name. This did not happen with Dr. Mendelson, who remained the same in informal interviews, even though I used his first name. Sharing power will lead to partnerships in education, in which the students will share privileges, but also responsibilities with the instructors (Jackson, 1966; Yukl, 1989).

The image of Dr. Mendelson brought back memories. When I began teaching, I had come from the research laboratory after finishing my master's degree in microbiology. I continued working in research for more than four years while I was "teaching" a ninth-grade general microbiology course. For three or four months I taught (and enjoyed) the evolution of the misconception of *spontaneous generation*, followed by its demise from accepted understanding with the development of scientific thinking and experimentation, such as Pasteur's experiments. At the end of one class, some students approached me and told me that they could not understand anything. It was "way over their heads." I was in shock. My first impulse was to say that it was their fault. But it was not. I understood that coming from the research laboratory and knowing and understanding the subject matter did not make me a good teacher. I also understood that students needed to struggle with concepts in order to learn and to challenge previous knowledge with new evidence. They need to become critical and know how to evaluate evidence. I realized that I could not do the learning for them. Since then I have learned to listen to the students. They are the best "thermometer" of my teaching. A high degree of shared care and friendship power, while lowering position power and recognizing various facets of expertise in my students, allowed me to develop as a teacher. I will always be thankful to those students who cared enough for me (and for themselves) to let me know and help me to improve as a teacher. Without their care I could easily have continued to use the power associated with my expertise and my position in order to disempower students.

Hedy Moscovici
Teacher Education Department
California State University–Dominguez Hills, Carson, California, USA

EDITORS

METALOGUE

Empowering Students as Co-Participants in a Scientific Discourse Community

KT: One of the most damaging parts of objectivism as it underpins much of the thinking of university scientists about college science teaching is a tendency to separate the knowing of science from the knower. The biology professors in this study were adamant that what they were to do was to plan a science content course, not a science methods course. For the most part, the students wanted to learn science to make them better science teachers. They were on the lookout for opportunities to connect what they were learning to their goals of becoming better science teachers. Much of the struggle described by Hedy in her chapter involves the efforts of the teachers to search for relevance and the insistence of some of the professors to make it science. The portrayal of science in her chapter, from the perspectives of the professors, is objectivist and at times elitist.

PG: The teacher that Hedy portrays, Dr. Mendelson, is etched in my mind, having watched him on videotapes while he tried to teach evolution in two-hour blocks. He reminded me of others whose teaching I have watched and of some of my own teachers. Dr. Mendelson also reminded me of Peter Taylor's impressions of Dr. Stern (Chapter 1). Like Ken, I believe it is the objectivist framework that molds the beliefs of college and university professors. I was caught in it for many years, until I started to develop a different lens for seeing what was happening in my classroom. Now I see the learners as trying to construct meaning, starting with their present experiences and expanding their understandings as they learn things that connect to prior understandings.

KT: If instructors do not make every effort to connect what they want students to learn to the primary discourses of the students, then there is a strong possibility that the students will not only not learn but they may learn to be helpless with respect to science. That is precisely what we should avoid at all costs. The last thing we would want is for a generation of teachers to have a learned helplessness with respect to science. On the contrary, if possible we would want prospective teachers to connect their language resources to the tasks selected for their science classes, and in the process become more science-like in the way they made sense of what they were learning. My ideal would be to see the emergence of a discourse that was increasingly science-like over the course of a semester of study. In contrast, what Hedy (Chapter 4) describes is an approach in which there was immediately an insistence on learning and using a vocabulary that often was inaccessible to the learners. The approach soon led to overload for many of the students, with the result that they felt disempowered.

PT: Ken, what you are saying about the relationship between language and knowledge is so important that it deserves a little more emphasis. There are two metaphors of mind that I have found very helpful when I think about teaching from a constructivist perspective. The first is the metaphor of *mind as embodied*, that is, the usual idea of the mind being located within the body (not just the brain) of the individual. This metaphor points us toward the educational metaphor of *learning as knowledge acquisition* (Sfard, 1998). Such a metaphor sits quite comfortably with many science teachers, who are concerned primarily with (the commodity of) students' conceptual understanding. They readily embrace (social) constructivism as a referent because of its emphasis on students constructing new concepts on the basis of what they know already. They design innovative teaching strategies to enable students to acquire scientifically correct concepts. But classroom discussion is justified largely as a means of enabling the powerful teacher to diagnose student understanding. Not surprisingly, classroom discussion is dominated by the voice of the question-asking, problem-posing teacher. Hedy's chapter prompts us to consider another metaphor—*mind as distributed in discursive space*—which points us to the metaphor of *learning as participation*. Students are learning to "become" a part of a community, indeed to construct "discursive resources" (such as new vocabulary, articulation of ideas, empathic challenging) in order to co-construct the community with the teacher and fellow students. The latter view of learning emphasizes the mutuality of educative relationships among teacher and students. A more democratic teacher would be inclined to share the locus of control of the classroom discourse in the interests of nurturing an environment in which students had plenty of opportunity to learn "to speak science." Now, to "get real" for a moment, we need to adopt a dialectical perspective on the desirability of college science professors making use, at different times and for different purposes, of both acquisition and participation metaphors of learning.

PG: When I watched the videotapes of the classroom, the students were obviously bored much of the time during extended lectures, even when there were slide presentations of interesting plants and animals. When Dr. Mendelson perfunctorily asked for questions at the end of a presentation, the students knew they did not know the content, but I sensed that they did not feel that they knew enough even to ask a question. They felt the power was not in their hands, and that if they asked a question, they might be ridiculed because the answer to the question was so obvious to Dr. Mendelson. Sometimes when Dr. Mendelson did answer questions, he did it in such a way that he introduced more vocabulary, so the students could still not comprehend the answer.

Empowering Professors as Co-Participants in a Professional Community

KT: The issue of collaboration between faculty in Education and Arts and Sciences also arose in Hedy's chapter. The critical issue for me is respect and trust. My experience has been that many faculty in departments and colleges of science do not respect the learning and thinking of science educators. The lack of respect gets reflected in an unwillingness to listen to what is being offered and to endeavor to learn from them. In the study described by Hedy there was a tendency of the biology professors to argue for the educators to stay away, to trust them to develop an appropriate course and to receive feedback from them on what worked and what did not. They did not want collaboration and did not accept it when it was offered. A challenge we all must address is how to build collaborative relationships on mutual trust and respect. Unlike the chapter by Abdullah Abbas and his colleagues (Chapter 7), the teachers in this study were not co-researchers. This is due to no fault of Hedy's. The teachers in this case were not willing to participate in research on their practices if educators suggested it, and they did not value research of this type as a means of improving teaching and learning.

PT: Perhaps the biology professors were constrained in their collaboration by the prevailing metaphor of *learning as acquisition*, which governed not only their teaching but their own learning. They did not feel inclined to want to acquire knowledge from others outside of their field. If, instead, they were open to the metaphor of their own *learning as co-participation* in building a professional community of teaching and learning, then educative relationships with educators as co-learners and co-researchers might have been possible. This suggestion raises the issue of authentic co-participation, in which both parties commit themselves to collaborative learning. This would rule out the prospect of the educators "sitting in judgment" on the Biologists' pedagogies. But it would not rule out the possibility of critical discourse arising once mutual trust had been well established. Of course, there remains the tricky question of how to initially engage the biology professors in an ethos of co-participation.

PG: There was another biology teacher in this group, Dr. Parker, whom Hedy also described. Dr. Parker did value educational research, and did display caring toward her students, but she was caught within the confines of the culture of objectivism. The power in the course development committee lay with Dr. Mendelson, the geneticist; a person focused on science content. He listened to the others on the committee, but made the critical decisions in the course.

BIBLIOGRAPHY

Aicken, F. (1991). *The nature of science* (2nd ed.). Portsmouth, NH: Heinemann.
Bruner, J. S. (1990). *Acts of meaning*. Cambridge, MA: Harvard University Press.
Caprio, M. W., McIntosh, W., & Koritz, H. (1989). Science education for the nonmajor: The problems. *Journal of College Science Teaching 18*, 424–426.
Eckensberger, L. H., & Meacham, J. A. (1984). The essentials of action theory: A framework for discussion. *Human Development 27*, 166–172.
Eisner, E. W., & Peshkin, A. (Eds.). (1990). *Qualitative inquiry in education: The continuing debate*. New York: Teachers College Press.

Erickson, F. (1986). Qualitative methods in research on teaching. In M. C. Wittrock (Ed.), *Handbook of research on teaching* (3rd ed., pp. 119–161). New York: Macmillan Publishing Company.

Foucault, M. (1979). *Discipline and punish: The birth of the prison.* New York: Vintage Books.

———. (1980). *Power/knowledge: Selected interviews & other writings 1972–1977.* New York: Pantheon Books.

Freire, P. (1990). *Pedagogy of the oppressed.* New York: The Continuum Publishing Co.

Gallagher, J. J. (Ed.). (1991). *Interpretive research in science education.* National Association for Research in Science Teaching (NARST) Monograph Number 4. Kansas State University, Manhattan, KS.

Giroux, H. A. (1989). Schooling as a form of cultural politics: Toward a pedagogy of and for differences. In H. A. Giroux & P. L. McLaren (Eds.), *Critical pedagogy, the state, and cultural struggle* (pp. 125–151). Albany: State University of New York Press.

Giroux, H. A., & Simon, R. (1989). Popular culture and critical pedagogy: Everyday life as a basis for curriculum knowledge. In H. A. Giroux & P. L. McLaren (Eds.), *Critical pedagogy, the state, and cultural struggle* (pp. 236–252). New York: State University of New York Press.

Grundy, S. (1989). *Curriculum: Product or praxis?* Philadelphia: The Falmer Press, Taylor & Francis Inc.

Guba, E. G., & Lincoln, Y. S. (1989). *Fourth generation evaluation.* Newbury Park, CA: Sage Publications, Inc.

Jackson, P. W. (1966). The way teaching is. In C. Hitchcock (Ed.), *The way teaching is: Report of the seminar on teaching* (pp. 7–27). Association for Supervision and Curriculum Development and the National Education Association, Washington, DC: National Education Association.

Karp, D. A., & Yoels, W. C. (1976). The college classroom: Some observations on the meaning of student participation. *Sociology and Social Research 60*(4), 421–439.

Latour, B. (1987). *Science in action: How to follow scientists and engineers through society.* Cambridge, MA: Harvard University Press.

Lemke, J. L. (1989). The language of science teaching. In C. Emihovich (Ed.), *Locating learning: Ethnographic perspectives on classroom research* (pp. 216–239). Norwood, NJ: Ablex Publishing Corporation.

———. (1990). *Talking science: Language, learning, and values.* Norwood, NJ: Ablex Publishing Corporation.

———. (1995). *Textual politics: Discourse and social dynamics.* Bristol, PA: Taylor & Francis Inc.

Macionis, J. J. (1993). *Sociology* (4th ed.). Englewood Cliffs, NJ: Prentice Hall.

McCrone, J. (1991). *The ape that spoke: Language and the evolution of the human mind.* New York: Avon Books.

Merriam, R. W. (1988). A function in trouble: Undergraduate science teaching in research universities. *Journal of College Science Teaching 18*(2), 102–106.

Moscovici, H. (1994). *An interpretive investigation of teaching and learning in a college biology course for prospective elementary and early childhood teachers.* Unpublished doctoral dissertation, The Florida State University, Tallahassee.

Moscovici, H., & Gilmer, P. J. (1996). Testing alternative assessment strategies—The ups and downs for science-teaching faculty, *Journal of College Science Teaching* *25*(5), 319–323.

National Research Council (U.S.). (1990). *Fulfilling the promise: Biology education in the nation's schools.* Washington, DC: National Academy Press.

Noddings, N. (1984). *Caring, a feminine approach to ethics and moral education.* Berkeley: University of California Press.

Patton, M. Q. (1983). *Qualitative evaluation methods.* Beverly Hills, CA: Sage Publications.

Piaget, J. (1967). *Six psychological studies.* New York: Random House.

Rutherford, J. F., & Ahlgren, A. (1990). *Science for all Americans.* New York: Oxford University Press.

Sfard, A. (1998). On two metaphors for learning and the dangers of choosing just one. *Educational Researcher 27*(2), 4–13.

Tobias, S. (1990). *They're not dumb, they're different: Stalking the second tier.* Tucson, AZ: Research Corporation.

Tobias, S., & Tomizuka, C. T. (1992). *Breaking the science barrier: How to explore and understand the sciences.* New York: College Entrance Examination Board.

Tobin, K. (1993). *Qualitative and quantitative approaches to research on learning environments.* Paper presented at the International Conference on Interpretive Research in Science Education, November 25–28, National Normal University of Taiwan, Taiwan.

Yukl, G. A. (1989). *Leadership in organizations* (2nd ed.). Englewood Cliffs, NJ: Prentice Hall.

CHAPTER 5

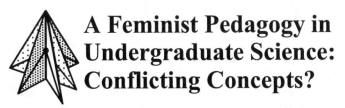

A Feminist Pedagogy in Undergraduate Science: Conflicting Concepts?

Kathryn Scantlebury

Feminist theorizing about the relationship between society, academe, and science has identified the dominant masculine image of science as a major factor in perpetuating gender-related inequities in undergraduate science education. As educators and scientists, if we accept and believe in slogans such as *less is more, science for all,* and *equity and excellence in education* that dominate documents, policy statements, and papers on science education reform, then in order to counteract the dominant masculine discourse of academic science, we must begin the *re-vision* of undergraduate science (National Science Foundation, 1996). By utilizing a feminist perspective and implementing feminist pedagogies in our science teaching, we can begin to address science's masculine image, the interwoven nature of gender and science, and the institutionalization of these concepts within society and academe (Keller, 1987).

I work in a unique environment for a feminist science educator, namely a chemistry and biochemistry department. Being a feminist scholar, I experience the alienation associated with my research, which is further exacerbated by being the only educational researcher in my department (Arpad, 1992). However, because of the high status afforded to science faculty from the university's administration and the K-12 education community, there is power and privilege associated with my position not given to my colleagues in education or women's studies. As a feminist, I am acutely aware of the added responsibility I have from this privilege to improve women's lives and challenge the social, political, and economic structures that reinforce subordination of underprivileged groups.

The Masculine Image of Science Education

Feminist critiques of scientific research have focused on the questions that scientists ask, the interpretation of those questions, the overall social context of science, and the way scientists practice their profession. In contrast, gender studies in science education focus on the sociocultural variables that impact on students' perceptions of undergraduate science teaching (Seymour & Hewitt, 1997), students' achievement (Ginorio, 1995; Tobias, 1990), intervention programs (Matyas & Malcolm, 1991), curricular issues (Barad, 1995; Brickhouse, 1994; Rosser, 1990, 1995), and the environment for women in academe (Hall & Sandler; 1984; Lewis, 1990; Morgan, 1996a; Thomas, 1990). Kelly (1985) separates the masculine image of science into four different components: that is numbers, packaging, classroom climate, and its presentation as a masculine preserve. When one considers who takes and who teaches science at the university level, and how university faculty generally teach science, universities are poor examples of equity in educational settings (Parker & Rennie, 1989; Rose, 1994; Scantlebury, 1994).

Science's strong masculine stereotypes can develop from undergraduate science's both formal and hidden curriculum. Science is portrayed as rational, logical, unemotional, and positivistic. These are characteristics associated stereotypically with the masculine gender, and they conflict with the stereotypic view of femininity (i.e., women are not rational or logical in thought, but subjective and allow emotions to rule their decisions) (Brickhouse, Carter, & Scantlebury, 1990). The "typical" formal curriculum for a science major is approximately fifty credit hours of college level science and mathematics courses. Students learn mostly through coursework and are often taught by professors who make little or no attempt to relate the material to the everyday world and cover material at a fast pace. In the other setting for learning science—teaching laboratories—experiments are often *cookbook* and bear little relationship or relevance to the lectures. By the end of their science experiences, students often have not posed a hypothesis, designed scientific experiments, or completed original research. Seymour and Hewitt describe the high student attrition rates from science, mathematics, and engineering (SME) as "the problem iceberg" (Seymour & Hewitt, 1997). Studies at the college level addressing gender bias have noted that in all subjects, female students receive less class time, instruction, and attention from professors compared with their male colleagues (Hall & Sandler, 1984).

The main reason cited by students who switched from SME to other fields is that the non-SME fields offered better education and greater intrinsic interest. Participants in Tobias's (1990) study of undergraduate science courses made similar observations regarding course content and pedagogy. Students' other reasons include rejecting the lifestyle of a career in science, being "turned off to science," unapproachability of the faculty, and poor teaching (Seymour, 1992). Students defined poor teaching as:

> Lecturing, inability or refusal to explain difficult ideas more than one way; reading from texts as a substitute for explanation; "silent teaching" (writing on the blackboard rather than explaining the material); emphasis in testing on rote learning rather than comprehension; lack of intellectual stimulation, challenge, or encouragement to think critically; failure to teach for knowledge transfer; dullness of material and style of presentation; failure to communicate enthusiasm for the subject; inflexibility or defensiveness in response to questions and discussion points; and unapproachable demeanor. (p. 236)

Male students tend to leave SME because of changing career goals. Women's reasons for leaving SME differ from those of men. Many female SME undergraduates switch because they feel their professors do not care about them. As Seymour notes, this issue is compounded by sociological influences on female students. Many female students in science cite a personal connection—a teacher, parent, or role model—as a reason for their interest and participation in SME (Eccles, 1989; Kahle & Meece, 1994). When women perceive that professors are not taking a personal interest in them, they find it difficult to maintain their enthusiasm and interest in the subject. Professors could begin to change their teaching practices by illustrating concepts with examples to which students can relate. Students may view this effort as an attempt by professors to relate science to the individual.

In addition, women students must confront the very subtle and consistent nature of gender inequities, often rendered invisible to the women, their peers, and the faculty by their very constancy (Spender, 1982). Thomas's (1990) study showed how female students in undergraduate physics saw themselves as the *Other* (Woolf, 1938). These students exist on the margins of the subject because their tutors and lecturers do not know how to interact with them. For many female students, the solution is to reject feminine values and attributes, and to accept the masculine ones so strongly associated with science. This acceptance of *male as norm* is also common outside the scientific community. Our society, and science in particular, is androcentric (Harding, 1991). Gender inequity is the *norm* and anything else is *not normal*. When one first deviates from the norm of gender inequity, the reaction from others, particularly those favored by the inequities, is typically swift and harsh (Sandler & Shoop, 1997). And, unless gender inequity is very blatant, such as in the case of sexual harassment, it is not perceived as a problem in the classroom by either faculty or students. Women in both Thomas's (1990) and Seymour's (1992) studies experienced subtle, and not so subtle, sexual harassment from their peers, and resentment from both faculty and peers of their presence in a traditional male arena.

In *revising* undergraduate science education, we need to consider (a) the patriarchal structure of academe and science; (b) the diversity of feminist perspectives; (c) the role of power and knowledge in defining a feminist pedagogy; and (d) the impediments to enacting a feminist praxis. In the following pages, I introduce the concept of feminist pedagogy, discuss the role of power and knowledge, and discuss the possible detours to achieving a feminist pedagogy that are created within the sciences by faculty, students, and the structures of the institutions.

Feminist Pedagogy and Feminist *Pedagogy*

What is "feminist pedagogy?" Gore (1993) italicizes this term to represent two different interpretations of the concept. *Feminist* pedagogy is a product from scholars within colleges of education and utilizes Western feminist epistemologies such as liberal, cultural, social, radical, postmodern, and post-structuralist in order to examine pedagogy. The foci for these *feminist* pedagogies vary depending upon the feminist discourse, but they all include political, critical, and praxis-oriented components (Weiner, 1994). Praxis defines the interrelationship between thought and action on those ideas, and it is a key component of a feminist epistemology. The implementation of a theoretical feminist perspective is summarized in the phrase "the personal is political." For feminists, different perspectives infer differing political and personal agendas, and these agendas influence praxis.

Liberal Feminism

Liberal feminism is a product of the Enlightenment or the Age of Reason period. A historical example of liberal feminist ideology is Mary Wollstonecraft's *The Vindication for the Rights of Woman* (Poston, 1975). This text is a call for educational equality, and a rebuttal of Rousseau's educational theories for women. Wollstonecraft ridiculed typical female education as superficial and as not preparing women for a productive life in society. Today, liberal feminists continue to argue for women's equal access to education and power as the means to establish societal equality. Pedagogical changes based on a liberal feminist approach include ensuring that women and girls receive the same educational opportunities as male students. However, Eccles (1989) has described this as the *deficit male* syndrome; that is, how can we "fix" women and girls to be more like men and boys so that they may succeed in science? Using a liberal feminist approach to feminist pedagogy, the underlying premises and the status quo of science and education are neither critiqued nor questioned.

Cultural Feminism

In contrast to liberal feminism, the premise of cultural feminism is that "women's political value system can be derived from traditional women's culture and applied to the public realm" (Donovan, 1985, p. 82). Leaders within this ideology believe that society needs women to enter public life to focus policy-making discussions about issues of family, children, and an improved quality of life for all. In *Herland*, Charlotte Perkins Gilman described a utopian feminist society based on matriarchal rule (Gilman, 1912). Historically, the ideas of cultural feminism supported the movement of women into teaching. Women teachers would "naturally" bring out the best in their students because teaching was viewed as an extension of the maternal role (Tyack & Hansot, 1990). However, this same ideology prevents women teachers from moving into managerial positions, such as principalships or superintendencies, because these positions are perceived as not fitting with women's caring and maternal nature. As a pedagogical approach, cultural feminism defines as feminine characteristics such as caring, collaboration, and cooperation. A science curriculum may focus on areas such as midwifery or nutrition, that is, subject areas traditionally ascribed to the feminine.

Social Feminism

Social feminism, developed from Marxist theory, argues that capitalism is the cause for women's oppression (Donovan, 1985). Specifically, social feminists examine the role of domestic labor, sex division of labor in the public sphere, the role of class, and praxis. Social feminists believe that education contributes to women's inequality and yet they do not see education as a vehicle to change the status quo of women's societal roles. Pedagogy based on social feminism may use environmental issues such as chemical waste dumping in Third World countries or silicon implants for women as examples of how capitalism may have a negative impact on women and their families.

Radical Feminism

Radical feminists identify society's patriarchal structure and androcentricity as the major challenges to women's equality. In education, scholars such as Spender (1982) and Delamont (1990) reflect that education is a patriarchal enterprise and, as such, any educational study is implicitly a study of gender issues. In this feminist paradigm, key terms are "patriarchal relations, domination and subordination, oppression and empowerment, women and girl-centeredness" (Weiner, 1994). Sue Rosser's work in the biological sciences exemplifies this approach (Rosser, 1990, 1995). However, Hildebrand (1995) notes that this perspective indicates that it is the curriculum and the teachers that need "fixing," that is, "education and science and the social discourse which create them are still left unchallenged" (p. 4).

Contemporary Feminisms

A major criticism regarding feminism is that, as a movement, it has focused traditionally on white, middle-class women. *Black feminists*, such as hooks (1994) and Collins (1991), articulate black feminist studies, while Rich (1986) broadens the discussion to include *lesbian feminism*. Finally, *postmodern feminism* and *poststructural feminism* have influenced feminists' views and thinking:

> Post-modernism is a historical category (namely, defining a post-modern era) and a systematic or ideal concept (namely a theoretical, analytic framework)...post-structural feminism seeks to analyze in more detail the working of patriarchy in all its manifestations—ideological, institutional,

organizational and subjective...social relations are viewed in terms of plurality and diversity rather than unity and consensus, enabling an articulation of alternative, more effective ways of thinking about or acting on issues of gender. (Weiner, 1994, pp. 62–63)

Along with other feminists, I view these different epistemologies as important for articulating the diversity. We need a variety of viewpoints based on the variety of human experiences (Harding, 1986). Using different feminist perspectives that influence the framing of educational research questions can potentially impact undergraduate science education.

Feminist *pedagogy*, on the other hand, focuses on instructional aspects, such as classroom and assessment practices, and is a product of women's studies programs. Feminist critiques of science and undergraduate science courses have suggested teaching strategies that would make science more gender-inclusive. For example, in laboratory courses, Rosser (1990) suggests that professors could increase and expand the type of empirical observations by using both qualitative and quantitative methods in data collection. Also, they could include personal experiences of students in discussions and pose gender as a facet of research questions. Lecture courses could encourage students to work cooperatively in groups and give essay assignments. And examinations could be changed from only multiple-choice questions to a mixture of multiple-choice, short answer, and problem solving. Barad (1995) *revised* an advanced physics course for undergraduate majors and challenged the notion that "physicists just want to have Phun." Other chapters in Rosser (1995) focus on case studies of curriculum and pedagogical revisions in mathematics and the physical, computer, and environmental sciences.

At this stage, the impact of feminist *pedagogy* is limited to curricular reforms and to changing classroom practices. In order for educators to break the hegemonic masculine discourse of university science courses, we must challenge the notion of science as a masculine preserve that is implicit within the status quo of established cultural practices of science and that sustains a press for the reproduction of gendered stereotypes. However, institutions' cultural practices are typically cyclical in nature, thereby reinforcing the status quo rather than encouraging re-conceptualization through divergent practices (Connell, 1987).

Feminist Pedagogy

What is needed in undergraduate science education reform, therefore, is an articulation of a *feminist pedagogy* that incorporates the three components of feminism, that is, political, critical, and praxis, as well as a blueprint for changing academe's patriarchal structure. It is important for me as a feminist teacher to understand the role of power, classroom dynamics, the impact of my curricular choices (both explicit and hidden), the role of women in my subject area, and the concept of gender.

Gender as a Linking Concept

Gender, in conjunction with ethnicity, class, and sexuality are components of peoples' identities (Kenway, Willis, Blackmore, & Rennie, 1997). Those identities are constantly changing, but gender is a linking concept on the social landscape. Therefore its definition should be articulated. Historically, when the term *gender* was introduced it was initially used as a synonym for "sex," particularly when related to women (Acker, 1992). To study gender, then, was to study women. The early studies about gender did not include men; it was easy to assume men did not have gender. However, Butler (1990) broadened the definition and noted that the concept was ambiguous among feminist scholars. Furthermore, hooks (1984) criticized a large portion of feminist scholarship because it essentially focuses on white, middle-class, heterosexual individuals and institutions, and ignores the issues of race, class, and sexuality. These distinctions of "men" and "women" ignore the variation of gender that exists and inherently tie a discussion of gender back to the different biological functions of male and female humans. The difficulty with "gender" as a category is that the simple interpretation of "masculine" and "feminine" does not deal adequately with the category's variation. The wider category definitions described by Lorber (1994) produce a definition of gender as more of a continuum. Gender as a continuum is social and relational, rather than categorical. In other words, gender exists only as a comparative quality (if someone is "less masculine" than another, he or she is also "more feminine" than that same individual, even if their biological sex is the same).

On the social landscape of academe, there are numerous gendered links to consider. For example, the links between faculty and students, administrators and faculty, students and administrators—all of these relations are gendered relations. The people involved in these relations are also influenced by their gendered identities and interpretations of gender roles. Consideration should be given to other links, such as those between curricular choices and the dominant discourse of the disciplines, between hierarchical structure and gendered nature of the university as an organization, and between teaching and research.

Power and Feminist Pedagogy

Another unifying concept in discussions of gender and science education is that of *power*. The issue of power is central to feminist practice (Morgan, 1996b). A feminist teacher struggles with the concepts of power and the usage of power in her/his class (Gabriel & Smithson, 1990). However, there is a power relationship between teacher and students that cannot be relinquished (Bayim, 1990). Why are feminists concerned about power? In Morgan's (1996b) words:

> Power in education is operationalized by creating, defining, encoding, transmitting and evaluating claims to knowledge....Encoding power includes control over the educational curriculum and control over the educational discourse which determines the actual language of the class-room and what are going to count as educationally relevant contributions to the conversation of education. Power over transmissions entails control over the conferring of pedagogical legitimacy on some forms of cognition, patterns of thought, and sets of values, and ignoring and devaluing others. Finally, mastery of the curriculum and demonstrated competence in the admissible forms of articulation are inextricably tied to control over the definitions of achievement. But, plainly, the people who encode, transmit, and define what will count as knowledge, the people who establish the rules of educational play in the classroom and the people who define and reward the winners, have the educational power. (p. 108)

Academics have long argued and struggled with defining power. Lipman-Blumen (1994) summarized the discussion by stating, "[M]ost definitions of power emphasize the ability to make others conform to one's wishes, often leaving relatively ambiguous the exact origins and nature of that capacity" (p. 109). In accepting this conceptualization of power, one could subordinate others and silence their voices. However, the concept of power as a negative entity is in direct conflict with feminist theory and praxis. Michel Foucault provides a different perspective of power (Gordon, 1980), as something that circulates, or rather something which only functions in the form of a chain....Power is employed and exercised through a net-like organization. And not only do individuals circulate between its threads; they are always in the positions of simultaneously undergoing and exercising this power...individuals are the vehicle of power, not its points of application. (p. 98)

From this perspective, power is conceived as a creative energy rather than a device for dominating others. Feminists can use their educational power to enact a feminist pedagogy. Yet, for academic women, there remains a paradox. Morgan (1996b) uses the metaphor of a *bearded mother* to describe the paradox that feminist academicians encounter, while claiming "the forms of rationality, the modes of cognition and the critical lucidity that has been seen to be the monopoly of bearded men with fully developed rational souls" (p. 125). Feminist teachers are placed in a contradiction because they are also expected to commit to an ethic of care, which is considered a motherly, and thus female, monopoly.

Feminist Pedagogy in Undergraduate Science

In her review of critical and feminist pedagogies, Gore (1993) notes that "feminist pedagogy discourse does not construct separate pedagogies according to contexts" (p. 19). For a feminist pedagogy to provide the impetus for changing undergraduate science, we do need to consider the context. Consequently, in keeping with Jane Roland Martin's (1994) concept of *gender-sensitivity*, of taking gender into account when it matters in an educational setting, we acknowledge that a hegemonic masculine discourse dominates undergraduate science. Therefore, one should consider gender. The question arises: What would be the characteristics of a context-specific *feminist pedagogy*?

From the science education community, Brickhouse (1998) outlines the impact on science from radical, liberal, socialist, and postmodern feminist perspectives. She discusses the dilemma that science educators and scientists face in teaching students about the nature of science because of the conflicting views on the issue from scientists, philosophers, and science educators. However, Brickhouse does not enter into a discussion regarding the impact of these feminisms on pedagogy or research in science education. Gore (1993) suggests that key aspects of feminist pedagogy are an emphasis on cooperative learning, empowering the student to gain knowledge from her/his experience, and *revising* courses and programs. In teaching undergraduate science, both changes in pedagogy and a revised curriculum that considers the culture and nature of science, especially with respect to its patriarchal structure and dominant discourse, are needed to reflect reforms suggested from feminist philosophers and science educators.

Extending Feminist Praxis

If there is any misleading concept, it is that of coeducation: that because women and men are sitting in the same classrooms, hearing the same lecture, reading the same books, performing the same laboratory experiments, they are receiving an equal education. They are not, first because the content of education itself validates men even as it invalidates women. Its very message is that men have been the shapers and thinkers of the world, and that this is only natural.
—Rich, *On Lies, Secrets, and Silence: Selected Prose*

The implementation of a feminist pedagogy is a political act. A first stage in designing a science course from a feminist perspective is to consider the various feminist theories, because the different theories have different implications for praxis. For example, a course developed around liberal feminism would not include a discussion of the patriarchal nature of science, while this issue is a key tenet for radical feminists. The course content, pedagogical practices, and assessment procedure should be reviewed through the perspective of a particular feminist theory.

Faculty, students, science departments, parents, and university administrators are stakeholders when faculty choose a feminist ideology and praxis. To teach from a feminist perspective is not without risk—both to the teacher and to the students. In this society, women and men do not have equal status. Students interpret gender roles and often adhere to those roles that are considered appropriate to their gender. In her descriptions of a Social Foundations course, Lewis (1990) documents the challenges that arise when implementing feminist praxis. During a discussion of male privilege and dominance, Lewis describes how her male students consciously (or unconsciously) revert to male privilege, while female students become concerned about isolating and ostracizing their male peers. During the course, Lewis noted that some women underwent transformations within their political stances that had an impact on their lives. A few of the men acknowledged that they had begun to *revise* their views. In discussing the classroom dynamics, Lewis begins to expose and discuss the impact of the hidden curriculum. However, as Martin (1994) warns:

> Consciousness raising is (not) any guarantee that a person will not succumb to a hidden curriculum. But still, one is in a better position to resist if one knows what is going on. Resistance to what one does not know is difficult, if not impossible. (p. 167)

Faculty who are gender-sensitive would be knowledgeable about the different learning experiences, both in and out of college, for women and men. They would appreciate that it is likely that by the time their students have reached college, women have received less of their teachers' attention than have men, and that the nature of that attention has been different. Seymour and Hewitt (1997) outline how women are often motivated extrinsically, succeeding at science to please faculty and/or parents, rather than themselves, as males are more likely to do. Gender-sensitive science faculty would be aware of these different patterns for female and male students. Aware of their competencies and not willing to overestimate their abilities, women students are more likely to seek faculty attention to continually check their work and receive encouragement regarding their aptitude for science. Women tend to answer when they are confident that they are correct. In contrast, their male peers, practicing risk-taking behaviors, will more often call out answers to questions and be confident that their answers are correct (Committee on Undergraduate Science Education, 1997).

In laboratory settings, female students, for various reasons, may assume the "domestic tasks" (e.g., collecting equipment, taking notes, and cleaning up). They may not have had the same opportunities as their male peers to use science equipment because, as the latter received more time, resources, and instruction, the female students probably sat quietly, completed their assignments, and waited their turn. In many situations, their turn never came. Therefore, as with anyone attempting a new task, female students may take longer to complete a laboratory activity or may need to ask the instructor for more help. Rosser (1990) also suggests that laboratory work and reports could include both quantitative and qualitative data and could take the form of both written and oral reports. She suggests encouraging students to include personal experiences in class discussions and to use gender as a variable in their research questions. These practices broaden students' conceptions of what is meant by "data" and give recognition to the radical feminist perspective that the *personal is political.*

A Cautionary Note

Feminist work values women's experience and thoughts and an ethic of care. However, it would be irresponsible not to warn female science faculty of the potential dangers, from their students, peers, and administrators, of enacting a feminist pedagogy. While recognizing students' differences, especially those linked to gender-related experiences, faculty must also continue to expect students to be responsible for their own learning. Although utilizing the metaphor of *bearded mothers* to describe the paradox, Morgan (1996b) is unable to offer us a solution to this problem. In an attempt to bring this issue to the foreground in my own teaching, I include several paragraphs from various essays by Adrienne Rich (1979) in the syllabus for my courses. When I introduce the syllabus, I discuss with the students my reasons for using these quotes.

Students' Rights

The following quote is taken from a convocation address Rich delivered to Douglass College. When discussing the quote in class, I remind my students that, at that time, Douglass College was an all-female college. However, similar to the experience that Lewis discusses, often male students are uncomfortable in reading the text, and female students take on a *caregiver* role toward them. When this dynamic occurs, it provides me with an opportunity to discuss many of the pedagogical issues I have mentioned in the preceding paragraphs.

> The first thing that I want to say to you who are students, is that you cannot afford to think of being here to receive an education; you will do much better to think of yourselves as being here to claim one. One of the dictionary definitions of the verb "to claim" is: "to take as the rightful owner; to assert in the face of possible contradiction." "To receive" is "to come into possession of; to act as receptacle or container for; to accept as authoritative or true." The difference is that between acting and being acted-upon, and for women it can literally mean the difference between life and death. (Rich, 1979, p. 231)

Using discussion groups and then reporting back to the whole class is a teaching strategy that I frequently use. I ask students to listen to one another in the groups. If one group of students is dominating the discussion, then I will assign students to report to the class. (In my circumstances, white male students are usually the dominant group over white females, as my university is predominantly white and I rarely have students from non-white groups in my classes.) However, after the first few class meetings students generally self-monitor on these issues. They are helped to do this by the feminist quotes on the syllabus and on my e-mail signature, by the posters showing only women scientists displayed in the laboratory in which I teach, and by my research agenda, which is advertised on my web page and publications, and which clearly articulates my feminist stance and beliefs.

At the end of each class session, I complete a personal crosscheck. With whom did I interact today? What conversations did I have with my students? Are there aspects of my teaching that are biased? If so, how can I change these? In amongst these concerns, I also consider what I know from personal experience and other researchers; that is, for some students any mention of feminist issues is excessive, and they react negatively. An extreme example of this occurred while I was teaching a graduate course on gender and education. A student attempted to use my classroom as a platform to pursue a confrontational agenda with me and the other students in the class. He wanted the class to read and discuss literature related to men's rights. The issue for me was not the subject but rather his source, namely a popular men's magazine that portrays women as sexual objects. Initially, I refused to allow the literature from this source to be distributed in my class but provided the student with other, reputable references that he could share with his colleagues. The student accused me of censorship. I negotiated to have several colleagues interested in men's rights to work with the student on his class presentation. He refused.

This situation became very traumatic for me, as it was a challenge to my feminist beliefs and principles. At the time, the men's rights movement dominated the public discourse. Yet the publication the student wanted to use would be inflammatory and insulting to his peers. I was unable to convince the student of the potential negative outcomes of his proposed action. Initially, my reaction in caring for the class and also the student was to refuse my permission. After several days of personal debate and seeking advice from colleagues, both feminist and not, my solution was to let the class decide. While keeping the student's identity confidential, I outlined the situation to the class. If the class agreed, in my absence (because I refused to participate in the discussion), the student could present the article to the group for discussion. I left the room while the class discussed the issue. Although the student was present, he did not speak while I was present, nor when I left the room. A few days later he dropped the class.

I often use my own class as an example with my students regarding feminist pedagogy and as I describe elsewhere "practice what I preach" (Scantlebury, 1994) with regard to the content taught and the assignments given. For example, a case study of the drug thalidomide provides an excellent explanation and rationale for why scientists use systematic procedures in naming chemical compounds and the important role of the Federal Drug Administration (FDA). In the '50s and early '60s, doctors prescribed thalidomide to pregnant women for morning sickness. The drug caused deformities in unborn babies that resulted in children with no arms and/or no legs. Although many factors contributed to the continued worldwide use of thalidomide after the link had been made between the drug and the birth defects, one problem was the naming system used to identify the compound. The drug was banned as thalidomide in the United States but was still being prescribed in other countries under different names. European doctors did not realize that the same drug was being repackaged by the drug company under another name. In the United States, the thalidomide case precipitated increased regulation of drug companies with regard to extensive testing of new drugs before giving approval for public use (Selinger, 1989). (Further discussions on the role of the FDA could involve how these laws affect patients with the AIDS virus.)

Also, I critique my syllabus. In doing so, I consider the following questions: What subtle messages are in the course materials? Regarding authorship, do gender and race balance the texts and papers? Is there variation in the assignments? Adrienne Rich's (1979) essay challenges both students and me to take responsibility for learning. Rich also challenges students and faculty to reconsider, from a broader perspective, the function of assessment and grading.

> If university education means anything beyond processing of human beings into expected roles, through credit hours, tests, and grades...it implies an ethical and intellectual contract between teacher and student. This contract must remain intuitive, dynamic, unwritten; but we must turn

to it again and again if learning is to be reclaimed from the depersonalizing and cheapening pressures of the present-day academic scene. (p. 231)

Included in my syllabi is a clearly defined grading policy
(Appendix 1). This quote provides the opportunity for my students and
me to discuss, and in some instances to implement a dynamic grading
contract. As part of that contract, grades in the course are either A, B or
"do it again." A student with a "do it again" receives an incomplete until
such time as s/he submits satisfactory work. However, as part of a
feminist pedagogical approach, I must be willing to change the criteria
while being cognizant of the potential consequences of such actions from
colleagues, administrators, and other class members. Students can choose
to complete most class assignments in a group. However, field
observations and a personal portfolio are individual efforts. We regularly
discuss students' progress toward completing these semester-long
assignments.

Similar to many of my teacher education colleagues, one course
assignment is a portfolio of each student's work. Working in groups of
three, the students grade a selection of their peers' portfolios. The
purpose of this exercise is twofold; first, students can learn from their
peers in viewing their work. Second, the experience of grading a
portfolio using a scoring rubric can highlight the rubric's flaws as well as
the advantages and disadvantages of the portfolio as an assessment tool.
Initially, we would also develop the rubric for the portfolio, but the
portfolio concept was often foreign to my students and this exercise for
this particular assignment was difficult and time-consuming. Instead we
now develop a rubric for judging students' posters on historical figures in
science. The scientists must belong to an underrepresented group, that is,
women and/or non-white ethnicity.

For both assignments, my grade on the assignments is one of a
group. However, because of my position, my power, and the
responsibility I have because I am the professor, I mediate if a student
requests changes to her/his grade. This rarely happens because after a
student has seen all the work from her/his colleagues s/he often realizes
that the grade given is fair.

Most preservice teacher education programs require students to
complete daily lesson plans and/or long-term plans such as units or
semester-long topics. As one part of the planning I require students to
identify equity issues. Initially, preservice teachers report that their

lessons have "no equity issues." This provides a starting point for a discussion on what students know, how they learn and how a student's gender, race, and socioeconomic status may influence her/his ability to learn science. Over the year, the preservice teachers begin to learn how students from different groups can be impacted by what they, as teachers, choose to teach and the strategies they select to do this.

My preservice teachers' ability and/or willingness to implement equitable teaching strategies is another aspect of the feminist pedagogy that I attempt to practice. The student teaching practicum is a major area where I perceive that I enact a feminist praxis. Consequently, the tools we use to evaluate students during their student teaching have items such as "calls on girls and boys equally," "does not use sexist language," and "uses nonstereotypic examples to illustrate concepts." My university position as the coordinator of the Secondary Science Education Program is a gatekeeper role. It is my responsibility to pass or fail students. For the student, successfully completing student teaching and other program requirements entitles her/him to the university's recommendation for certification. As I have described elsewhere, a major goal of the program is to graduate science teachers who are gender-sensitive (Scantlebury, 1994).

Why do I consider these to be feminist strategies? I consciously attempt to design assignments that require different skills from students and highlight issues related to underrepresented groups in science, especially women. Additionally, where possible I try to share the power that I have as a university professor. I recognize the role of power in the micropolitics of my class and the program that I oversee. Adrienne Rich (1979) challenged the faculty of Douglass College to "work against, as well as with, in ourselves, in our students, in the content of the curriculum, in the structure of the institution, in the society at large" (p. 240). By challenging my students' conceptions of gender and science I consciously attempt to enact a feminist praxis in my teaching.

We cannot discuss the reform of undergraduate science education without recognizing that the dominant discourse of academe is one of hegemonic masculinity. Unless teachers and students actively work to counteract this discourse, it pervades and rules through the classes, laboratories, lectures and informal learning settings within the university. Given the strong masculine image of science and engineering, the hegemonic masculine discourse that dominates universities is strongest in science, engineering, and mathematics departments.

Summary

The hidden curriculum affects all participants in the education system: students, faculty, teaching assistants, and parents. As teachers attempting to enact a feminist pedagogy, we need to acknowledge this difference and actively work to counteract such biases in our classes.

Do the students in our science classes perceive that the female students learn that their opinions do not matter, while the male students learn that female students are taking up some of the space, time, and energy that used to belong to them? Do the students learn that science is a process and accessible to all students, or do we maintain the concept and the belief that science is a subject for the elite, the chosen few?

As professors, we can use our positions within the university to further feminist causes, to practice feminist pedagogies, and to challenge and change each level of the patriarchal structure. We are all at different stages of the journey toward a *feminist pedagogy*, and our implementation of this concept will vary depending on our political viewpoints. I believe that science classrooms can be feminist within the patriarchal structures of the society/academe/culture of science.

History provides hope for feminist scholars such as myself. After a lifetime of working for women's emancipation, Susan B. Anthony's last publicly spoken words on that subject, three weeks prior to her death, were that "failure is impossible" (Barry, 1988). Similar to the struggle that Susan B. Anthony championed, a feminist pedagogy in science is a lifelong undertaking.

Kathryn Scantlebury
Department of Chemistry & Biochemistry
University of Delaware, Newark, Delaware, USA

EDITORS

METALOGUE

Science as Masculine: A Gender-Sensitive Perspective

KT: Like Penny (Chapter 17), Kate is located in a department in which there are few women. Unlike Penny, Kate's primary appointment is in chemistry education rather than chemistry. Her situation in the academy provides her with a somewhat unique opportunity to speak about the nature of science. On this occasion she describes science as masculine and draws our attention to the difficulty that many females have in accessing and appropriating a discourse that is masculine in character. In her discussion of feminist perspectives on science Kate draws on many sources, including Barton (1997), whose present work involves teaching science to homeless children in New York. Barton makes a strong point that has arisen in several chapters in this book, and that is that scientific discourse cannot be separated from those who know and do science. If a person is to become a learner of science or a doer of science then it seems axiomatic that s/he must use his/her primary discourse (of the home) to make sense of science or the activities in which s/he is to engage in the process of learning science. To the extent that scientific discourse is associated with communities that are masculine, white, and middle class, it can be anticipated that individuals who are not white, middle class, or masculine will encounter difficulties in learning science. When science is perceived as objective truths about the universe, it is difficult to see how it can be taught as a social, cultural, political, and gendered activity in which humans have engaged for particular purposes that are connected to the historical contexts pertaining at the time.

PT: The traditional stereotypical image of (Western) science as the source of objective truths is problematic in a number of ways. First, it leads us to believe that science appears to transcend all local cultures, to be immune from and to create an immunity against local influence; thus, for "scientific" we implicitly read "a-cultural." Second, its official reporting genre of third-(im)person(al) passive voice speaking the dialect of propositional deductive logic obscures from view the epistemologies of practice of scientists, especially the creative processes of discovery and the sociopolitical processes of justification; for "scientific reason" we implicitly read "cold logic." Third, over centuries the memory of the constructed nature of this image has faded from view, leaving a seemingly timeless and impervious sense of naturalness; for "scientific knowledge" we implicitly read "mirror of Nature." Little wonder that professors of science suspended within the webs of this historically dominant mythology favor a teacher-centered pedagogy of cool authoritarian and decontextualized transmissionism. It is the uncritical pedagogical reproduction of this traditional image that is being contested by (a diversity of) contemporary theories of (social) constructivism and feminism. From these perspectives, science can be understood as a culturally framed (discursive) activity whose knowledge claims arise from and are legitimated by social and political processes.

Feminist theory adds considerable power to this perspective by reminding us that the traditional stereotypical image of (Western) science is *gendered*; it reflects the image of its masculinist (mostly male) designers, those with a predilection for an excessively rationalist (or *separate*) way of knowing and acting. Feminist theory also reminds us that science has a (not-so-well-recognized) feminine image arising from the feminine epistemologies of practice of leading women scientists. It stands to reason, therefore, that professors of science should be developing gender-sensitive and gender-inclusive epistemologies of teaching practice if they are to counter the inequities afforded by the traditional stereotypical image of science.

PG: I do believe that there are other ways of knowing, and that these are generally ignored in traditional, male-dominated science. To open up the field to others not traditionally in science (such as females, minorities, and handicapped individuals), we need to learn to see alternative ways of viewing science. I think that Kate is opening her students' eyes to alternative views of science. Kate's students are prospective secondary science teachers, who will influence their students once they start their teaching careers. One thing that college and university science faculty may not have considered is that by influencing teachers we impact their students, many of whom will eventually come to the university. The cycle is thus continuous.

Countering Excessive Masculinity via Critical Co-Participation

KT: There are very many feminist groups and, as Gore has pointed out (emphasized by Kate), these groups often do not acknowledge one another's research and scholarly thinking. There is not a coherent set of implications that we should keep in mind when we plan and enact college science curricula. However, one might consider not only the disadvantages of being a woman in science and choose examples to illustrate those disadvantages, but also the advantages of being a male in doing science and engaging in this course. By bringing such issues to consciousness we create objects for reflection and possible change. In a sense we also are developing the open discourse advocated by Mark (Chapter 16). Open discussions of the nature of science as it is represented in this course can facilitate individuals creating objects about which students can reflect and compare to their own goals and situations. Discussing science as a political act in general is one dimension of what might be done, but discussing science as it pertains to the practices in this community allow learners to adapt the discourse to meet their own goals.

PT: Yes, Kate points out that a range of feminist theories abounds, but that one of the important linking concepts, one that is often misconstrued, is that of *gender*. But it is well to realize that the gender categories of masculine and feminine should not necessarily be equated with the biological (sex) categories of male and female. In terms of their preferred ways of knowing as learners in a science classroom, males and females can be located along a masculine–feminine continuum (although we'd expect the majority of males to be skewed toward the masculine end). Although an excessively masculine pedagogy is likely to cater mostly to male students, it is also likely to cater to some (strongly masculine) females and vice versa for a feminist pedagogy. Thus, it is important that we differentiate between a *feminist pedagogy* that aims (radically?) to cater largely to the interests of those (traditionally disadvantaged) females (and some males) who cluster near the feminine end, and a *gender-inclusive pedagogy* that aims (utopically?) to cater to the entire masculine–feminine spectrum. Given the hegemony of the masculinist image of science, it is understandable that countering feminist pedagogies have been constructed in order to promote the interests of traditionally disadvantaged (feminine) students. Perhaps the next step is toward a gender-inclusive pedagogy that aims to promote a pluralist image of science and engage all students in enacting a range of epistemologies of scientific practice. I envision this transformative practice (praxis) taking place both inside the classroom/laboratory and outside in the broader community, where science already is being practiced professionally or where (social? environmental?) problems exist that (students of) science can address (see, for example, Kielborn & Gilmer, 1999).

KT: The issue of hegemonic masculinity reminds me of Bourdieu's *habitus*. Identifying aspects of a discourse by discussing it in ways that have been advocated by Kate and other authors in this book will only identify those characteristics that are accessible to the oral parts of discourse. In an earlier discussion I alluded to a "between the ears" perspective of knowing. The parts of discourse that are accessible to speaking and writing are just a tip of an iceberg of what we know and can do. The habitus is a major part of what might be referred to as hegemony. A habitus mediates interactions within a community and supports the customs and taboos. My point in raising this concept is to address the idea that speaking about it will create a context in which the changes that matter can be identified and effected. I do not think this is the case. The major parts of habitus always will be beyond language, and one comes to know of the habitus by participating in the community as one who knows and can do. When this occurs, a participant will know the habitus and of course the hegemony—but the knowing will not be communicated by speaking or writing, only by doing in a sense that the participants know what to do and what not to do in the many contexts of practice. Thus, praxis is realized by co-participation in the community, not by intellectualizing practice through reflection on action.

PT: Ken, I take it that you mean that the practice of discussing critically the dominant masculinist image of science and its consequentially restrictive (harmful?) effects on learning is a necessary but insufficient (intellectual) activity. That for professors to teach science in transformative ways they should aim to work changes in the habitus of their science classrooms, of their departments/institutions, and of society at large. The challenge, then, is to act in pedagogically thoughtful ways that engage students in both doing science and talking about the doing. I envision that a gender-inclusive professor of science would aim to foster open and critical classroom discourse for exploring, from the standpoint of students' firsthand science, learning experiences in a range of contexts, the advantages and disadvantages of various permutations of masculinist and feminist ways of knowing and acting scientifically.

BIBLIOGRAPHY

Acker, J. (1992). Gendered institutions: From sex roles to gendered institutions, *Contemporary Sociology 21*, 565–569.

Arpad, S. (1992). The personal cost of the feminist knowledge explosion. In C. Kramarae & D. Spender (Eds.), *The knowledge explosions: Generations of feminist scholarship* (pp. 333–339). New York: Teachers College Press.

Barad, K. (1995). A feminist approach to teaching quantum physics. In S. Rosser (Ed.), *Teaching the majority: Breaking the gender barrier in science, mathematics and engineering* (pp. 43–78). New York: Teachers College Press.

Barton, A.C. (1997). Liberatory science education: Weaving connections between feminist theory and science education. *Curriculum Inquiry 27*, 141–163.

Barry, K. (1988). *Susan B. Anthony: A biography of a singular feminist*. New York: New York University Press.

Bayim, N. (1990). The feminist teacher of literature: Feminist or teacher? In S. Gabriel & I. Smithson (Eds.), *Gender in the classroom: Power and pedagogy* (pp. 60–77). Urbana: University of Illinois Press.

Bourdieu, P. (   ).

Brickhouse, N. (1994). Bringing in the outsiders: Reshaping the sciences of the future. *Journal of Curriculum Studies 26*(4), 401–416.

———. (1998). Feminism(s) and science education. In B. J. Fraser & K. Tobin (Eds.), *International handbook of science education* (pp. 1067–1082). The Netherlands: Kluwer Academic Publishers.

Brickhouse, N., Carter, C., & Scantlebury, K. (1990). Women in chemistry: Shifting the equilibrium towards success. *Journal of Chemical Education 67*, 116–118.

Butler, J. (1990). *Gender trouble*. New York: Routledge.

Collins, P. H. (1991). *Black feminist thought: Knowledge, consciousness and the politics of empowerment*. New York: Routledge.

Committee on Undergraduate Science Education (1997). *Science teaching reconsidered: A handbook*. Washington, DC: National Academy Press.

Connell, R. W. (1987). *Gender and power*. Stanford, CA: Stanford University Press.

Delamont, S. (1990). *Sex roles and the school*. New York: Routledge.

Donovan, J. (1985). *Feminist theory: The intellectual traditions of American feminism*. New York: Fred Unger Publishing Company.

Eccles, J. (1989). Bringing young women to math and science. In M. Crawford & M. Gentry (Eds.), *Gender and thought: Psychological perspectives* (pp. 36–58). New York: Springer-Verlag.

Gabriel, S., & Smithson, I. (Eds.). (1990). *Gender in the classroom: Power and pedagogy*. Urbana: University of Illinois Press.

Gilman, C.P. (1912). *Herland*. New York: Pantheon Books.

Ginorio, A. (1995). *Warming the climate for women in academic science.* Washington, DC: Association of American Colleges and Universities Program on the Status and Education of Women.

Gordon, M. (Ed.). (1980). *Power/knowledge: Selected interviews and other writings, 1972–1977.* New York: Pantheon Books.

Gore, J. (1993). *The struggles for pedagogies: Critical and feminist discourse as regimes of truth.* London: Routledge.

Hall, R., & Sandler, B. (1984). *The classroom climate: A chilly one for women?* Washington, DC: Project on the Status and Education of Women, American Association of Colleges.

Harding, S. (1986). *The science question in feminism.* Ithaca, NY: Cornell University Press.

————. (1991). *Whose science? Whose knowledge? Thinking from women's lives.* Ithaca, NY: Cornell University Press.

Hildebrand, G. (1995, April). Re/Viewing gender and science education via multiple frames of reference. Paper presented at the annual meeting of the National Association of Research in Science Teaching, San Francisco, CA.

hooks, b. (1994). *Teaching to transgress: Education as the practice of freedom.* New York: Routledge.

Kahle, J. B., & Meece, J. (1994). Research on gender issues in the classroom. In D. Gabel (Ed.), *Handbook of research in science teaching and learning* (pp. 542–576). Washington, DC: National Science Teachers Association.

Keller, E. F. (1987). Women scientist and feminist critiques of science. In S. Graubard (Ed.), *Daedalus, Learning about women: Gender, politics and power* (pp. 77–92). Cambridge, MA: American Academy of Sciences.

Kelly, A. (1985). The construction of masculine science. *British Journal of Sociology of Education 6,* 133–153.

Kenway, J., & Willis, S., with Blackmore, J., & Rennie, L. (1997). *Answering back: Girls, boys and feminism in schools.* St. Leonards, Australia: Allen & Unwin.

Kielborn, T. L., & Gilmer, P. G. (Eds.). (1999). *Meaningful science: Teachers doing inquiry + teaching science.* [Monograph] Tallahassee, FL: SouthEastern Regional Vision for Education.

Lewis, M. (1990). Interrupting patriarchy: Politics, resistance and transformation in the feminist classroom. *Harvard Educational Review 60*(4), 467–488.

Lipman-Blumen, J. (1994). The existential bases of power relationships: The gender role case. In L Radtke & H. Stam (Eds.), *Power/Gender: Social relations in theory and practice* (pp. 108–135). Thousand Oaks, CA: Sage Publications.

Lorber, J. (1994). *Paradoxes of gender.* New Haven, CT: Yale University Press.

Martin, J. R. (1994). *Changing the educational landscape: Philosophy, women and the curriculum.* New York: Routledge.

Matyas, M., & Malcolm, S. (1991). *Investing in human potential: Science and engineering at the crossroads.* Washington, DC: American Association for the Advancement of Science.

Morgan, K. (1996a). Describing the emperor's new clothes: Three myths of educational (in)equity. In A. Diller, B. Houston, K. P. Morgan, & A. Ayim (Eds.), *The gender question in education: Theory, pedagogy and politics* (pp. 105–123). Boulder, CO: Westview Press.

———. (1996b). The perils and paradoxes of the bearded mothers. In A. Diller, B. Houston, K. P. Morgan, & A. Ayim (Eds.), *The gender question in education: Theory, pedagogy and politics* (pp. 124–134). Boulder, CO: Westview Press.

National Science Foundation (1996). *Shaping the future: New expectations for undergraduate education in science, mathematics, engineering and technology.* Washington, DC: National Science Foundation (NSF 96–139).

Parker, L., & Rennie, L. (1989). Gender issues in science education with special reference to teacher education. In *Discipline review of teacher education in mathematics and science (Vol. 3)*, Department of Education and Employment, Commonwealth of Australia (pp. 230–247). Canberra: Australian Government Publishing Service.

Poston, C. (Ed.). (1975). *A vindication of the rights of woman: Mary Wollstonecraft.* New York: Norton & Company.

Rich, A. C. (1979). *On lies, secrets, and silence: Selected prose, 1966–1978.* London: Norton.

———. (1986). Compulsory heterosexuality and lesbian existence. In *Blood, bread and poetry: Selected prose 1979–1985.* New York: Norton.

Rose, H. (1994). *Love, power and knowledge: Towards a feminist transformation of the sciences.* Bloomington: Indiana University Press.

Rosser, S. (1990). *Female friendly science.* Elmsford, NY: Pergamon Press.

———. (Ed.). (1995). *Teaching the majority: Breaking the gender barrier in science, mathematics and engineering.* New York: Teachers College Press.

Sandler, B., & Shoop, R. (Eds.). (1997). *Sexual harassment on campus: A guide for administrators, faculty, and students.* Needham Heights, MA: Allyn and Bacon.

Scantlebury, K. (1994). Emphasizing gender issues in the undergraduate preparation of science teachers: Practicing what we preach. *Journal of Women and Minorities in Science and Engineering 1*(2), 153–164.

Selinger, B. (1989). *Chemistry in the marketplace.* Orlando, FL: Harcourt Brace Jovanovich.

Seymour, E. (1992). The "problem iceberg" in science, mathematics, and engineering education: Student explanations for high attrition rates. *Journal of College Science Teaching* February, 230–238.

Seymour, E., & Hewitt, N. (1997). *Talking about leaving: Why undergraduates leave the sciences.* Boulder, CO: Westview Press.

Spender, D. (1982). *Invisible women: The schooling scandal.* London: Writers and Readers Publishing Cooperative Society.

Thomas, K. (1990). *Gender and the subject in higher education.* London: Open University Press.

Tobias, S. (1990). *They're not dumb, they're different: Stalking the second tier.* Tucson, AZ: Research Corporation.

Tyack, D., & Hansot, E. (1990). *Learning together: A history of coeducation in American public schools.* New Haven, CT: Yale University Press.

Weiner, G. (1994). *Feminism in education: An introduction.* London: Open University Press.

Woolf, V. (1938). *Three guineas.* New York: Harcourt Brace Jovanovich.

APPENDIX

Excerpt from the Syllabus of the Course: Teaching Science in the Secondary Schools

Assessment

If university education means anything beyond processing of human beings into expected roles, through credit hours, tests, and grades....it implies an ethical and intellectual contract between teacher and student. This contract must remain intuitive, dynamic, unwritten; but we must turn to it again and again if learning is to be reclaimed from the depersonalizing and cheapening pressures of the present-day academic scene. (Rich, A.C. [1979]. Claiming an education in *On lies, secrets, and silence: Selected prose, 1966–1978*. [p 231]. New York: W.W. Norton)

Grades for this course will be based on the following:

1. Field observation.
2. Journal—weekly reflective journal response to one or more of the following: class discussions, readings, speakers, issues on education, observations of other teaching settings (e.g., other courses, field placement, informal learning settings).
3. Homework will be assigned, as appropriate.
4. Teaching plans and execution.
5. Prompt and regular attendance is a requirement of the course.

Portfolio	25%
Planning a curriculum[1]	20%
Microteaching	20%
Field Observations/Assignment	20%
Homework[2]	5%
PDOs	5%
History of science	5%

NOTES

1. Curriculum planning includes a unit plan, a unit assessment plan, a resource unit, and atopic plan.
2. Homework includes, but is not limited to, biweekly journal, evaluation of textbooks, professional development opportunities. (Examples of PDOs include attending professional conferences, participating in teacher workshops, judging at science fairs, attending science or science education seminars on campus, assisting in-service teachers in their school programs, or reviewing professional journal articles.)

CHAPTER 6

 ## College Physics Teaching: From Boundary Work to Border Crossing and Community Building

Wolff-Michael Roth and Kenneth Tobin

> *To say that two communities have trouble getting along because the words they use are so hard to translate into each other is just to say that the linguistic behavior of inhabitants of one community may, like the rest of their behavior, be hard for inhabitants of the other community to predict.*
> —Rorty, *Contingency, Irony, and Solidarity*

Miller, a physics professor with more than twenty years of experience, lectures in this physics course, which he had specially designed for his audience, preservice elementary teachers, as part of a large project for enhancing teacher preparation. While lecturing on Newton's second law, Miller questions: "What makes this thing move? If I want to make it move, what do I have to do?" Several students call out: "Force."

Satisfied that the students have provided the expected answer, Miller continues:

I'd have to touch it and if I do that it moves. That is the outside object that he's referring to and he gives a name to how you touch a ball and he calls it force. So, if we put a force on there we can make the thing move. If we don't put a force on there, we can't make it move and that's basically the second law. I'm going to write down F for force. But this is just a quantitative way for saying effective outside objects and does it matter what way I push it? If I push it this way, which way does it go? What happens if I push it that way? It will go that way? So, force I have to say how hard I push it and what direction I push it because the thing's going to move in different directions. I put a line over a quantity when I mean that I have to tell you two things: How big it is and what direction it is. Now the ball moved, we all see that. What changed? We have position, velocity, and acceleration. When I touched it, what did I do? I made this comment earlier, that you cannot change velocity and position unless you have some acceleration for a time, just like you can't make any money unless you work for any time, no matter how much a person pays you. The question is: What property of the body does this outside object change? Are the properties that we're interested in velocity, acceleration, and position?

A student calls out: "Velocity."
Miller responds:

The truth is, it changes acceleration and continues, then once the acceleration exists, the velocity will develop and the position will change. It almost will happen at the same time, but they don't really happen at the same time. If we could really record the motion, we could find that it follows this pattern of the falling ball. You notice this when you go in your car. You step on the gas. Does the car instantly go to forty miles an hour? No. It takes some time to get to forty miles an hour. What changes is that the force creates an acceleration and immediately after that acceleration is created, you get a velocity and a change in position. So, that's the order in which things are related. So, the amount of acceleration, he said, was proportional to how big the force was, and that makes sense. If I hit this thing gently it goes slowly. What happens if I hit it hard? It goes faster.

In many respects this excerpt from one of Miller's lectures is typical of those we observed. It features several aspects that we identified as making it difficult for the students to know physics in the way Miller intended. On a global level, it was mostly Miller who spoke. He not only presented information, but he also asked questions that he answered himself. When students called out frequently one-word contributions, as in our vignette, Miller did not ascertain whether they actually were using words in the way he did to describe objects and events in canonical ways. But even when discrepancies were evident, such as when students called out "velocity" whereas Miller expected "acceleration," he did not pursue the issue but continued presenting subject matter content.

Language is ambiguous, even if conversation participants share a good deal of common ground, experiences, and common sense. Ambiguities increase when speakers and listeners are from different discourse communities, as were Miller and his students. Thus, when Miller asked, "What makes things move?" he did so within a framework where a force (F), velocity (v), acceleration (a), and position (x) stand in a particular mathematical relationship expressed in three equations:

$$F = ma \qquad a = \frac{dv}{dt} \qquad v = \frac{dx}{dt}$$

However, for someone who does not know these relationships, such as the students in this class, the question, "What makes this thing move?" and the ratified answer, "Force," also make sense in a pre-Newtonian discourse in which motive forces are central. Thus, if these students used Aristotelian descriptions and explanations of motion (the extensive literature of misconceptions in physics has shown that this is the case for the majority of students at the college level, even with engineering and science backgrounds), they could interpret Miller's lecture as fitting their own ways of talking, despite deep-seated differences. That is, these lectures allowed incompatible explanations to go undetected until they surfaced as student errors at some later point (usually in an exam, or in our interviews). In the present case, one of these errors became evident as a student called out "velocity" (compatible with the pre-Newtonian concept of motive force) when Miller was expecting the Newtonian concept of acceleration.

Miller introduced so many new ways of representing that the students' task of following was virtually impossible. For example, in the excerpt, Miller mentioned a notation for representing the magnitude and direction of a quantity without providing students with any opportunities to engage in this practice themselves. In a matter of one sentence, he wrote and uttered, "I put a line over a quantity when I mean that I have to tell you two things: How big it is and what direction it is." From then on, he appeared to assume that his audience would competently follow when he switched between the two different representations.

In an effort to provide students with an opportunity to make sense of his physics talk, Miller frequently translated into a vernacular or tried to establish analogies with other phenomena familiar to students. However, in this translation, he often misrepresented the physics he wanted to teach. In the above episode, Miller talked about the shift in the onset of acceleration, change in velocity, and displacement ("Once the acceleration exists, the velocity will develop and the position will change. It almost will happen at the same time, but they don't really happen at the same time.") Such a shift is not compatible with Miller's canonical framework represented in the three equations, for the latter's strength comes from the use of calculus and infinitesimal changes (instantaneous changes of position and velocity with time). In the same way, the translations into vernacular allowed ways of hearing that are incompatible with Newtonian physics. Miller's suggestion, "If I hit this thing gently it goes slowly. What happens if I hit it hard? It goes faster," could possibly be heard as suggesting a link between force and velocity (goes slowly, goes faster) rather than the intended one between force and acceleration, according to Newton's second law.

We engaged in several detailed studies of Miller's teaching because of the problems in the teaching and learning context that emerged as the course was enacted (Roth, Tobin, & Shaw, 1997; Roth & Tobin, 1996). Miller is a leading theoretical physicist and volunteered to be involved in the development and teaching of a new course for prospective elementary teachers. He is an exemplar of what funding agencies, such as the National Science Foundation, have in mind when they require arts and sciences faculty to teach such courses as a requisite for funding: a leading scientist who wants to teach prospective elementary teachers, volunteering to do so, chairing a committee to help plan a course, and then teaching the course in successive years. Since he was not successful, by his own account and that of students, we felt that we should study the enacted curriculum and learn from what happened (e.g., Roth & Tobin, 1996).

In our studies, we identified tremendous discontinuities between Miller's discursive and representational practices on the one hand, and those with which students were familiar on the other. The same kinds of discontinuity appear to have occurred in meetings between the scientists and science educators when they talked about the directions to be taken in design and implementation of a new science course for elementary education majors. However, to bring about change, we had to be able to engage with Miller in a conversation without making the same mistakes just described: We needed to develop a discourse about teaching college level physics that is accessible to both professors like Miller and science education reformers like ourselves. We had to begin building this new discourse from a common experience and develop it so that it became useful for designing and enacting change.

In this chapter, we attempt to come to grips with the challenge of assisting college teachers to improve their teaching practices through several iterations of a dialectic between observation and experience on the one hand and theoretical considerations on the other. We begin by looking at the way Miller accounts for his own teaching in terms of a more traditional view of knowing and learning. Subsequently, we offer a different view of knowing and learning as changing participation in changing practices. Using this perspective as a tool, we return to Miller's teaching to disclose possible constraints on learning in his class. We identify several points of departure for a journey of change and then provide the autobiographically inspired account of change in physics teaching of a professor (Ashmore) who shared a number of biographical details with Miller.

A Traditional View of Knowing and Learning

To bring about change in college physics teaching, we need to have a good understanding of the theoretical descriptions of knowing and learning that professors typically bring to class, and how these descriptions relate to the enacted curriculum.

Miller's discourse about knowing and learning physics was based on his experiences with physics majors, his own learning of physics, his days as a college basketball player, his coaching experience with children's basketball teams, everyday notions of learning, and his ontology. The curriculum he enacted in the course for preservice elementary teachers was quite consistent with his reflections about teaching and learning. He suggested that understandings would emerge over time and that students would be exposed to ideas in this course that would mature over time and possibly prove to be useful in the future when they were teaching. These beliefs may have led him to be less concerned with assisting students to enact physics (doing, talking, and using inscriptions) during the course and to provide activities designed to build understandings. Overarching these beliefs of the teacher is a belief set about the power of the teacher over students. In essence Miller did not see a need to negotiate with students about the enacted curriculum, nor did he appear to take seriously their notions of what ought to have been taught. Expertise in the content was perceived as the most significant source of power, and since he knew physics and the students did not, it made sense to him that he ought to have the power to make the critical curricular decisions.

Miller was an expert in physics and taught from the perspective of one thoroughly familiar with the discursive and representational practices of the field. Miller suggested that physics teachers ought to be in control and present subject matter content in chunks that could be digested by students. He taught from the perspective of one who knows science and set a stage designed to make the salient aspects of physics visible to his students. His lectures and demonstrations were a part of a stage act in which a content master revealed to students the essential steps in coming to know (cf. Roth & Tobin, 1996). He controlled the selection of activities, the order in which content was introduced, and the pace of the activities. As coach, he knew that his students would not grasp every detail and develop understandings during the lesson. Instead, he believed that at some future time, understandings of physics would click into place.

One of Miller's core resources for explaining his teaching was the claim that knowledge of physics existed in a knower-independent universe. He commented:

> The laws of physics are not mine. They are all of ours. And all of us, which now number into millions, and billions, who have ever tried them came to the same conclusions. Students are not able to decide if something is right or wrong, because things behave in a certain way, and it is not a matter of opinion. I cannot tell them or force them to accept anything, but I can tell them what everyone else believes is the case. So they can accept it, or not accept it.

The beliefs expressed in this excerpt are consistent with Miller's actions in controlling content selection in the curriculum. The excerpt also indicates that for Miller, scientific truths are beyond question. Accordingly, he often presented science as certain and unchanging. Most of the time the content was associated with exercises from worksheets or on questions assigned for homework. As did many of his colleagues and consistent with his ontological claims, Miller frequently expected students to infer the laws of nature either from direct observation or by manipulating various inscriptions. In Miller's world, these inscriptions referred to natural phenomena in a self-same way, showing the same structure. His sense was confirmed by his experience with physics majors who, after hard work and with the passage of time, usually arrive at the correct way of knowing physics. Other parts of his lectures were designed to reveal the truths about nature:

> Any experience in the area of science is knowing what is true and what is not true....So science is just recording these truths. I think it is very important to have the experience of knowing something is true. So my goal was to show the students that certain things are true. I tried to take

them through a process because it makes no sense if I just tell them that it's true.

Miller assumed that students would understand what he intended to communicate in talk, gesture, and inscription, and that they would follow his transformations of inscriptions from some initial phenomenon to a representation that is common in the physics community. For example, Miller rolled a ball down an inclined plane. This inclined plane was adjusted so that distances traveled by the object came out as integer numbers (1, 4, 9, 16 ft) that were exactly the squared values of the time that had passed (1, 2, 3, 4 s). After adding a zero in each of these sets (i.e., 0, 1, 4, 9, 16 ft; 0, 1, 2, 3, 4 s), Miller transformed these sets into a data table, from which he generated new tables and graphs. In and through his inscriptions, numbers, tables, plots, best-fit curves, and sketches, he expected students to see the laws of kinematics and dynamics as part of the fundamental structure of the world. That is, he expected students to discover and understand the laws of motion through the tables with the data sets he had generated from the rolling ball. To make students see, Miller employed inscriptions, physics talk, and talk about inscriptions. Unlike scientists who use conversations to make sense of inscriptions, Miller's presentation assumed the self-evidence of talk and inscription. In his view the meanings were there to be heard, seen, and experienced.

In Miller's view, goals on the one hand and knowing and learning on the other are independent. Miller endorsed the view that the course was to emphasize physics content and not how to teach physics ("I am teaching you content so you can have a strong background in science, so that you will be comfortable to teach science"). He therefore focused on content and its potential suitability for prospective elementary teachers irrespective of their own goals for the course. What he had not anticipated was the extent to which the students would enter the course with expectations that everything they learned would be related directly to their professional goals. (The research on situated cognition actually suggests a tight linkage between cognition, knowing and learning, and the goals and means to achieve them.) The following comment from Miller indicates his perception that goals with such an orientation were inappropriate.

> I don't know what the goals of the students were, and maybe it was to teach science to their students. But they cannot be the goals of this course.

> If the students are coming in with those goals, that needs to be changed.
> Now to bring things in that relate to the teaching of schoolchildren should
> not harm the course in any way, but the goal is not to teach you how to
> teach elementary children, it is to teach science.

Miller was not responsive to the professional goals of the students to an appreciable extent, and therefore may have missed important opportunities to help his students to learn and understand. Rather than seeing the students' goal orientations as means to the end of having prospective teachers learn science content, he perceived their goals as a challenge to persuade them to accept his point of view.

Knowing and Learning as Participation in Practice

Miller's ways of explaining knowing and learning are widely shared. Despite mounting evidence that students do not learn, and in fact are turned away from science, many teachers still theorize and enact their teaching by telling about science. However, new and different ways of thinking about knowing and learning have recently evolved that require different ways of teaching. A small number of studies have shown that these new ways of thinking about knowing and learning are fruitful for designing learning environments that foster a greater degree of understanding than more traditional environments allow. In this section, we present the most promising of these views. (There are others, such as conceptual change and use of analogies, each with their own problems. For a thorough review see Duit, in press.)

We understand knowing a subject as participation in the activities of a community with its specific recognizable practices. This is an anti-representationalist view of knowing. Rather than thinking of knowledge as something represented (such as a concept), that is constructed and held somewhere in the mind, we talk about competent participation. Knowing physics therefore means to participate in talking about relevant objects and events in the ways physicists do, using acknowledged words, sentences, gestures, inscriptions, and so forth. Learning physics, rather than being the transfer or construction of some mental content, is regarded as increasing participation in the life world of those whose conversational objects are linked to the field of physics.

Language, Cognition, and Being-in-the-World

Philosophers of language such as Heidegger and Wittgenstein argued that language is not simply one among many forms of representation used to read out individual thoughts from one cognitive system to another. Rather, both Heidegger and Wittgenstein view language as central to our being. Language constitutes our world, though it is not prior to the world. Rather, language and the world of our experience co-evolve; in other words, language is deeply grounded in experience, but experience is mediated by language. So there is no escape from language, no world behind the words; there is no getting to the edge and beyond words (Dreyfus, 1992). Language is central to any human form of life.

Our everyday language grounds all the other specialized idioms we use in our work (e.g., as a mathematician, business manager, science teacher, or science education researcher). Becoming a scientist or a scientifically literate person requires an individual to develop the competence to participate to an increasing extent in the ongoing discourses of scientific communities. For example, graduate students in physics become competent physicists through their increasing participation in a form of life, laboratory life: The longer they participate, the more they are able to competently participate in the conversation of the recognized masters, their thesis advisers, mentors, and so forth (Traweek, 1988).

What happens as individuals participate in a form of life is also a feature of Bakhtin's work. Here, words are characterized as culturally shared and distributed cognitive artifacts. They do not exist inside or outside of individual consciousness; language lies "on the borderline between oneself and the other," and "word in language is half someone else's." The word becomes "one's own" only when "the speaker populates it with his own intention, his own accent, when he appropriates the word, adapting it to his own semantic and expressive intention" (Bakhtin, 1981, p. 293).

Why is participating in language as a form of life so important? Our hunch is that much as with perceptual acts (seeing), our competence in using a language has to be experientially grounded. That is, we cannot learn a language or more simply, an individual word, without using it in a variety of different situations. We have to get a feel for what it can do for us: We have to learn a language through use. This is at the root of Wittgenstein's (1968) advice that words have no meanings in themselves; they achieve meaning only through their use in particular language games.

Boundary Work and Border Crossing

Inscriptions. Sociologists of scientific knowledge have observed scientists' almost obsessive preoccupation with inscriptions, that is, visual or verbal representations of nature, extracted from the laboratory, cleaned, re-drawn, transformed, and finally displayed to support scientific text (Latour, 1987, 1993). The meanings of these inscriptions (like those of other scientific and technological artifacts) are constructed as collaborators achieve two nested convergences. First, the correspondence between inscription and natural objects is the result of making convergences between language and experiment such that the distinctions between words and the world are reified. The second convergence is established as scientific collaborators assure one another that they are talking the same language (Gooding, 1992; Rorty, 1989).

Conscription devices. When inscriptions are constructed in a collaborative effort, they also serve to organize workers, the work process, and the language workers deploy during the task. Because these inscriptions enlist the participation of those who will employ them (since users must engage in generating, editing, and correcting them), they have been termed conscription devices (Henderson, 1991). By referring to an inscription as a whole or in part, social actors can engage each other, focus each other's attention and communication. In this way inscription devices can serve as mnemonics for past stages in a conversation that develops over and about the inscription. These conscription devices are so important to scientific and technological conversations that scientists and engineers will stop a meeting to fetch a design drawing, render a more or less faithful facsimile on the whiteboard, or render a diagram in gesture. As such, conscription devices are inscriptions with the function of coordinating and constraining the activities of two or more actors.

Boundary/coherence objects. The same inscriptions, when used in different circumstances, frequently give rise to significantly different discourses. Inscriptions can therefore be used to distinguish between communities, and therefore are *boundary objects*; or they can be used to coordinate activities between different communities, and therefore are *coherence objects* (Star, 1989; Star & Griesemer, 1989). When they are used as coherence objects, inscriptions fulfill an important function in the coordination of work across communities of practice. Star and Griesemer provide a detailed analysis of how a museum curator coordinated the practices of hunters, lay botanists, and government employees by means of collection protocols that allowed him to establish an impressive record of the Californian fauna at the University of California–Berkeley's Museum of Vertebrate Zoology. In a similar way, one can conceptualize the coordination of work in an airport operations room in terms of the boundary objects used there, a complex sheet. A complex sheet contains a matrix that maps incoming and outgoing planes, and includes cells for the transfer of people and baggage (Suchman & Trigg, 1991). By means of the complex sheet, the activities of a variety of people are coordinated such as to guarantee the appropriate routing of passengers and baggage.

When they are used in the second sense, inscriptions do boundary work and create distinctions. In this way, scientists draw clear lines between themselves and those they claim to do non-science, such as evolutionists, sociologists, or parapsychologists (Gieryn, 1996; Gieryn, Bevins, & Zehr, 1985; Pinch, 1979). In their article *Setting Boundaries between Science and Law*, Solomon and Hackett (1996) describe how the ability to read and evaluate particular inscriptions (whole articles, data in various forms within articles) was used in a court case by some to distinguish valid and good science from bad science, good scientific evidence from unscientific testimony, and established fact from unsettled speculations. In this view, only peers (who peruse inscriptions in a particular way) can evaluate scientific research and therefore draw the distinction between science and non-science. Closer to education, the competency with which individuals use particular inscriptions (mathematical equations, graphs, diagrams, etc.) is used to draw distinctions about who enters particular programs, who is to be preferred during hiring processes, and so forth.

Boundary objects can be used to define power. Practices of representation and representations of practice make secure particular forms of authority, and therefore determine relationships of power and knowledge. In this sense, Western culture encases historical and institutional structures that both privilege and exclude particular readings, aesthetics, authority, voices, representations, and forms of participation (Giroux, 1992; Star, 1991). Teachers therefore have to begin the unlearning of their own privilege to be able to listen to students and speak in ways to promote their learning.

Border Crossing. When inscriptions are used as conscription devices and coherence objects, they become sites where individuals can cross preexisting borders. They can become openings through which outsiders become insiders. A number of studies have illustrated such border crossings. In one study, the design drawings in architectural studios played the role of conscription device in the conversations between studio masters and students (Schön, 1987). As they engaged each other in conversations about the learners' first drawings, studio masters and students increasingly learned to understand each others' concerns, construct the meanings of particular representations (words, drawings, gestures), and use practical and conceptual tools in the same way. Therefore, they began to participate in a common language game and form of life (Wittgenstein, 1968). In the same way, elementary and high school physics students increasingly participated in canonical forms of discourse as they communicated over and argued with different inscriptions such as diagrams, animated computer-based representations, or semantic networks (Roth, 1995a, 1996; Roth & Roychoudhury, 1994). Inscriptions when used as conscription devices and coherence objects can therefore be used to allow teachers and students to write, speak, and listen in a language in which meaning becomes multiaccentual and dispersed and resists permanent closure. This is a language in which one speaks *with* rather than exclusively for others (Giroux, 1992).

We agree with Giroux (1992) that educators should engage in a border pedagogy that decenters relationships of knowledge and power, and foregrounds ways in which knowledge can be remapped, reterritorialized, and decentered in the wider interests of rewriting the borders and coordinates to increase the extent to which all students can participate authentically. We consider inscriptions (tables, texts, graphs, diagrams, etc.) as sites where students and teachers negotiate their different uses and readings and engage in becoming increasingly literate in their initially exclusive discourses. Teachers and their students become border-crossers, mediated in this work through the use of conscription devices and coherence objects. In this effort, borderlands are being created in which the very production and acquisition of knowledge is being used by students to rewrite their own histories, identities, and learning possibilities (Giroux, 1992).

Language, Science Learning, and Community Building

In the context of learning physics specifically, and science more generally, we view becoming scientifically literate as increasing competence in an ongoing discourse of a scientific community (e.g., the conservation discourse of an organization such as the World Wildlife Fund, in public rallies against nuclear plants, or as a politician who has to make decisions related to the Centers for Disease Control). That is, increasing the extent to which authentic participation occurs in a form of life that is science-like. Arguing some point, defending one's experimental choices, designing an experiment in a collective so that different groups can test different hypotheses or follow different routes of argument would constitute some of the activities that we consider appropriate for students to engage in to learn science. In the course of such activities, students would learn incidentally the idiom of science, in itself embedded within the idiom of everyday-life English. The overall purpose for conducting activities that one can rationally defend then couches all other concerns, one of which would be to argue one's case most effectively. Students would look up words, not for the purpose of memorizing another word, but because they want to make a particular point.

In the past, researchers and teachers frequently assumed that when children were allowed to invent and use their own scientific and mathematical language games, they would automatically take ownership (e.g., Cobb et al., 1991; Lampert, 1986; Roth, 1995b). One of our studies allowed us to separate ownership and invention of a language game (McGinn, Roth, Boutonné, & Woszcyna, 1995). We now look not simply at whether activities allow students to invent and deploy their language games, but whether activity structure and artifacts in the students' and teachers' experience allow them to build a common language game. Teachers then face the challenge to design the students' learning environments in such a way that they can create and deploy language games that make sense to all, which is the same challenge that designers of computer systems face when they create new workplaces (Ehn & Kyng, 1991). Teachers are the play-makers who set the stage for participation in language games by finding and supporting ways in which students and teachers can, as a collective, develop new ways of talking science that are sensible to students and legitimate for the teacher. The starting point has to be a language game sensible to both. Ordinarily, everyday language can be regarded as common ground. Starting with this common ground, new and more viable language games can be constructed by the collectivity, students and teachers alike. Ehn and Kyng suggest that in this effort, what is important is not if the artifacts that serve as conversational topics mirror "real things," but whether they encourage interaction and reflection. These new language games enable students to see the world in new ways; for language games and the world mutually constitute each other. And because students co-participate in establishing and maintaining these new forms of talking, they are, as Bakhtin (1981) would say, populating the language games with their own intentions.

A Practice Perspective on Miller's Teaching

New perspectives assist us to see and understand events in different ways. They reveal previously inaccessible aspects, and necessitate different consequences to be drawn. As we pointed out above, Miller's discursive resources for explaining his teaching and the enacted curriculum we observed were relatively consistent. However, the evidence gathered from students' notebooks, interviews, and classroom actions suggests that there existed an impasse. Despite Miller's good intentions, students did not come to command any appreciable competence in doing physics, that is, talking about, explaining, or representing physical phenomena. The critique in this section is not intended to indict Miller as a person. Rather, critique allows us to identify possible points of departure for transformation in teaching practices.

Miller had his discursive resources for making sense of the situation: the problem lay in the characteristics of these students. He described the elementary education majors as capable, but not as well educated as physics majors ("I guarantee that the students with a 3.5 average that are going into elementary education are not as well educated as a lot of people who are going into chemistry and physics with a 2.7 average.") Physics majors were more like aggressive scholars who thought through problems and endeavored to understand. Elementary education majors were resistant to what he called real thinking and reluctant to deal with uncertainty.

Our own epistemology of practice throws a different light on students' lack of physics competence. We will focus here on two issues in the lectures, inscriptions and language, because a reconsideration of their deployment can actually lead to a transformation in Miller's practice of teaching physics.

Myth of Discovery

The world is no longer understood as inherently structured. What and how we see is intricately related to the language, inscriptions, and tools we command. Therefore, the world looks different to physicists and students because of the differences in language, inscriptions, and tools they use, and the competence with which they are used (Roth, McRobbie, Lucas, & Boutonné, 1997). From this perspective, Miller's attempts to show students the truths about nature are ill fated not because his students are unwilling or less educated, but because their resources and competence lead them to different conclusions. In the context of assessment, conclusions that differ from canonical physics mean lower grades. Although Miller also wanted to emphasize science as a process, he was thwarted in this goal by students' drive to obtain his answers and truths.

Miller expected students to infer the correct structure of the world. However, since structure is not regarded as an inherent property of the world, but as originating from interactions with it (seeing, describing, inscribing, applying tools), physics teachers no longer can expect students to arrive at canonical answers on their own. When left to themselves to make sense of experience, students will arrive at answers, but these are likely to differ from canonical constructions. Teaching strategies are then needed to deal with this diversity of talking about, acting toward, and describing phenomena, and to help students select those practices that are intelligible, plausible, and fruitful for dealing with a variety of interesting phenomena in the domain of physics.

Generating Boundaries

Miller, in the scientist's usual fashion, made extensive use of inscriptions in his teaching. These were his actual tools that revealed the patterns of nature as he saw them. He formulated some real world phenomena in terms of numbers—such as two coordinated lists of position-time measures—converted them into tables, and then transformed them through a cascade of inscriptions into velocity- and acceleration-time tables, plots, and best fits (regression lines).[1] Consistent with various analyses of inscription use (Latour, 1987; Lemke, 1998), Miller displayed these inscriptions as part of a rhetorical effort to convince the audience that Aristotle had it all wrong and Galileo got it right.[2] However, because his audience did not have enough competence to follow the transformations, or to deconstruct the use of these inscriptions in this particular way, the inscriptions actually served as boundary objects. The boundary objects reflect a chasm between Miller, a competent user, who transformed experience into various phenomenologically different objects. The audience became aware that a performance was occurring in which they could not take part. The nature of these inscriptions as boundary objects was especially evidenced in the different discourses that emerged. In Miller's case, inscriptions were integrated with his other actions and talk, to make arguments about physical phenomena. In the students' case, their talk focused on the construction of the inscription itself. Miller employed the inscription-technologies to master nature that illustrated his power and distinguished him from students.

Border Crossings

Miller experienced considerable difficulties in bridging physics discourse and everyday ways of speaking. As can be expected, his physics register, including the various mathematical inscriptions and associated discourse, conformed to the norms of his community. However, when he tried to make his understandings more approachable by changing into everyday talk, the accuracy of physics and mathematics was very often lost. As Miller translated from canonical physics talk to a vernacular he believed accessible to students, the physics became watered down or disappeared completely. For example, in the opening excerpt from the lecture, Miller suggested a delay between the moment a force creates an acceleration and the moment one can observe velocity and change in position. Here, the order was such that first acceleration was imparted, then velocity developed, with the consequence of a changing position. These views are similar to the medieval notion of impetus, or impressed force, still used by the younger Galileo. There, projectiles could be moved only by a motive force impressed by the projector (Galileo, 1960).

As he crossed the boundary to employ an everyday register, Miller used the same Aristotelian descriptive language that science education researchers associate with misconceptions, alternative frameworks, or naive conceptions. The point here is that crossing the borders between physics and everyday registers is not easy, even for competent speakers in both domains such as Miller. Because our everyday registers actually function as the root system to which all other discourses default when it comes to making sense, crossing into the domain-specific discourses of any field is a difficult undertaking.

Controlling the Discourse

When instructors like Miller maintain unilateral control of the dialogue, particularly when students submit (as did the students in our studies), they are unable to diagnose students' current competence. At the same time, their students are unable to publicly test their understandings or explore the instructor's meanings. Miller's use of inscriptions and discourse that were inaccessible to students in fact became instruments for creating borders and for instituting his position of power. His representational practices instituted his claims to objectivity, universality, and consensus.

Miller's control over the discourse also allowed him to pursue his own goals, to teach pure physics, and disregard the goals students brought to the class. The preservice teachers wanted to learn physics in ways that would allow them to teach it later. The gap between Miller and his students is like that which existed between traditional software designers and end users of their products; the designers' discourse was one of pure design, which led to products that did not fit in with the users' discourses, which were concerned with the everyday requirements of their workplaces (Ehn & Kyng, 1991). As goals provide meaningful frameworks to contextualize what we do and learn (Lave, 1988), Miller probably missed many teaching opportunities in his desire to keep physics discourse separate from possible pedagogical issues for teaching physics in elementary schools.

Politics of Representation

In the process of teaching physics, Miller's presentations disregarded students' experiences and their ways of talking about phenomena. Miller did not acknowledge their everyday experience, built on a foundation of Aristotelian ideas, as capital for developing canonical science. The controlled laboratory experiment, particularly as represented in various mathematical inscriptions (equations, tables, and graphs), was used by Miller as evidence to support truth, Newton, and implicitly all modern thought. This opposition between everyday experience and scientific thought was repeatedly expressed by the careful positioning of chalkboard displays and coincidentally, the location of experiments. Newton, canonical physics, and the truth, as Miller saw it, were recorded on the front blackboard. Aristotle's views and medieval thought, which corresponded to students' own ways of understanding, were noted on a blackboard off to the side. This created a stark contrast between students' experiences and what Miller declared to be truth. What students had learned, often intuitively, from their life experiences, and from some of the activities conducted in the class, was consistent with Aristotelian thought. Students had conducted experiments in which the time it took different coins to drop to the ground was measured. Although repeated and averaged, consecutive experiments demonstrated differences between coins. Knowing that these were not the expected (correct!) results, one student group made a direct comparison, which again showed a difference. Miller also simultaneously dropped a tennis ball and a basketball from a building to show that objects fell at the same rate. Again, there was a difference in favor of the heavier ball, and a piece of evidence in favor of Aristotle. In all these instances, Miller asked students to disregard the differences and appealed to measurement error, uncontrolled conditions, friction, and other influences that interfered with seeing what was supposed to be seen; and he used an array of techniques including mathematical discourse and representations to persuade students to accept the scientific discourse.

Not all speakers in a conversation have the same authority. Mehan (1993) described how the reports of school psychologists and nurses were more privileged than the voices of parents in regard to decision making about disability. Miller's presentations gained their authority in a similar way, by the very nature of their constructions. His inscriptions, language, and descriptions have privileged status because they are (to students) ambiguous, contain technical terms that have specialized uses, and are difficult to understand. His students do not challenge his claims because they perceive their knowledge to be inadequate, and they cannot learn easily from the way Miller presents information, grounds assertions, and represents nature in language, gesture, and inscriptions. Miller's presentations received further authority because of his institutionally designated position of authority and gatekeeper for the physics community. In this sense, Miller's representations are not value-free but have a political nature. Representational technologies are intrinsically political, not only in that they embody the ideologies and agendas of those that create and use them, but in the less obvious and more pervasive sense that they constitute active elements in the organization of the relationships of people to each other and with their environment (Akrich, 1987).

Points of Departure: Crossing Borders

In the context of systemic change in college and university physics teaching, we have to think about two types of existing borders. On the one hand, there are those that we showed to exist between physics teachers (such as Miller) and their students, particularly those who do not major in the field. On the other hand, we also have to consider crossing and reframing borders that currently exist between physics teachers and reformers (often situated in science education departments). These two concerns are fundamentally related to the creation of new discourse communities in physics and physics education. To engage in the necessary border pedagogy, we can start by identifying sites such as particular inscriptions that all constituents can consider as coherence objects, which, in the work of reconfiguring boundaries, become conscription devices and mediation devices for the emergence of new discourses.

Border Pedagogy: From Vernacular to Physics

What can Miller do so that students in his classes begin to embark on a journey during which they will shift and cross the boundaries between the communities of the home and physics? In terms of knowledge transmission, lectures appear to have limitations when compared with books and other information sources. Errors are more likely to occur in the real-time of a lecture than in books that have been edited. The rate of information dissemination in lectures seldom adjusts to all students' rates of learning. In our practice perspective, the co-presence of physics teachers and their students affords learning opportunities that far exceed those of information transmission that could be left to reading. Co-presence affords co-participation in the practices of talking, inscribing, and operating on worldly objects. Co-participating allows students and teachers to interact with all the benefits of communication: the constant correction of passing theories about the words, gestures, and inscriptions of others to account for their mumbles, stumbles, malapropisms, metaphors, tics, seizures, psychotic symptoms, egregious stupidity, strokes of genius, and the like (Rorty, 1989).

To avoid learning binds, interactions between practitioners in a field (such as Miller) and newcomers (such as elementary education majors) are necessary to deal with the ambiguities, vagueness, and inexpressibilities of communication. Teachers and students both have to be willing to engage in the work of crossing and reconstructing boundaries. Corrections of misunderstandings depend on students' and teachers' abilities and willingness to interactively search for a convergence of meaning.

Rather than trying to transfer information and knowledge into the minds of students by means of lectures and demonstrations, a practice perspective focuses on participation. The question now becomes, To what extent do physics lectures, seminars, or demonstrations allow students and teachers to co-participate in talking about, gesturing over, inscribing, and handling objects and phenomena? Much as we would not expect someone to become a basketball player or fluent speaker of a foreign language by watching others, we no longer expect that students learn physics practices by observing professors.

Miller's sports-related resources for talking about coaching may be a starting point for a common discourse. Miller would probably not expect that one can learn basketball by watching others play. His coaching metaphor of teaching and learning could play a central role in changing his practice of teaching; it is in terms of this metaphor that Schön (1987) described how newcomers and old-timers in a variety of fields learn to cross over apparently unbridgeable communication gaps (like those that we demonstrated in our studies of Miller's teaching) to a seeming convergence of meaning. How do these coaches do it?

Because students and coaches bring to the learning situation their different experiences and ways of seeing, there need to be some concrete things over and about which they can converse. These things can be artifacts the students previously produced or the performance of some activity. Their ensuing dialogue then incorporates three features: It takes place in the context of students' artifacts or performance; it makes use of actions, words, and gestures; and it depends on the reciprocal reflection on meaning. With each of their interventions, coaches experiment in communication, testing their own diagnoses of students' practices and the effectiveness of their own communication.

Border Pedagogy: From Physics to Physics Education

In the same way as we proposed the beginning of a conversation between physicists and their students, we have to begin a conversation between physicists and physics educators. Despite our critical perspective on physics teaching earlier in this chapter, we do not consider our discourse as a master narrative that physics teachers have to adopt. Rather, we see our critique as one example of how difficulties of teaching and learning can be brought into a different light. Such a critique could but does not have to become a starting point for reform. What we need is to begin a conversation between physicists and physics educators that ultimately leads to increased student participation in physics discourse and better physics teaching at the elementary school level. We think of conversations that can lead to the design of learning environments that allow students to develop more competent physics practices than those they develop to date. The Swedish designers of computer-based work environments understood that if they were to be successful, they had to abandon a high-horse approach of imposing their discourse on the potential end users; they understood that they had to develop a common discourse such that it could handle in sufficient complexity the intricacies of working in or designing computer-based publishing (Ehn, 1992). Our question then has to be, What are common artifacts that could serve as points of departure for creating discourses between teachers like Miller and ourselves? The following autobiographically inspired account of teaching science practices illustrates that change is possible. This is not a master narrative about how change *ought* to happen, but simply one account of how change *may* come about.

Changing Ways of Teaching Physics

Ashmore watches the videotape from his physics class for elementary preservice and inservice teachers earlier this week. He has been teaching physics and physical science for ten years now, but only recently began videotaping his own lessons. Over the past few months, he has learned a lot about himself and about students in his physics

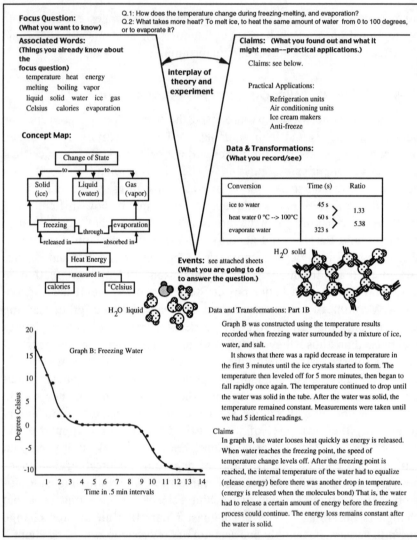

Focus Question:
(What you want to know)

Q.1: How does the temperature change during freezing-melting, and evaporation?
Q.2: What takes more heat? To melt ice, to heat the same amount of water from 0 to 100 degrees, or to evaporate it?

Associated Words:
(Things you already know about the focus question)

temperature heat energy
melting boiling vapor
liquid solid water ice gas
Celsius calories evaporation

Concept Map:

Change of State
— to — — to —
Solid (ice)   Liquid (water)   Gas (vapor)

freezing   evaporation
through
released in   absorbed in

Heat Energy
measured in
calories   °Celsius

H_2O liquid

interplay of theory and experiment

Claims: (What you found out and what it might mean--practical applications.)

Claims: see below.

Practical Applications:
Refrigeration units
Air conditioning units
Ice cream makers
Anti-freeze

Data & Transformations:
(What you record/see)

Conversion	Time (s)	Ratio
ice to water	45 s	1.33
heat water 0 °C --> 100°C	60 s	5.38
evaporate water	323 s	

Events: see attached sheets
(What you are going to do to answer the question.)

H_2O solid

Data and Transformations: Part 1B

Graph B was constructed using the temperature results recorded when freezing water surrounded by a mixture of ice, water, and salt.

It shows that there was a rapid decrease in temperature in the first 3 minutes until the ice crystals started to form. The temperature then leveled off for 5 more minutes, then began to fall rapidly once again. The temperature continued to drop until the water was solid in the tube. After the water was solid, the temperature remained constant. Measurements were taken until we had 5 identical readings.

Claims

In graph B, the water looses heat quickly as energy is released. When water reaches the freezing point, the speed of temperature change levels off. After the freezing point is reached, the internal temperature of the water had to equalize (release energy) before there was another drop in temperature. (energy is released when the molecules bond) That is, the water had to release a certain amount of energy before the freezing process could continue. The energy loss remains constant after the water is solid.

Graph B: Freezing Water

(y-axis: Degrees Celsius, from -10 to 20)
(x-axis: Time in .5 min intervals, 1 to 14)

Figure 1. Excerpt from an Activity Report Completed by Three Preservice Teachers in a College Physics Course

classes, which he taught without lectures. The course consisted almost exclusively of collaborative small group investigations and problem solving. The investigations centered around *focus questions* to which students sought answers experimentally. These activities were conducted by the students with minimal procedural guidance. During these sessions,

Ashmore used concept maps to summarize the key ideas of each of the relevant chapters. Most of the time, students were in small groups and therefore got to talk physics a lot. They experimented together, worked on their reports, and prepared collaborative concept maps to express the theoretical background of their laboratory experiments and to represent their learning during these laboratory experiments. He made students responsible for reading *Conceptual Physics* (Hewitt, 1989) and provided opportunities to engage in and to talk physics in the context of the activities he designed.

At the moment, Ashmore is struggling with the tension between students' making sense of phenomena in ways that are not consistent with his own understandings, and assisting them to adopt different ways of talking about and interpreting phenomena without falling into a mode of telling. The video shows him approaching a group of three elementary teachers sitting in front of a graph they had produced from the temperature of a vial of water after it was put into a mixture of ice, salt, and water. This is the same group that had submitted the report that lay next to him (Figure 1).

Ashmore: What I am particularly interested in is the interpretation that you give to this part (points to the flat in the freezing curve, Graph B in Figure 1). Because down here (flat part) it stays flat, then drops off. Why doesn't it go up immediately as soon as you started heating here in Graph A?

Carla: When the temperature is rising (gestures along graph where temperature goes up), then there has to be a change in kinetic energy of molecules.

Jill: Because it takes energy to change state, right here (points to flat part of temperature curve) it's changing state to ice (moves finger along line and stops at downward bend) here.

Ashmore: Where does the energy come from?

Gina: The energy comes from water.

Carla: The ice around (looks at Jill) the ice-water-salt mixture..

Jill: Yes, the salt-ice-water mixture.

Ashmore: In this case (moves finger along flat part of curve), do you lose or gain energy? Where does it go to?

Jill: The energy is being lost.

Carla:	Into the ice.
Gina:	The energy is being lost here (points to flat part) because of change of state and more energy is being used.
Carla:	All the energy that is being lost is not showing up here on the temperature because it is not being transferred.
Jill:	Is being lost because of change of state.

As he watches the tape and listens to the conversation between students and himself, he is quite pleased with the physics talk he is hearing and the success of his hunch about the graphs being sites for sense-making conversations. Students have begun to engage in conversations about the objects and phenomena he prepared for them. As they interact with each other, he visits each group and participates in the conversations. Sometimes, as in the episode he just watched, he asks questions to engage students in committing to sense making. At other times, he simply lends a hand, or provides students with some information that will help them along in their investigation. He thinks that students still have some way to go until they begin to pose a lot of questions themselves that will help them to generate plausible interpretations for the inscriptions they have generated. The episode also reifies his beliefs about graphs as devices to support sense making as students talk about, point to, and gesture with reference to them.

He had not always viewed students' work, and particularly the graphs and diagrams they generated during their work, in the same way. Through his training as a physicist and his interests in epistemology he had come to believe that structure is out there in the world. All one had to do was look well enough at the world around, and one could discover the laws that govern the world. When he began teaching, he wanted his students to discover the world for themselves. In this way, his approach to teaching was much more insidious than anything Miller had done. Miller at least told his students what he wanted them to learn while, Ashmore asked his students to find out for themselves what had taken scientists more than 2,000 years to put together. He had also believed, as did Miller, that the structure of nature and graphs were self-same. Failure to read this structure from graphs was attributed to problems in students' motivation or cognitive ability.

Ashmore's views changed considerably when he began to videotape his classes and took the time to watch and try to understand what was happening. As he began to watch the videotapes to examine in detail what students actually did, his experience was much like that of his colleague Schoenfeld:[3] What he saw was nothing like what he expected, and nothing like what he experienced as a teacher. In class, teachers are concerned with keeping the lessons going; making sure that those students who need help are attended to; and when equipment breaks down, teachers assist students to get back on track. As outcomes, they see only what students produce, on homework, lab reports, or tests and examinations. With the videotapes, things changed. Ashmore began to realize that his students did not see the same things he saw when they looked at the same objects and events. They seemed to act in and describe different worlds. Because they brought their own understandings to interpret the world around them, they came to different conclusions than Ashmore, who had been enculturated to particular ways of seeing and talking about the world in his physics community.

As he observed his interactions with students, Ashmore realized how much he dominated the interactions, cut off students, and evidently did not listen to what they wanted to communicate. When he read Schön's account of the different teacher–student interactions in an architectural practicum, Ashmore realized that all his teaching ultimately would fail if he did not listen to students with the same intent as he expected his students to listen to him. He just had to find situations in which he and his students could establish a conversation and make sense of each other's talk. Ashmore realized that situations that had some concrete referent available could make it easier to repair conversational troubles. He recognized that, for example, graphs and other forms of representations would be ideal settings for learning. Students could engage each other to come to agreements about the meanings of inscriptions, and he could engage students in conversations in the course of which they could find out about each other's meanings.

Ashmore thought about the similarities in Miller's and his own biography. Both had been highly successful athletes, and coaching had become to both an important metaphor for thinking about teaching and learning. However, unlike Miller, Ashmore thought that coaching was essentially a dialogue in which students made sense of what the coach said after they engaged in sporting activities. Whatever a coach could say beforehand made little sense unless there was a prior experience that the athlete could use to make sense of the coach's talk. It was as if you had to begin to engage in an activity before you knew what it was about. Now Ashmore knows that Schön described coaching in an architectural design studio, master classes of music, and interactions between expert and novice psychoanalysts. In any case, for coaching to be successful, coach and athlete/student had to be able to understand each other and therefore to listen to each other. He remembered one particular year in his athletic career when the communication between him and his coach had been disrupted, with dire consequences. After a year of success in which he had been runner-up in the world championships came a slump in performance that continued for several years until Ashmore and his coach reestablished an open and trusting relationship. Miller, with his experiences as athlete and coach, should be able to relate to this way of making sense of teaching and learning.

In the class Ashmore currently teaches, students' feedback seems to support his new realization that graphs are not representations of nature, but the outcome of scientific practices that, as any other practices, are appropriated as individuals participate in relevant activities. Some of the comments in the students' reflections read, "I am also learning more with each lab about correctly interpreting data recorded in graphs" (Jill); "With each lab I learn a little more about graphing and interpreting data" (Carla); and "It was good to be able to talk out the interpretations of the graphs, especially for someone like myself who has not had the experience with analyzing numerical data in this format" (Mary).

Increasingly Ashmore had come to realize that one of the key facilitating factors for his students' learning was the fact that they began to think about possible ways of conducting the investigations with their own students. The students in this class appeared to learn more and were better motivated because they related physics to their own teaching: "This would be an interesting experiment to develop for use in a primary classroom. The main problem would be in working with a heat source. Primary children and teachers (me) aren't too safe around hot flames but perhaps a safer hot plate could be used" (Lois); "This year, when we do experiments within the class, I will be able to relate to the children more, from their perspective" (Gina); and "The concepts of freezing and evaporation can be introduced to very young children, and it is important as a teacher to have a good grasp of what is actually going on so that as much information as possible is available for teaching" (Christina). As they thought about and asked for advice in these matters, they reflected on and became increasingly aware of their own understandings that had been in agreement with the scientific canon: "I cannot believe that I have lived so many years misunderstanding such a simple process as boiling water. I always thought that the bubbles in boiling water were air bubbles. But there you go, I was mistaken" (Christina).

After viewing the video, Ashmore turned to the report Carla, Gina, and Jill had submitted (partially reproduced in Figure 1). He was quite pleased to recognize how far the three had come. Initially, they were filled, like most of their peers, with immense anxieties, afraid to interpret and make commitments about what they observed. Now they had begun to integrate their own prior experiences ("Predicting with success is always gratifying, and figuring out why results are different than expected is also exciting, especially in a group") with the new ones in this lab, and they integrated what they did with their readings. He could see how his interactions with them appeared to have settled out in the students' interpretations and claims.

There are still things to be done. Ashmore sees that the three students included degrees Celsius as a measure of heat energy (Figure 1). He recalls having read about the conceptual problems children have in which they do not seem to differentiate heat and temperature. He decides to address the issue of heat and temperature in a whole-class conversation, in part, because other students in the class appear to have developed a more differentiated discourse that distinguished between heat and temperature. Ashmore wants to make use of this existing expertise to assist more students in developing a differentiated discourse. Given the molecular-level drawings of water in the solid, liquid, and gas states many of them had included in their reports, he considers introducing the notion of entropy, which scientists use to describe the degree of order in a system.

The notion of learning by becoming a member of a community as one participates in practices was instantly meaningful to Ashmore. He remembered the time when he first entered the physics labs where he did his graduate work. Initially, everything appeared foreign, especially what the other graduate students, post-docs, and the professors were talking about. Although they tried to explain what they were doing and saying, there appeared to be a border between him and his colleagues. With time, however, as he spent more time in the labs, he became more and more part of the daily routines. He became a member of the community and found himself on the other side of what had appeared to be an impassable boundary. He remembers that a lot of becoming part of the community involved the use of various kinds of inscriptions, printouts from a multichannel analyzer, curves drawn by a plotter connected to apparatus, and computer programs and images.

A few weeks later, Ashmore received the students' anonymous evaluations of the physics course. Almost all students shared in the assessment that they had learned a lot of physics, decreased their anxieties toward the subject, generated new ideas for teaching physics to children, and developed more positive attitudes about teaching physics in elementary schools. The following is one of these evaluations:

> I liked the way Ashmore approaches this physics course, particularly his acceptance and use of the cooperative learning model. Because of this model, we were given the opportunity to converse and discuss concepts, labs, etc. He is a very caring individual and this is so evident in his feedback, insistence that there are no dumb questions, and his concern that we feel more positive about physics. Many of us entered this classroom looking at the floor, little confidence and low self-esteem about the course content. I no longer look at the floor, I almost feel like I can cope with these concepts. Especially the way he connects and applies these concepts to everyday practical situations. He has helped me make meaning and sense of much of the material, not only through teaching, but through hands-on, collaborative activities. Ashmore is my idea of a good teacher; he is more a partner in learning (a collaborator) rather than a teacher (lecturer). As I look at the course objectives, I see that we met them. I feel I have developed a sense of meaning (relevant and purposeful) of momentum, force, velocity, energy, heat, temperature. I now have a better attitude toward physics. [Anonymous student]

Conclusion

At the time of our research, we perceived a considerable discontinuity between the two communities represented in this study. Our analyses of the discourse that characterized the physics course designed to meet the needs of preservice elementary teachers underscores the challenge of these students' physics learning. To be successful in mediating learning, college physics teachers and members of the physics community need to provide for situations in which they can build a common discourse with learners. This common discourse needs to be in terms of the language already accessible to students, from which they can develop canonical physics talk. Miller's present focus on learning powerful ideas and making visible the processes underpinning the development of these ideas floundered because the learners had few discursive resources to connect with Miller. They also had few opportunities to co-participate with him in doing and talking physics. Co-participation appears to be an important key to successful learning, meaningful interaction during which the teacher and learners cooperate to create a discourse that can later be applied to the teaching of science in elementary schools. Co-participation in physics discourse appears to be the kind of experience that allowed students to learn so much in Ashmore's class. In addition, Ashmore's students took every opportunity among themselves and with him to talk about possible ways of conducting the investigations with, and explaining the phenomena to, the children they were teaching. In this interplay between their own goals, becoming better science teachers, and learning physics themselves, their learning was often quite incidental. With their subject matter competence also grew their confidence in themselves as learners, with few hesitations to accept their own prior ideas as different from and less fruitful than the ones they learned in this class.

It should be possible to assist college physics teachers to understand the limitations of older learning and teaching paradigms. Learners are not simply computing devices that need to be stuffed with information, and with languages that help us interface with the world, especially with other people. Our being, language, and the worlds we inhabit and experience are fundamentally interconnected. Once we take such a perspective, our teaching changes. We can set up opportunities for students to create life worlds, to create the worlds they experience and inhabit. Science no longer is a subject, but a useful way of life, with an idiom that makes useful distinctions that our everyday idiom does not make. Words make differences, they disclose objects and imbue them with their familiar aspects. Words disclose, and in this disclosure, they appropriate the world. As language, they are aspects of a life form that we appropriate and become part of through participation. Physics teaching should make differences. But these differences should not be those that create boundaries. Physics teaching should not be boundary work that establishes discourse communities that no longer can converse with each other. Physics teaching should make a difference.

Ashmore changed his views about knowing and learning and, as a consequence, made a difference in the lives of students. He allowed them to cross the borders that his elementary education majors had experienced, between themselves and knowing physics. Miller too had made a start in recognizing that he was responsible in that his teaching did not achieve its intended goals (see opening quotation). If he is prepared to view himself as a learner with respect to his experience, then further gains can be anticipated to the benefit of his students who, in turn, will go forward into their classrooms and provide youngsters with a sound start on a journey toward scientific literacy. As learners, we too could engage together with Miller to explore new ways of describing and framing teaching and learning, and draw implications for teaching. But we know from previous research that while change of perspectives is possible, it does not come easily and involves great commitment to change. It had taken Ashmore years to change, and it still takes continuous reflections about his actions on the basis of the videotapes of his classes. In Miller's situation, modifying his teaching to adapt it to preservice elementary teachers' needs may still require a long and arduous process. However, a modification is needed lest the chasm between him and the students be widened.

Wolff-Michael Roth
Applied Cognitive Science
University of Victoria, Victoria, British Columbia, Canada

Kenneth Tobin
Graduate School of Education
University of Pennsylvania, Philadelphia, Pennsylvania, USA

EDITORS

METALOGUE

Re-Conceptualization via Discourses about Learning

KT: In the chapter that I coauthored with Michael Roth we experienced the helplessness of being a learner in the kinds of physics classes taught by Miller. As you can read from the story of Ashmore, both Michael and I appreciated the position taken by Miller. We had been just like that in so many ways. I will never forget the day when I was interviewing two of the best students from Miller's class. So that they would not be caught without any images of what the class was, like we had a videotape segment to allow them to reconstruct a lecture about a ball rolling down an inclined plane. After I had asked a mature-aged woman student to tell me what she knew about the physics associated with the ball rolling down the slope, there was a very long period of silence. Then she began to cry. "I know nothing," she sobbed. Fighting back my own tears, I encouraged her to begin by telling me a story of what happened. As the narrative began she gained in confidence and created an oral text that became an object for her own critique. During the interview she used her narrative as a way to relearn physics. In the process she also made visible a number of non-Newtonian ideas—kernels for someone like me to use as a basis for more teaching of physics. My main points here are that I would not like to have my better students feel that they had wasted their time in a whole course, as she had proclaimed. Second, giving her an opportunity to use resources associated with her primary discourse enabled her to build a bridge back into the less familiar discourse of physics.

PT: As I sat in the classes of the various college science professors who make up the textual character of Dr. Stern (Chapter 1), I too recalled my own none-too-joyous learning experiences over the years at the hands of various far-from-exemplary science teachers. The main difference now is that I know helplessness is not attributable simply to some deficiency in my own makeup. I am empowered sufficiently with educational insight to identify bad pedagogy when I "smell" it. But I also have developed advanced learning skills that enable me to hold a critical self-reflective discourse with myself about the quality of my conceptual understanding and the nature of (emotional, conceptual) impediments to its development. So I empathize with the very many students who continue to be disempowered in science classes by inadequate teaching that can, in the absence of any discourse about the nature of learning, promote "slow silent death" by a thousand expository lectures.

PG: A few years ago I cotaught a physical science course with Miller, and I can hear clearly Miller's voice and discourse reflected in Tobin and Roth's chapter. Changing our teaching involves a change in our belief systems, and it takes time to reflect and learn and change. Just today I had a discussion with Miller and feel like he is starting to move on the pathway toward self-reconceptualization. His being involved in the science education community and on dissertation committees in science education are causing him to reflect. It would be interesting to teach another course with him now, and see if he does teach differently than he did a few years ago when he started on this journey in science education. I know I teach radically differently ever since teaching the honors general chemistry course that I describe in this book (Chapter 17).

Alternative Ways of Portraying
Ethically Nonexemplary Pedagogy

KT: There has been a tendency for readers of our research with Miller to see what we were writing as teacher bashing. I can see their point but want to maintain that we hold no disrespect for Miller. In fact I spoke with him on the telephone just this week. We are good friends. My point is rather simple. We must learn from our practices. No matter how well we think we are teaching we can always improve. But what should we do differently? In this chapter we set out to address that question in a serious way by trying to provide a rationale for why Miller taught as he did. Although Miller's ways of thinking about teaching and learning are markedly different from those of Mark Campbell Williams or Penny J. Gilmer in this book (Chapters 16 and 17, respectively) they are representative of the thinking that underpins most of the approaches to teaching at the undergraduate level of college.

PT: Ken, you have raised again (see Metalogue of Chapter 2) the thorny ethical question of how to deal with pedagogy that is far from exemplary, particularly when one encounters it as a researcher. Should the researcher's agenda (serving the broader educational community, perhaps) assume a privileged moral status and permit him to press on, documenting the case and conducting validity checks regardless of the consequences? As we know from experience, teachers can become quite offended by what they perceive to be unfair or prejudicial (interpretive) judgments by researchers. Are the warrants of friendship or educative intent sufficient to justify potentially offensive research of this kind? There is no simple answer, of course, and so we need to tread carefully. Indeed, some feminist perspectives would urge us to act always with an ethic of care as the guiding principle in our (educative) relationships with teachers (no matter how villainous the teacher?). In researching the character of Dr. Stern, I found a way through the ethical maze by constructing a *phenomenological account* of the impact of the teaching on my own sensibilities/sensitivities, which I had finely tuned to be empathic to struggling learners, and I constructed a *composite character* for whom credibility checks were not needed (Chapter 1). That has proved to be a promising line of inquiry. Complementing this approach is the *autobiographical account* of "Ashmore," an exemplary professor of college physics. Perhaps the transformation of college science teaching hinges on the future proliferation of autobiographical accounts such as this, written by professors of science (in collaboration with skilled science education researchers). Doubtless, their authenticity makes for compelling reading.

KT: Michael is a fine example of what can be accomplished as a teacher–researcher. Also a physicist he (Ashmore) undertook studies in science education and then returned to high school science classes, where he engaged in one of the more spectacular programs of research in science teaching and learning ever to be undertaken by a teacher. His willingness to always return to classrooms, roll up his sleeves and learn about teaching by teaching and *co-teaching* with others is an example from which we all can learn.

PT: Of course, the practice of co-teaching is not unproblematic, as can be seen clearly in Sabitra's chapter on *team teaching* (Chapter 11).

PG: One thing that really impresses me about Michael's approach to his educational research is that he transcribes the audiotape on the same evening that he taught the class, and he develops research ideas from concentrating on the discourse and questions in his classroom before he sees his students the next day. I think one reason that Michael is so productive is that he doesn't lose a moment to what he can learn and research.

NOTES

1. We provided elsewhere a fine-grained analysis of these transformations through the entire cascade of inscriptions, and Miller's construction of a money-earning analogy (Roth, Tobin, & Shaw, 1997).
2. Unfortunately, Miller did not represent the historical development of understanding about motion on the incline. His lecture constituted a pedagogical reconstruction of the events, which deleted much of Galileo's struggle to arrive at the equations that he, Miller, arrived at in a matter of a lesson.
3. Alan Schoenfeld (1987) provides an interesting account of the changes in his understanding of problem solving in mathematics when he started to watch what students actually did rather than what he thought they did.

BIBLIOGRAPHY

Akrich, M. (1987). How can technical objects be described? Paper presented at the Second Workshop on Social and Historical Studies of Technology, Twente University.

Bakhtin, M. (1981). *The dialogic imagination*. Austin: University of Texas Press.

Cobb, P., Wood, T., Yackel, E., Nicholls, J., Wheatley, G., Trigatti, B., & Perlwitz, M. (1991). Problem-centered mathematics projects. *Journal of Research in Mathematics Education 22*, 3–29.

Dreyfus, H. L. (1992). *What computers still can't do: A critique of artificial reason*. Cambridge, MA: MIT Press.

Duit, R. (in press) Conceptual change in science education. In M. Carretero, W. Schnotz, & S. Vosniadou (Eds.), *Conceptual change*. Mahwah, NJ: Lawrence Erlbaum Associates.

Ehn, P. (1992). Scandinavian design: On participation and skill. In P. S. Adler & T. A. Winograd (Eds.), *Usability: Turning technologies into tools* (pp. 96–132). New York: Oxford University Press.

Ehn, P., & Kyng, M. (1991). Cardboard computers: Mocking-it-up or hands-on the future. In J. Greenbaum & M. Kyng (Eds.), *Design at work: Cooperative design of computer systems* (pp. 169–195). Hillsdale, NJ: Lawrence Erlbaum Associates.

Galileo, G. (1960). *On motion and on mechanics.* Madison: University of Wisconsin Press.

Gieryn, T. (1996). Policing STS: A boundary-work souvenir from the Smithsonian exhibition on science in American life. *Science, Technology, & Human Values 21,* 100–115.

Gieryn, T. F., Bevins, G. M., & Zehr, S. C. (1985). Professionalization of American scientists: Public science in the creation/evolution trials. *American Sociological Review 50,* 392–409.

Giroux, H. (1992). *Border crossings: Cultural workers and the politics of education.* New York: Routledge.

Gooding, D. (1992). Putting agency back into experiment. In A. Pickering (Ed.), *Science as practice and culture* (pp. 65–112). Chicago: University of Chicago Press.

Henderson, K. (1991). Flexible sketches and inflexible databases: Visual communication, conscription devices, and boundary objects in design engineering. *Science, Technology, & Human Values 16,* 448–473.

Hewitt, P. G. (1989). *Conceptual physics, 6th Ed.* Glenview, IL: Scott, Foresman.

Lampert, M. (1986). Knowing, doing, and teaching multiplication. *Cognition and Instruction 3,* 305–342.

Latour, B. (1987). *Science in action: How to follow scientists and engineers through society.* Milton Keynes, UK: Open University Press.

———. (1993). *La Clef de Berlin et Autres Leáons d'un Amateur de Sciences.* Paris: Editions la Découverte.

Lave, J. (1988). *Cognition in practice: mind, mathematics and culture in everyday life.* Cambridge, England: Cambridge University Press.

Lemke, J. L. (1998). Multiplying meaning: visual and verbal semiotics in scientific text. In J. R. Martin (Ed.), *Scientific discourse* (pp. 87–113). New York: Longmans.

McGinn, M. K., Roth, W.-M., Boutonné, S., & Woszczyna, C. (1995). The transformation of individual and collective knowledge in elementary science classrooms that are organized as knowledge-building communities. *Research in Science Education 25,* 163–189.

Mehan, H. (1993). Beneath the skin and between the ears: A case study in the politics of representation. In S. Chaiklin & J. Lave (Eds.), *Understanding practice: perspectives on activity and context* (pp. 241–268). Cambridge, England: Cambridge University Press.

Pinch, T. (1979). Normal explanations of the paranormal: The demarcation problem in fraud and parapsychology. *Social Studies of Science 9,* 329–348.

Rorty, R. (1989). *Contingency, irony, and solidarity.* Cambridge, England: Cambridge University Press.

Roth, W.-M. (1995a). Affordances of computers in teacher-student interactions: The case of interactive physics. *Journal of Research in Science Teaching 32,* 329–347.

———. (1995b). Inventors, copycats, and everyone else: The emergence of shared (arti)facts and concepts as defining aspects of classroom communities. *Science Education 79,* 475–502.

————. (1996). Thinking with hands, eyes, and signs: multimodal science talk in a grade 6/7 unit on simple machines. *Interactive Learning Environments 4*, 170–187.

Roth, W.-M., McRobbie, C., Lucas, K. B., & Boutonné, S. (1997). The local production of order in traditional science laboratories: A phenomenological analysis. *Learning and Instruction 7*, 107–136.

Roth, W.-M., & Roychoudhury, A. (1994). Science discourse through collaborative concept mapping: New perspectives for the science teacher. *International Journal for Science Education 16*, 437–455.

Roth, W.-M., & Tobin, K. (1996). Aristotle and natural observation versus Galileo and scientific experiment: An analysis of lectures in physics for elementary teachers in terms of discourse and inscriptions. *Journal of Research in Science Teaching 33*, 135–157.

Roth, W.-M., Tobin, K., & Shaw, K. (1997). Cascades of inscriptions and the representation of nature: How numbers, tables, graphs, and money come to represent a rolling ball. *International Journal of Science Education 19*, 1075–1091.

Schoenfeld, A. H. (1987). Confessions of an accidental theorist. *For the Learning of Mathematics 7*, 30–38.

Schön, D. A. (1987). *Educating the reflective practitioner*. San Francisco: Jossey-Bass.

Solomon, S. M., & Hackett, E. J. (1996). Setting boundaries between science and law: Lessons from *Daubert v. Merrell Dow Pharmaceuticals, Inc. Science, Technology & Human Values 21*, 131–156.

Star, S. L. (1989). Layered space, formal representations and long-distance control: The politics of information. *Fundamenta Scientiae 10*, 125–154.

————. (1991). Power, technology and the phenomenology of conventions: On being allergic to onions. In J. Law (Ed.), *A sociology of monsters: Essays on power, technology and domination* (pp. 26–56). London and New York: Routledge.

Star, S. L., & Griesemer, J. R. (1989). Institutional ecology, translations and boundary objects: Amateurs and professionals in Berkeley's Museum of Vertebrate Zoology, 1907–39. *Social Studies of Science 19*, 387–420.

Suchman, L. A., & Trigg, R. H. (1991). Understanding practice: video as a medium for reflection and design. In J. Greenbaum & M. Kyng (Eds.), *Design at work: cooperative design of computer systems* (pp. 65–89). Hillsdale: Lawrence Erlbaum Associates.

Traweek, S. (1988). *Beamtimes and lifetimes: The world of high energy physicists*. Cambridge, MA: MIT Press.

Wittgenstein, L. (1968). *Philosophical investigations*. Oxford: Basil Blackwell.

SECTION II

 **Pushing the Envelope**

CHAPTER 7

Promoting Active Learning in a University Chemistry Class: Metaphors as Referents for Teachers' Roles and Actions

Abdullah O. Abbas, Kenneth A. Goldsby, and Penny J. Gilmer

University faculty who teach required science courses for prospective teachers generally use traditional approaches to teaching in which learning is conceptualized in terms of a transmission model. Thus, prospective teachers learn science in learning environments that encourage memorization of science information and that place relatively little emphasis on understanding knowledge and how to use it in daily life. It is not surprising, therefore, that many new teachers of science create learning environments in their own classes that are very similar to the environments they experienced as students of science.

More than 400 national reports (Hurd, 1994) published in the United States during the 1980s and 1990s (e.g., National Research Council, 1996; National Science Teachers Association, 1992; Rutherford & Ahlgren, 1990) have called for reform in science teaching and learning, particularly in classes for prospective teachers. Shortcomings of science education have been traced back to the elementary school level; so the preparation of prospective elementary school teachers to teach science is a major concern. These national reports call for major changes in what, when, and how we teach science, many recommending more constructivist teaching approaches (Glasersfeld, 1989) in science classes for prospective elementary school teachers. There is a prevailing view that prospective teachers need to experience learning environments in which meaningful learning takes place.

A recent study of exemplary science teachers has linked success at improving student interest in science with providing a context in which meaningful learning can take place. Such teachers, as noted by Collette & Chiappetta (1994), modify "the context within which the subject matter is taught so that it connects with what students know and is perceived to be relevant, yet in a manner that relates to the curriculum" (p. 76). They endeavor to make their teaching appealing by creating an environment that is rich in interesting activities and topics. Such an environment motivates students to be actively involved in the learning process, using prior experience and knowledge to construct meanings in new situations.

This chapter draws on the results of a study designed to examine the classroom actions of an exemplary university chemistry teacher of prospective elementary school teachers and the way that actions relate to his beliefs, goals, and teaching roles. The central purpose of the study is to understand the metaphors that Mark (a pseudonym) uses as referents to conceptualize his roles and frame his actions and interactions and, thereby, to generate an exciting environment for learning. We use pseudonyms for the students throughout the chapter to protect the anonymity of the participants.

The significance of this case study concerns the ways in which a university chemistry teacher and his students conceptualize their own roles and those of others in the class, and how the teacher's varying roles mediate the construction and maintenance of an environment that encourages and stimulates students to learn science actively.

Roles of the Authors

Abdullah Abbas conducted his doctoral dissertation research on Mark's teaching (Abbas, 1997). Abdullah was particularly interested in studting Mark because he wanted to improve his own teaching in his native country of Yemen, where he is a professor of science education. Abdullah chose to study Mark's teaching because Mark was an exemplary teacher, as evidenced in prior studies by Brush (1993) and Duffy (1993).

The second author, Kenneth Goldsby, is Mark, the teacher whom Abdullah studied. We chose a pseudonym so Ken himself does not get fixed in time as he was during the study. Ken was involved formatively throughout the study, and also served as the "outside" representative on

Abdullah's College of Education dissertation committee. While Mark is frozen in time, Ken continues to learn, develop, and evolve in his beliefs and philosophy on the practice of teaching.

The third author, Penny J. Gilmer, was one of two principal investigators with Ken Tobin for the National Science Foundation grant that supported the development of this course. Penny was active in working with teams of scientists, K-12 teachers, and science educators to develop three new interdisciplinary science courses, including the physical science course described here, designed for prospective elementary school teachers. She also was Abdullah's major professor.

An Interpretive Research Approach

We employed an interpretive research design, described by Erickson (1986) and Guba & Lincoln (1989), in the study. Erickson (1986) defines interpretive research as "the immediate and local meanings of actions, as defined from the [participants'] points of view" (p. 119). The basic assumption of this type of research is that the researcher and her or his subjective experience provide the lens through which the research develops (Brush, 1993). The main sources of data for this study are field notes, transcript analysis of interviews with the teacher and students, and analyses of videotaped excerpts. We employ additional data sources to ensure that the inferences we constructed are consistent with the variety of data.

Constructivism (Glasersfeld, 1989) and co-participation (Lemke, 1995; Schön, 1985) are the main theoretical frameworks of the study. The basic tenet of constructivism is that knowledge is personally constructed but socially mediated (Tobin & Tippins, 1993). Knowledge is always the result of constructive activity by an individual while he or she exists in a cultural sense. This epistemology enables us to make sense of how students learn and how the teacher works with them to create learning environments that enhance their learning.

Within a constructivist framework, participants become empowered and act and reflect on their actions. Through interaction with others, the students can negotiate meanings of actions to arrive at a consensus on what has been learned. Co-participation implies, as stated by Tobin (1997), "the presence of a shared language that can be accessed by all participants to communicate with one another such that meaningful learning occurs" (p. 369). The teachers who know science assist students to learn by engaging in activities in which co-participation occurs (Roth, 1995). In such situations, students are empowered to engage actively in the process of knowledge construction and have the autonomy to ask questions when they have problems (Tobin, 1997).

An interpretive approach to studying science teaching and learning, as suggested by Guba & Lincoln (1989), requires that the study takes place in the natural location or place, such as a classroom or a laboratory, where the events occur that we wish to interpret. We conducted this study in a university physical science course designed especially for prospective elementary school teachers. Our focus is on the actions of Mark, the chemistry teacher. His reading and reflection on the audiotape and videotaped transcriptions provide what Guba & Lincoln (1989) refer to as "member checks." The researcher's prolonged and intensive engagement (Guba & Lincoln, 1989) in the class during the whole semester added to the credibility of the study.

The following vignette is a reflection on the class taught by Mark in which he utilizes dry ice and liquid nitrogen in a class activity. It is based on field notes, an analysis of the videotape and interviews, and a review of students' journals. The vignette provides insights into the learning environment that Mark endeavored to create in order to maximize students' participation in science learning.

Smashing Up the Ball Against the Wall Caught Me off Guard

It was Friday. Students and teachers were wearing heavy clothes because it was very cold. Two teachers and the two teaching assistants, Chris and me, plus twenty-one students, all females, were in the class. Students were arranged in five groups, each group with four or five students. The physics teacher, Adams, finished his presentation about error in measurement at about noon.

The chemistry professor, Mark, began talking about temperature and explained the connection between temperature and the concepts of density, volume, and mass, which were explained by Adams in the first hour of the class. Then he gave a short presentation, for approximately fifteen minutes, about temperature, Fahrenheit scale, Celsius scale, Kelvin scale, and the boiling and freezing points of water in each scale. He used the chalkboard to draw three thermometers representing Celsius (^oC), Fahrenheit (^oF) and Kelvin (K). For each scale the students were *shown both the boiling and freezing points of water.*

At about 12:15 P.M., Mark asked the students to start activities with liquid nitrogen and dry ice (solid carbon dioxide). He clarified to the students how liquid nitrogen and dry ice are very cold, their temperatures being 189.8°C (-320°F) and -87.8°C (-109°F), respectively. Then he poured some liquid nitrogen into his hand, but it turned to gas as quickly as it reached his hand. He explained to them how this demonstration indicates that the hand temperature is very high compared to the temperature of the liquid nitrogen. The students became excited after this demonstration. When he showed them the dry ice, they were enthusiastic to see how these small pellets turn from a solid phase to carbon dioxide gas without getting their hands or the table wet. Mark described the different uses of dry ice in our daily lives, such as keeping drugs or foods refrigerated when they are sent by mail.

After a short break, students came back to the class eager to start their activities. Before starting, Mark showed them how the volume of nitrogen gas can be measured by putting an amount of liquid nitrogen into a test tube, then immediately putting a balloon over the top of the test tube. The balloon began to inflate due to the liquid becoming a gas. Then Mark asked them to start group work and write down their observations and comments about what they observed. He gave each group some liquid nitrogen in a Styrofoam cup while the teaching assistant distributed balloons and some dry ice to each of the groups. Each student worked in her group to perform the activity.

On the first occasion when balloons were being immersed in the liquid nitrogen, some students moved away and put their hands over their ears. They expected the balloons to burst. Students showed their excitement as they worked, especially when Mark, who previously had immersed a racquetball in liquid nitrogen, threw it against the wall and it shattered into many pieces. Some students searched the floor of the room looking for pieces of the ball. One student, Nora, commented on these actions in an interview as:

> That was neat. I liked playing with that. That was fun, to see [the ball] smash up against the wall, and the balloons with the air, and then the dry ice that we used and just threw on the table. I like doing crazy things like that. I didn't expect him to throw the ball up against the wall. That caught me off guard.

Aisha wrote in her journal the following description of what her group did during the dry ice/liquid nitrogen/balloon activity in her group:

> We experimented with small pellets of dry ice.....We stuck one pellet of dry ice in the balloon and tied it up. The dry ice begins to blow the balloon up slowly and then eventually almost appears to stop until we pick it back up and shake it and bring the inside to motion. We also did some experiments with the liquid nitrogen. [Mark] stuck a racquetball into the liquid nitrogen and it immediately turned into a glass-like state. [He] then threw the ball against the wall and it shattered into many pieces. We stuck our balloons with the dry ice in it into our cup filled with liquid nitrogen. The balloon seemed to collapse and [shriveled] up when coming in contact with the nitrogen.

While work in the activity continued, students discussed with each other within groups what they observed and sought help from Mark when they needed assistance. Mark moved from one group to another to assist students with their work, participate in discussion, and help them to understand what to do.

One student asked a good question, "What will happen if we blow a balloon up using regular air and put it into the liquid nitrogen?" Mark asked each student to predict and discuss within each group what would happen in this example, and to record their predictions before putting balloons with air into the liquid nitrogen. When they finished recording their predictions they inflated their balloons with air. Aisha wrote in her journal about her predictions and observations as follows:

> My prediction about blowing the balloon up with regular air instead of using the gas given off from dry ice is that it will burst under the pressure of the density of air inside the regular balloon. My guess is that it will pop. The air balloon shriveled quickly after being immersed into the liquid nitrogen then went back to its normal state as it was before being blown up. Then, we tried the dry ice once again and it seemed to drip condensation.

After a while they were asked to look for whatever they could see inside the balloons filled with air, such as liquid, after putting them in the liquid nitrogen. The teachers tried to help students to observe liquid inside the balloons. Some students could see a liquid, while some of them could not. Huda, for example, wrote in her journal, "I saw a liquid in the balloon for a second, but then it disappeared as the balloon got its shape back. I have no idea what it is." But Aisha wrote in her journal,

"Although we were asked to see something inside of the air-filled balloon many times, I never really saw anything worth noting. Nothing was inside of the balloon." Before ending the class, students within each group discussed their observations and wrote what they thought in their journals.

Teaching for Active Learning

Beliefs about how students learn science can have a direct influence on the classroom roles of science teachers (Tobin, 1990a). A belief, as noted by Tobin & LaMaster (1995), is "knowledge that is viable in that it enables an individual to meet [his or] her goals in specific circumstances" (p. 226). When a science teacher has a belief that students learn science by listening to his or her lectures, for example, this teacher tends to use a transmission-absorption model for teaching and learning. Within this model, students pay less attention to meaningful learning since the teacher's goal is to cover the syllabus of the course and prepare students to pass the tests.

However, Mark's approach to teaching science is dynamic. He encourages students to engage actively in discussions and activities intended to promote their interest in chemistry learning. This approach is consistent with his goals of creating learner-centered classrooms and maximizing the participation of his students who are non-science majors and frequently have negative feelings about science (Brush, 1993). Using different strategies for teaching indicates that Mark is always on the lookout for ways to stimulate active and varied student participation. Chemistry classes are rich with exciting demonstrations and hands-on activities, such as the dry ice, liquid nitrogen, and balloons described in the vignette, which Mark selects to stimulate students' active engagement in the learning process.

In an interview, Mark emphasizes his beliefs about the importance of active learning in class of prospective teachers:

> I understand the constructivist's view that we cannot transfer knowledge intact to our students, but initially I thought that I could at least transfer my enthusiasm about science intact to someone else. I know now that I cannot do that, or at least I can only do it for a little while. The students have to construct their own interests. My role is to facilitate that process. I have to help the student find some reason that this material is interesting

or relevant. I can't transfer my interest or my sense of relevance to other students. They have to construct that for themselves.

Underlying Mark's beliefs about learners who are actively involved in science learning is a set of beliefs about the nature of learning and knowledge that he referred to as constructivism. As a referent for teaching, constructivism can shape the direction of a teacher's actions and his role as a science teacher in the classroom. Mark believes that to have students learn science actively means to create an environment in which alternative ways of teaching and learning can facilitate students' learning processes. If we consider learning from a social constructivist perspective, learners are no longer passive agents, but they engage in interactive discussions or in small group problem-solving activities. Accordingly, the teacher's role changes from a giver of knowledge to a mediator of learning (Tobin & Tippins, 1993). The science teacher, as Tobin and Tippins (1993) note, "takes account of what the students know, maximizes social interactions between learners such that they can negotiate meaning, and provides a variety of sensory experiences from which learning is built" (p. 10).

Mark's main goal is to maximize the participation of his students. He is always looking for ways to stimulate active and varied ways of participation among them. His classroom actions aim to stimulate students to participate and learn actively through doing science in which they can construct their own understandings of science concepts. It is clear that he works hard in chemistry classes to move from teacher-centered to learner-centered approaches of teaching and learning where alternative ways of science teaching and learning could be utilized. These findings are consistent with those of two studies conducted in the past few years about this course (Brush, 1993; Duffy, 1993).

It seems that Mark's beliefs about science teaching and learning evolved during his work as a science faculty member and a college science teacher. He comments on how his beliefs about science teaching and learning evolved during his work as a college science teacher in the following way:

I had heard a great deal about constructivism before beginning these courses. I think we all know that you learn by doing. Show me something and I'll remember it for a minute; let me do it and I'll remember forever. So I thought constructivism was just another way of saying that the best way to learn something was to do it. That is why I wanted to get on board [in this project] because I knew that is true from my own experience with laboratory research. The best teaching we do, the best learning environment we create for students, is when we bring them into our laboratory and have them do science in a mentoring relationship.

To achieve the goal of setting up exciting environments for learning in the classroom, Mark used metaphors to conceptualize his roles in creating such environments. Mark was able to switch his actions based on which of the constituent metaphors he used as a referent to frame his actions and interactions, and thereby, to create an exciting environment for learning.

"Learning is an Exciting Trip"

An assumption underlying this study is that many of Mark's beliefs about teaching and learning are metaphorical. We propose that, as Mark reflects on his actions and considers the various roles that he might adopt, he makes sense of his roles by the use of metaphors. Mark's personal epistemology is considered relevant in deciding whether or not a particular role is appropriate for use in his science teaching. If the role is consistent with his beliefs, the decision might be to adopt the role but, if not, the role might be considered inappropriate (Tobin, 1990b). Data in the study suggest that actions in Mark's classroom are metaphorical and consistent with his belief that students can learn science meaningfully when they enjoy the learning process.

Teaching practices in Mark's classes for the last few years (Duffy, 1993; Brush, 1993) indicate that Mark, as a constructivist teacher, developed his own metaphor to serve as a referent for his teaching practices. It is clear that his actions are embedded in the metaphor that "learning is an exciting trip." Mark used his beliefs, which are associated with constructivism, as a referent for actions in his classes and for the metaphors in which he embedded his actions. To be consistent with his beliefs that students should enjoy while learning science meaningfully, Mark uses the "trip" metaphor to construct a vision of what science classes for prospective teachers could be like. Such a metaphor is evident in interviews and in actions and interactions in Mark's classroom.

Mark describes actions and interactions in his classroom in terms of this "trip" metaphor, in which he conceptualizes his role as the trip driver, or a tour guide, and students' role as the travelers. In an interview,

Mark describes the way in which he uses the metaphor of a trip to justify his approach in teaching and learning:

> Learning is like a trip, and the teacher is the driver. [The driver points to a building] "And this is the Capitol and it was built in 1856...Let's go over here, I want to show you the legislative building, two bodies."...The students are in the back saying, "What is that building?" You go, "That's a neat building, do you want to go see it?" Or, "That building is not very interesting. Do you really want to go there? Maybe we can come back to it if we have time." This is a quickly thrown together metaphor for what I think the teacher should be doing. The first time I taught the chemistry course [an early version of the physical science course] I thought I should empower the student. I thought, "It's their course...I'll let them drive. I'll let them decide where we go." Sounds great in theory, but it did not work very well. Students are not used to being empowered. They don't know where to go or how to get there.

Underlying the trip metaphor is a more constructivist epistemology in which every person, including the driver, is involved actively in the trip and associated activities. Another important referent used by Mark is enjoyment. He believes that learning chemistry, especially for elementary education majors, should be enjoyable. The driver endeavors to create a high-quality atmosphere in which travelers enjoy learning about things and places along the way in their trip. This is consistent with Mark's actions in classes for prospective teachers (Duffy, 1993; Brush, 1993) to create exciting environments of learning through the use of different demonstrations and learning activities.

In a similar study, Ritchie (1994) described how a teacher used a metaphor of "teacher as travel agent" to transform her teaching to better agree with constructivism. This teacher used the metaphor as a referent for actions in her classroom. After a period of time, the teacher taught in a more routine manner and was less reliant on using the metaphor to guide her teaching practices. Using the trip metaphor allows Mark to focus on facilitating student learning through the use of instructional strategies to maximize the participation of all students in learning activities.

The trip driver enjoys the trip when he feels that other participants in the trip do so. Mark, too, teaches science with interest when he feels that students in the classroom become excited and engaged actively in learning activities. It is obvious that Mark becomes very excited about what is going on in the classroom, especially when students are fully involved in the events. We observed this situation in many classes during the semester, such as the classes on temperature, on covalent bonding, and on reduction and oxidation. Mark describes his feeling when instructional strategies work effectively to maximize students' interest and participation:

> I put a lot of time and energy into it [the physical science course], especially when I am teaching my part. I take it personally and I take it hard when things don't work, and I feel great when things do work.

Students' Prior Knowledge

The main goal for Mark as a trip driver is that all participants in the trip enjoy learning and knowing new and exciting places and things along the way of the trip. Mark encourages students to participate actively in the trip program using their prior knowledge and experience to construct new knowledge through watching, listening, asking questions, arguing with others, and reading carefully the brochures and posters about their journey. Each participant has his or her own prior knowledge and ideas about places and things on the trip. The driver can use his experience to help participants construct new knowledge about things and places on the trip.

Within this metaphor, students are active agents on the trip to achieve the goal of learning with understanding about topics and concepts in the course. Students use their prior knowledge built by home, TV, high school, and other organizations in the community to construct new knowledge about science concepts. Learning by construction implies a change in prior knowledge, where change can mean replacement, addition, or modification of extant knowledge (Cobern, 1993). Mark commented on the importance of students to conduct a fruitful trip of learning as follows:

> The teacher is in charge, the teacher sets the tone of the class...but at the same time, the students are very important. So the students can say, "Let's go here, I want to see this," but the teacher drives them. They say, "All right, let's go here and spend an hour." [The teacher may reply] "That's a good idea," or, "Don't go there," or, "Don't stay very long," or "No, we are not going over there, we just don't have time, we can't stop there." That is the teacher's job, but the teacher does his or her job best if he or she does it with the input of the class.

Students as Co-Participants

Since Mark's main goal as a teacher is to maximize the participation of his students, using the metaphor of a trip as a referent for actions encourages the class community to develop interaction in the classroom. The trip community creates a shared language that permits all participants to co-participate. Co-participation implies that each of the participants shares a language and can understand what is happening to the extent that there is freedom to participate and learn with understanding (Schön, 1985). The language is negotiated and is constantly evolving as learning occurs. Mark stimulates co-participation among participants through the use of different instructional strategies. Students learned chemical concepts with understanding because they feel that they have the power to talk, ask and answer questions, and express their ideas. Tobin & Tippins (1996) emphasize the importance of creating a discourse that is shared among participants in science classes: "In a community in which co-participation is occurring there are interactions among participants in which negotiation and consensus building are apparent and learners are empowered to participate and learn because of their ability to use a shared language" (p. 715).

The data suggest that the way in which Mark perceives his roles in the classroom is embedded in the metaphor that "learning is an exciting trip."

Changing Teaching Roles

Through the use of the trip metaphor as a referent for actions in the classroom, Mark could constrain his roles and students' roles in science teaching and learning. In terms of the teacher's actions, students could also construct metaphors for their roles as learners to constrain their actions and to mediate those of their teacher (Tobin & Tippins, 1996). Mark emphasized his role in the classroom as a driver for a trip:

> I think my role in the classroom is to first of all [to be] the driver. I'm the expert; the person who says what we are going to do and when we are going to do it. The teacher is the person who makes decisions "on the fly" about how things are going in the classroom. That is my role as teacher.

Within the metaphor of a trip and to be consistent with his beliefs about teaching and learning, Mark considers himself as a driver for the trip in which the participants want to enjoy and learn. Mark views himself as no more than one of the other participants in the trip, but with more experience about the route. Since he is always driving in this way, he is an expert traveler with more knowledge about exciting things and places along the way. In each trip, the driver learns more about activities that maximize participants' enjoyment and involvement, and the activities that do not work.

In science teaching and learning, Mark feels that some activities and demonstrations—such as the dry ice, liquid nitrogen, and balloon activity described in the vignette above—are effective in the classroom, but other activities are not. Mark expresses his frustration about how the fluorescent detergent demonstration does not work in stimulating students in the classroom as follows:

> I think [the fluorescent detergent demonstration] is just really neat, and I talk about how and why these fluorescent dyes are added to detergent, and how it is like a bluing agent. The dyes absorb UV light and emit high energy visible light, which masks the yellow color in white clothes. I think it is so cool. I've done it three times in three different courses and it hasn't worked yet. I mean it is a cool little demonstration, but it just doesn't work. I've got it in my notes that this doesn't work, but each year I still try it because I think it is so cool. I think it is one of the neatest simple things you can do in the classroom, but it doesn't work for the students. They basically understand it, but it doesn't get them excited. It doesn't get them thinking.

When the goal of a trip driver is to create environments in which participants enjoy their trip, the driver needs to use other roles to maximize travelers' interest. Such a driver can change his role from a driver to a tour guide or a controller, for example, based on contexts and situations in the trip. Mark is able to change his actions in the classroom as the context of learning changes. The metaphor of a travel agent or a trip driver encompasses managerial roles as well as aspects of constructivist learning theory, as noted by Ritchie (1994):

> The travel agent metaphor eliminates the need for several metaphors, from which to select or "switch" to, depending on the role requirement. The travel agent [as well as the trip driver] teacher encourages students to explore new routes as well as visiting well known destinations by establishing a supportive environment based on mutual respect and trust. The link between the teacher as a travel agent and constructivism helps validate the use of the metaphor in this context. (p. 296)

In class Mark shifts from one role to another within the trip metaphor as a referent for actions. In the trip of learning, Mark is able to switch his role from a driver to a tour guide, an entertainer, a learner, and a controller in order to create learning environments in which students learn science with interest. Change of teachers' roles in the classroom was predicted by Glasersfeld (1988), when he noted that the teacher's role "will no longer be to dispense 'truth' but rather to help and guide the student in the conceptual organization of certain areas of experience." Mark's actions shift according to the role he uses as a referent for a specific action in a particular situation to facilitate learning.

For instance, in a class to discuss a question in the last quiz about "why isopropyl alcohol is very soluble in water, but it is also an effective degreaser," Mark uses several roles within the trip metaphor to help students learn chemistry concepts with understanding. The text of discussion in the class (Appendix) showed how Mark shifts from one role to another to stimulate student–teacher and student–student interactions to learn about dying and bleaching concepts.

Teacher as Guide

Mark is a guide or a facilitator when he helps students to understand the difference between the concepts of dying and bleaching (see Appendix). For example, Mark uses the "red rover, red rover, please send someone over" game to help students understand the difference between water–water and water–grease interactions. It was clear that Mark guides the students in this discourse to relate what they are learning to their previous knowledge and to their daily experiences with chemicals, such as soap, Vaseline, and waterproof mascara. To facilitate the learning of science it is essential that the teacher "infuses science into the classroom community by mediating between the languages of the child and of science" (Tobin, 1997).

Teacher as Entertainer

Mark is an entertainer when he tells the students about what happened to his hair as a result of dying and bleaching when he was a graduate student. Mark as a learner is clear when he tells them, "I will find out why hair turns green, chemically." In addition, he pays attention to students' ideas and uses simple language to share discussion in the class. Furthermore, he is a controller when he initiates the discussion by saying, "One more time with the quiz, and we'll go back to talk about electrolyte activity," and ends it with, "Let's take advantage of the twenty minutes we have left…"

Engaged Students

Mark's actions of shifting from one role to another stimulate students to engage actively in the learning process. They are excited to hear Mark's hair story, which encourages them to participate in the discussion and talk about their experiences regarding dying and bleaching. Mark encourages students to use their own language to become involved and participate in discussing, arguing, and asking questions. Such involvement encourages students to think how to relate what they are learning to their daily life experiences. A student, for example, talks about how her father used kerosene to clean grease on his hands after working on his car. Another student talks about how her sister's hair became green as a result of swimming daily in a pool.

Students feel that they can learn with understanding through talking, arguing, asking questions, and expressing their misunderstandings. A student in the class, Muna, comments on Mark's roles in the classroom: "I like [Mark] a lot more.…He is just more down to earth. He is funny; he explains it a little more in common terms."

Mark is an enthusiastic learner who takes an interest in the students and their learning. Use of the trip metaphor helps him to build a congenial classroom atmosphere where students feel secure to speak before their peers. Mark is an approachable teacher who does not intimidate. He should be commended for reaching out to the students in a friendly, nonthreatening manner. Students such as Muna do take an interest in chemistry because they like Mark. Students appreciate the way he speaks in the classroom, by using simple terms and everyday examples that they can understand.

Conclusion

Using constructivism as a referent to develop science teaching and learning in college level classes is receiving increasing consideration. A constructivist epistemology calls for a reconceptualization of what a science teacher is and what he or she does in the classroom (Herron & Eubanks, 1996). This epistemology can be used as an alternative referent to allow teachers to frame problems in different ways and ultimately to obtain different alternative solutions (Tobin & LaMaster, 1995). Using constructivism (rather than objectivism) as a referent for teaching helps a science teacher to think of appropriate ways to conceptualize his or her roles in the classroom. Metaphor is a way that the teacher can conceptualize his or her roles in representing knowledge of teaching and learning. The term "teacher's role" refers to how a teacher considers his or her position when he or she is teaching a particular concept in a particular context.

A science teacher in college classes can construct his/her own metaphors to describe aspects of teaching and learning. Using metaphors can help teachers think about their roles and students' roles while teaching science. Tobin & Tippins (1996) report three significant aspects of metaphors in terms of potential applications to science teacher education. First, metaphors can be used as a way to describe teaching. Second, metaphors can be used as a referent to constrain teacher and student actions in the classroom. Third, metaphors can be used as a generative tool to build new knowledge. Using metaphors as referents to understand teaching and learning has the potential to change what happens in classrooms.

Mark used constructivism as an alternative referent to describe science teaching and learning in his classes. Accordingly, he developed a new metaphor that was consistent with his beliefs to provide a rationale for teaching science in a different way. Mark described his teaching role in terms of the teacher as a trip driver or a tour guide who helps and encourages travelers to enjoy knowing things and places on their trip. Mark utilizes this metaphor as a referent for his actions in classes to create a student-centered approach to teaching and learning in the classroom.

Mark's main goal as a teacher is to maximize the participation of his students, and he is always looking for ways to stimulate active and varied involvement of all participants in the trip. To be consistent with his beliefs and goals that prospective teachers should enjoy their journey of learning chemistry, Mark, the driver in the journey, uses the metaphors of controller, facilitator, learner, and entertainer as referents to create pleasant learning environments. He is able to switch his actions based on which of the constituent metaphors he uses as a referent to frame his actions and interactions and, thereby, is able to create an exciting environment for learning.

Abdullah O. Abbas
Faculty of Education,
Sanaa University, Sanaa, Yemen

Kenneth A. Goldsby
Department of Chemistry,
Florida State University, Tallahassee, Florida, USA

Penny J. Gilmer
Department of Chemistry and Biochemistry,
Florida State University, Tallahassee, Florida, USA

EDITORS

METALOGUE

Educators and Scientists Collaborating
to Improve Teaching and Learning

KT: Abdullah, a doctoral student at the time, collaborated closely with Ken Goldsby, the teacher in the study and a member of Abdullah's doctoral committee, to study Goldsby's teaching of chemistry over several years. A new approach to chemistry teaching was being implemented in a series of courses for prospective elementary school teachers. Penny Gilmer also was a chemist in the same university as Goldsby, and she was engaged in research on Goldsby's teaching and also her own professional practices. It is a welcome but somewhat unusual practice for college teachers to view teaching and learning as objects for research. If this practice could become more widespread, studies of college science teaching and learning, from within, could be seeds for the reform of college science teaching. Other chapters in this section report on these types of studies.

PG: I remember well how Abdullah and I would listen to videotapes and audiotapes of Ken Goldsby teaching (while I sometimes explained some of the chemistry to Abdullah), read and reread the transcripts of the classes, listened to audiotapes of interviews with freshman American females, and sifted through the NUD*IST [Non-numerical Unstructured Data Indexing Searching and Theorizing] analyses of the voluminous amounts of data to tease apart the lessons that we learned. Our chapter here is but one of them. It is amazing to me how Abdullah made this transformation of cultures and language to study the complex language of science in a university classroom.

Teaching Against the Grain

KT: Mark, the teacher in the study, was renowned for his teaching ability. However, he had a great deal to learn in terms of rethinking his roles from a constructivist perspective. Initially, he was skeptical as he thought about what he believed about learning and then what he believed as a bench chemist. The reflexive linkages between two somewhat disparate communities—educators and chemists—created a conflict for him. His approach to teaching was somewhat unique to a handful of professors in the chemistry department, and throughout the College of Arts and Sciences the trend was to teach through better transmission of information to increasing class sizes. In addition, the administration within the university favored good teaching but not at the expense of research in science. There was pressure not to allocate too much time to research in science education. So, in many respects, this chapter and others in this book by practicing scientists (Chapters 5, 6, 8, 9 and 17) are against the grain. The habitus of universities seems to support objectivism as a way of thinking about science and science teaching.

PG: Abdullah chose to study Kenneth Goldsby, my junior colleague in chemistry, as Ken Goldsby is a champion teacher who really cares about his students' learning. Ken Goldsby took a chance to be involved in this study of his own classroom, while he was still an assistant professor. His and my inorganic chemistry colleagues might have looked unfavorably at his promotion to associate professor and granting of his tenure because they might say he could have been concentrating on his chemistry research instead of looking at his own teaching. He knew that it was important to be a part of this community of scientists and science educators involved in research in science education. He knew that he could learn from this study of his own classroom. Also he knew that what we learned from his classroom might influence others' teaching. Kenneth Goldsby has high ideals, and he chose to participate.

KT: Mark knew a lot about good teaching in a somewhat intuitive way. He was not afraid to ask when he did not know something and he expected his students to do the same. When he heard about constructivism he certainly looked for consistencies and inconsistencies in his teaching through that theoretical frame. But when I think of Mark, I think of a person who tried to bridge the gaps between teachers and learners. I remember watching a videotape shown by Abbas and wondering where was Mark in the picture. I could hear his voice from time to time but could not see him at all. Then I got it. He was that kid in the blue jeans. Wearing his baseball hat "back-to-front." Mark was indistinguishable from his students in that lesson. It characterised Mark in many respects. He was a with-it teacher whose epistemology of action was appropriate. He cared a great deal about the students and wanted them to learn chemistry.

PG: I went to visit Mark's classroom for the first day of the semester one year when he was teaching physical science. I remember how he captivated the students. He used a strobe light and could vary the rate of his strobe. He had a fan going and asked the students why he could make the fan appear to stop moving. There were gasps of amazement and sounds of excitement in their voices and movements. He warned them not to put their finger into the blades to test it. I remember he had a tie on that day, and how it moved in jerky movements that really caught the students' attention. This is an example of how Mark got his students excited about learning. That was just an appetizer in a series of experiments in one day in his classroom. Mark exudes excitement and interest for his students.

Using New Metaphors to Connect With Students

PT: Yes, that sense of excitement is evident in the account of Mark's teaching. What strikes me as being highly unusual about Mark as a professor of chemistry is his concern for creating a classroom environment that provides rich learning experiences for students. He expects students in his classroom to learn deeply about chemistry and learning, unlike Dr. Stern (Chapter 1), who regards the classroom as a place of preparation (the assemblage of facts) for learning elsewhere. Using social constructivism as a referent, Mark is keen to engage students emotionally and intellectually in "moving toward" making sense from a chemist's viewpoint of phenomena that appear in their everyday lives. Their learning is situated initially in the world of familiar experiences. Through guided inquiry (e.g., predict-observe-explain strategy) and co-participation in an initially highly accessible discourse, Mark skillfully urges his students to adopt the discourse of a chemist and, in so doing, to co-construct chemistry concepts. Other science teachers interested in adopting social constructivism as a referent for their own teaching practices will find Mark's use of metaphor a very powerful and accessible strategy for understanding anew the relationship between their teaching roles and their students' learning roles.

KT: Metaphor is a powerful way to organize actions. Hence by conceptualizing the *teacher as provocateur,* teacher and student roles can be linked as a coherent set to be enacted in accordance with the metaphor as a referent.

PT: Metaphor is a powerful tool that can free us from dogma, especially the dogma of thinking dualistically (see Chapter 9). Rather than believing that it is necessary to reject objectivism and replace it (ideologically) with constructivism, I believe that it is more helpful to think about how we might add to our pedagogical repertoire metaphors based on alternative referents. The power of metaphor is that it can be used to think creatively about the relationship between teaching and learning. There might be times when it is appropriate to think (metaphorically) about teaching as though it involves transmitting objective facts and about learning as absorption, especially when learning efficiently by rote recall is required. At other times, we might choose to teach that facts should be regarded (metaphorically) as human constructs that are situated historically in the development of (Western) scientific ideas, and that learning is best considered as a process of constructing fallible understandings, legitimated by community agreement.

BIBLIOGRAPHY

Abbas, A. O. (1997). *The teacher's role in college level classes for non-science majors: A constructivist approach for teaching prospective science teachers.* Unpublished Ph.D. dissertation, The Florida State University, Tallahassee, FL.

Brush, S. S. (1993). *A case study of learning chemistry in a college physical science course developed for prospective elementary teachers.* Unpublished Ph.D. dissertation, The Florida State University, Tallahassee, FL.

Cobern, W. W. (1993). Contextual constructivism: The impact of culture on the learning and teaching of science. In K. G. Tobin (Ed.), *The practice of constructivism in science education* (pp. 51–69). Hillsdale, NJ: Lawrence Erlbaum Associates.

Collette, A. T., & Chiappetta, E. L. (1994). *Science instruction in the middle and secondary schools* (3rd ed.). New York: Merrill.

Duffy, D. (1993). *Fostering productive learning environments in college chemistry for prospective elementary teachers.* Unpublished master's thesis, Florida State University, Tallahassee, FL.

Erickson, F. (1986). Qualitative methods in research on teaching. In M. C. Wittrock (Ed.), *Handbook of research on teaching* (3rd ed.) (pp. 119–159). New York: Macmillan.

Glasersfeld, E. von. (1988). *Environment and communication.* A paper presented at the Sixth International Congress on Mathematics Education, Budapest.

————. (1989). Cognition, construction of knowledge, and teaching. *Synthese 80* (1), 121–140.

Guba, E. G., & Lincoln, Y. S. (1989). *Fourth generation evaluation.* Newbury Park, CA: Sage Publications.

Herron, J. D., & Eubanks, I. D. (1996). *The chemistry classroom: Formulas for successful teaching.* Washington, DC: American Chemical Society.

Hurd, P. D. (1994). New minds for a new age: Prologue to modernizing the science curriculum, *Science Education 78*, 103–116.

Lemke, J. L. (1995). *Textual politics: Discourse and social dynamics.* London: Taylor & Francis.

National Research Council (U.S.). (1996). *National science education standards.* Washington, DC: National Academy Press.

National Science Teachers Association (1992). *Scope, sequence and coordination of secondary school science: The content core.* Washington, DC.

Ritchie, S. M. (1994). Metaphor as a tool for constructivist science teaching. *International Journal of Science Education 16*(3), 293–303.

Roth, W.-M. (1995). *Authentic school science: Knowing and learning in open-inquiry science laboratories.* Dordrecht, The Netherlands: Kluwer Academic Publishers.

Rutherford, F. J., & Ahlgren, A. (1990). *Science for all Americans.* New York: Oxford University Press.

Schön, D. (1985). *The design studio: An exploration of its traditions and potentials.* London: RIBA Publications.

Tobin, K. (1990a). Teacher mind frames and science learning. In K. Tobin, J. B. Kahle, & B. J. Fraser (Eds.), *Windows into science classrooms: Problems associated with higher-level cognitive learning* (pp. 33–91). New York: The Falmer Press.

————. (1990b). Changing metaphors and beliefs: A master switch for teaching? *Theory into Practice 29*(2), 122–127.

————. (1997). The teaching and learning of elementary science. In G. D. Phye (Ed.), *A handbook of academic learning: Construction of knowledge* (pp. 369–403). San Diego: Academic Press.

Tobin, K., & LaMaster, S. U. (1995). Relationships between metaphors, beliefs, and actions in a context of science curriculum change. *Journal of Research in Science Teaching 32*(3), 225–242.

Tobin, K & Tippins, D. (1993). Constructivism as a referent for teaching and learning. In K. Tobin (Ed.), *The practice of constructivism in science education* (pp. 3–21). Hillsdale, NJ: Lawrence Erlbaum Associates.

————. (1996). Metaphors as seeds for conceptual change and the improvement of science teaching. *Science Education 80*(6), 711–730.

APPENDIX

 A Text of Classroom Discourse Between Mark and the Students

Mark:	One more time with the quiz, then let's go back to talk about the electrolyte activity. The question about this molecule [Mark points to the structure of an isopropanol molecule on the board and continus discussing the answer]....So isopropanol doesn't really act like a soap. And you know that. When you use isopropanol, it doesn't suds up and it doesn't make a foam. But it is a good degreaser and people often use it as a, as a, a...
Student 1:	Astringent.
Mark:	Astringent. Thank you. Isopropanol is the cheapest astringent that you can use. [You know] it's soluble in water because rubbing alcohol is a mixture of isopropanol and water...[Mark continues using the chalkboard to draw a structure clarifying the relationship between isopropanol, water, and degreaser.] There was an implication that the grease would go to the carbon, and that's not really right, the grease goes to the nonpolar part of the molecule. It is this whole part right here that's nonpolar. Does that make sense? The combination and application of the...
Student 2:	What part is the nonpolar part?
Mark:	The part that is nonpolar. The part that has carbon-hydrogen bonds...[Mark goes back to the board to clarify which part is polar and which is nonpolar in the structure of the molecule.]
Student 3:	So doesn't a carbon have something in the middle...I mean...because they're the only ones that can...

Mark:	Maybe only one that…if the grease went into the water, it moved here to break up the strong water–water interactions. Therefore there's no grease because the interaction between the hydrocarbon and the water is very weak, compared to the interaction between the water and water. Imagine you…played red rover, red rover, please send someone over, and you line up and hold hands. What's the rule to that game?
Students:	You have to try to break the barriers.
Mark:	If you break through you're free. If you don't break through, you're captured. When you call someone over to play, do you pick the big kid or the small kid?
Student 4:	The small kid.
Mark:	You pick the small kid because the small kid doesn't have the energy to break that interaction. But water molecules have a very tight hold on each other. The only thing that's going to break it up is something which has a strong interaction with water, like a polar molecule, like this OH group, like an ion, that's something that's going to break up water…
Student 5:	So if the outer edges were blocked because they were larger molecules…they have to bond with O and H?
Mark:	Yeah, that's right. This part is a hydrogen bond…
Student 5:	How do you know that grease is attracted to the hydrocarbon?
Mark:	The question is how do you know that grease is attracted to this part of the molecule. I would think that it's common knowledge. Give me an example. I mean, can you think about something you do know that you've done that involves grease being attracted to molecules like that?
Student 6:	The thing when we make soap?
Mark:	Yeah, the soap, you don't know which end is doing it. So let's pick something which is just a hydrocarbon.
Student 7:	Is it because they're both nonpolar?
Mark:	Okay, that's why it happens, but it doesn't answer the question, "How do I know that works?"

Student 1:   How about water, I mean you put grease in water, you know like if you're boiling, getting ready to do spaghetti noodles...

Mark: The grease floats on the top. So you know the grease doesn't go underwater. That tells you it doesn't interact with water, grease interacts with hydrocarbon. Let's think about this example: Did you ever have grease on your hands working on a car or something like that? What do you use to clean them off? Ever used kerosene?

Student 8:   My dad used to do it.

Mark: You're painting with an oil-based paint. What do you clean the brushes with?

Student 4:   Mineral spirits.

Mark: Mineral spirits. Mineral spirits is hydrocarbon. It's like gasoline. Let's use the mascara example, when you are trying to take waterproof mascara off, which is greasy, that's why it is not soluble in water, it's a grease. Water doesn't do a very good job. Somebody mentioned Vaseline, which is kind of hard to get out. When I was a kid, when I was in high school, they had a fifties party. Everybody slicked their hair back. I didn't have anything to do that, so I used Vaseline. [Students are laughing.] Do you know how long it takes to wash Vaseline out of your hair? If you wash your hair three times a day, it takes exactly twenty days to get the Vaseline out of your hair. [More laughing in the class] For a party in college, I wanted to go as a beach bum, so I dyed my hair blond, or tried to. I bought the lightest blond dye you could buy, put it in my hair. It didn't lighten it at all. So I rooted around the medicine cabinet and got hydrogen peroxide. I dipped my hair in it very carefully, [but the hydrogen peroxide] didn't touch it. [Students became excited] What am I going to do? I am a chemistry major in graduate school. I know how to solve all these problems. I take the towel, I soak the towel in hydrogen peroxide, wrap my hair with it and take a nap. Wake up two hours later, my hair turned orange. Flaming Richie Cunningham orange, [Students are very excited] which kept getting lighter and lighter over the next month.

Student 6:   Did you shave your head? [Students laugh]

Mark: What I did was, I went to this party, and heard from at least five different people, "I had a cousin who did that once and all his hair fell out," or, "all her hair fell out." That's when I learned what it takes to make hydrogen peroxide a good oxidizing agent. Every bit of what we're talking about now you're going to understand in terms of the simple chemistry in this class: acid-base chemistry reactions. Then I ended up at the time being a red hair for a couple of redox (reduction-oxidation) weeks. What I did was, I dyed my hair black. When I bought the blond dye, I bought the black dye also, I dyed my hair black after the party, so a couple of months later, when the dye started to wear off, I started having black roots with black and orange hair. That's back before people did that on purpose, so then I got a really short haircut. The reason why I am telling you the things I did is I am trying to think of things you might have done, use the makeup, clean the grease off your hands, because I think you can make sense of things you had experience...

Student 9:   Okay, let's say I wanted to dye my hair blond, and I applied everything, and I came out with green hair, what went wrong?

Student 10:   My friend did it, she had blond hair, and she wanted to dye her hair black, and it turned green.

Student 9:   You're supposed to do it strand by strand.

Mark: Let me tell you what I learned about this.

Student 3:   What if you sprayed lemon juice in your hair, what color would it turn? [Students were giving their answers all at once.]

Mark: Every point I'm trying to get across, every bit of this, you can understand in terms of the chemistry that we're talking about.

Student 10:   Why if you have blond hair and a light color hair, and you swim like in the summer and you have chlorine, why does your hair turn green?

Mark: This is so cool. Let's talk about this. You'll get a short answer and then with a promise we'll come back to this.

	When a person dyes their hair blond, what do we call that process? What do they do to their hair?
Students:	Bleaching.
Mark:	Bleaching. Dying is something different from bleaching. When you dye something you put a different color over it. When you bleach something you take the color out. When you want to bleach your clothing to take out stains, to take out color, or when you accidentally, or in my case when you are washing the colored clothes and you forget that you don't put the Clorox in those. [Students laugh.] What are you doing to them? You're also bleaching, right? What do you add to clothes to bleach?
Students:	Bleach.
Mark:	You add bleach, you add Clorox. What do you add to a pool to kill the microbes?
Students:	Chlorine.
Mark:	Chlorine. Same thing, the chlorine that you add to the pool is chemically the same thing as the Clorox that you add to your clothes. They are both oxidizing agents, oxidizing agents are molecules that rip electrons out of things....When you combine an oxidizing agent with an acid it becomes a better oxidizing agent and you bleach your hair better.
Student 5:	My mom used to tell me when I was little, I had a lot longer hair, she used to say that if you wet your hair down with a hose before you put something...
Student 9:	I think if you cut it up into...
Mark:	I think you're making up this green stuff. [Students argue that it is possible that your hair can turn green.] Who here would be willing to do that? I would like to see this.
Student 10:	I have pictures of when I was little, I can bring to you, with green hair.
Mark:	And you didn't dye it green because of a party.
Student 11:	If you swim every day and you don't rinse your hair, you don't rinse the chlorine out of your hair [students give their explanations all at the same time].

Student 9:   It's not green like her jacket (pointing to a student's jacket)

Student 12:   It's like a green tint.

Student 13:   My sister was platinum blond and she was a swimmer, she swam every single day of the week, her hair turned green.

Student 3:   I used to be a lifeguard, and little kids used to come to the pool and their hair used to be that color, I mean their hair was green.

Mark: I could deal with it, I will find out why hair turns green, chemically....I will find out chemically why hair turns green. If you guys work as hard as you can to apply the level of enthusiasm and interest that you applied to this conversation [students are laughing], then on Monday we'll do makeovers [students still laughing]. Aren't you impressed with what I know about makeovers, there's a term I don't get to use in my speech every day; let's take advantage of the twenty minutes we have left.

CHAPTER 8

 **Effective Strategies for Active Learning in the Small Chemistry Classroom or Laboratory**

Rosalind Humerick

Most of us in reviewing our educational experiences can dredge up horror stories of former teachers who were competent in their academic fields yet, in our view, did not know the first thing about teaching and relating to a class of students. Many college instructors, including myself, have had no formal teacher preparation and have learned to sink or swim on the job. We might acquire some sort of feel for what the students think of us as teachers, but most of us are not really aware of what we do in the classroom and the effect that can have on student learning. Student and faculty evaluations may provide some insight into our strengths and weaknesses, but I believe that for real improvement more systematic evaluative information is needed.

I recently took part in an action research study designed to evaluate the teaching and learning of science in the small community college where I work (Gibson & Humerick, 1997). This was the first time that I had taken part in an educational research project. Prior to the study, I was not consciously aware of what I did in the classroom or laboratory and how that had been influenced by my former teachers. Through the study, I found that I became more aware of my own educational history and much more familiar with what students believe will help them learn and where that learning takes place. This knowledge has, and is influencing, the way I teach today.

Collins and Spiegel (1995) refer to the work of Kurt Lewin (1946), who defined action research as

...a spiral of circles of research that each begin with a description of what
is occurring in the "field of action" followed by an action plan. The
movement from the field of action to the action plan requires discussion,
negotiation, exploration of opportunities, assessment of possibilities, and
examination of constraints. The action plan is followed by an action step
that is continuously monitored. Learning, discussing, reflecting, understanding, rethinking, and replanning occurs during the action and monitoring.
The final arc in the circle of research is an evaluation of the effect of the
plan and action on the field of action. This evaluation in turn leads to a
new action plan and the cycle of research begins anew. (Collins &
Spiegel, 1995, p. 117)

I would like to propose that all teachers take a look at their own
teaching. Teachers must learn to grow, and self-examination, although a
somewhat daunting, even frightening experience, is well worth the effort.
So much of what we do in the classroom every day is done instinctively
without us actually asking ourselves what we are trying to achieve. All
teachers can improve through self-examination, and becoming a teacher-
researcher is an ideal method for this examination.

An Action Research Study

The study that I describe briefly in this chapter was part of a long-
range project designed to examine the preparation of science and
mathematics teachers in Florida (Elmesky, Muire, Griffiths, Taylor &
Tobin, 1996; Tobin & Muire, 1997). As many elementary school
teachers take their science and mathematics content courses at the
undergraduate level at community colleges, an important question to be
asked is, "How are science and mathematics taught at this level?"

I first heard about the study when I was approached by the dean of
general education at my college. A colleague, biology instructor Lib
Gibson, had been asked to participate, and my dean thought that I might
be interested. I was, and Lib and I agreed to collaborate on the project.

During a meeting with Ken Tobin and the project staff from
Florida State University held in Tallahassee, Lib and I learned about the
long-range project and our role as community college regional
evaluators. Florida is divided into six educational regions, and the study
had a community college evaluator from each region. We were
introduced to the idea of interpretive research and were given a hands-on
session using qualitative analysis computer software. We discussed data
collection and evaluation planning and gained experience in interviewing

university students who had taken their mathematics and science courses at community colleges. Additionally we enrolled in a distance learning course offered by Ken Tobin. This course was designed to enable us to become proficient in interpretative research. We were guided by Guba and Lincoln's *Fourth Generation Evaluation* (1989).

Our study looked at the teaching and learning of college science, with a particular emphasis on the teaching and learning of chemistry. We generated data from both faculty and students, and our analysis focused on understanding beliefs and attitudes about teaching and learning. Lib and I were influenced enormously by Peter Taylor's visit to our college. We were in the beginning stages and had gathered data from mathematics and science teachers and students. Peter helped us to narrow our focus and encouraged us to concentrate on my chemistry teaching. He explained the term *action research* to us and really inspired me to become an action-researcher.

Autobiography of a Teacher

Before focusing on colleagues and students, I asked myself what had influenced my own teaching style, and what were its main characteristics. I reviewed my own past teachers and found that I had modeled myself on those whom I deemed to be "successful" teachers. I was influenced strongly by those whom I met early in my educational career. Two immediate things came to mind: For me to be a successful learner, I needed (1) a teacher who understood the content material and (2) a teacher who generated a calm, caring attitude and did not intimidate me or other students. During my high school years, I had been exposed to a teacher who did not know the content matter and another who seemed to take great pleasure in denigrating students. Neither of these teachers was in the area of chemistry—in this area I had two very positive role models, one a high school teacher, the other a college professor.

I have vivid memories of my first day of class with the teacher who denigrated students. She introduced herself by saying, "I am known as the dragon, and I don't like children." At a subsequent parents' evening she told my parents after a long pause, "Rosalind, well, at least she doesn't cry when I shout at her." We used to sit in our seats quaking with fear till she arrived, and then during class we would be paralyzed when asked to answer questions, in case we got them wrong. I used to live for Thursdays, when I didn't have to see her. One year I did so badly that I was sent to a lower division class—unfortunately, I started to achieve and next term was back in her class.

My applied mathematics teacher did not intimidate me. However, she used to burst into tears when she was not able to solve problems. During the first year of the course it didn't matter too much, as I was able to work them out for myself; in fact, she would often ask me for my solution. The second year was a disaster—I felt lost and frustrated and her tears did not help the situation. In the end it was her students who suffered, as we were not adequately prepared to take our national examinations.

Both of the chemistry teachers who influenced me were exceptional at explaining difficult topics in a comprehensible way, and both conveyed a caring attitude such that if you didn't understand something then it was their problem as well as yours. This is not to say that they did not demand excellence from their students; they did, but you felt as though they were on your side and were genuinely interested in your learning. Their influence today on what I as a teacher of science do in my classroom came out very strongly in the results of our research on my chemistry classes.

At high school, I was taught chemistry in small classes (ranging from one to thirty students) where there was a high degree of student–teacher interaction during class, where textbooks were seldom used except to assign homework problems, and where laboratory facilities were poor in terms of capital resources but rich in terms of skills and techniques learned. I found that when I went on to further my education I excelled just as well as those who had studied chemistry, where many more resources were available.

From the moment I started my college teaching career, I have endeavored to present to students a teacher who explains concepts in everyday language, who demystifies chemistry, and who really cares that they learn successfully. I want my classroom to be a place where active learning takes place, where students' minds are engaged for most of the fifty-minute lecture period—consequently, I try to incorporate a lot of problem solving in class and ask questions continually in an attempt to get students to "see" connections between concepts and be able to reason things out. I also do not want students to be afraid to ask questions during class or to be afraid to get the wrong answer. I want to create a team where we all help each other.

I correlate lectures with laboratories and make sure that the two are "in sync" with each other. One advantage that a small school has over a large school is that I teach both the lecture and laboratory section to the same students. My early teaching experience was as a graduate teaching assistant at a large four-year university where 1,500 students took a freshman chemistry laboratory each quarter. The approach we used was the so-called "cookbook" method, where students were told the theory behind the laboratory before they did it and were given the necessary mathematical equations. For example, a typical laboratory on density would be: Let's measure the mass and volume of an object and calculate the density using the equation "density equals mass divided by volume." During my first year of college teaching, I used this same method, which is characteristic of many available commercial laboratory manuals, and mimicked the laboratories that I had done as a student.

Five years ago, my chemistry colleague and I decided to try a different approach to laboratories. We found a commercial laboratory manual that used simple equipment (a necessity for us) but also used an inquiry approach where, through logical reasoning, students proposed hypotheses that they then went on to test (Abraham & Pavelich, 1991). A "guided inquiry" laboratory about density would still have students measure the mass and volume of objects made out of the same material. But the main difference is that students are asked to find qualitative and often quantitative relationships between data, and the term "density" would not be mentioned in the laboratory manual. Graphical techniques are often suggested as a method for finding a linear relationship between variables. I have had lots of success with this approach. I found the comments about the laboratory interesting when written by students who had taken both an introductory chemistry course, where we use the so-called cookbook approach, and the general chemistry course, where we use the inquiry approach.

Generation and Analysis of Data

The study focused mainly on two of my chemistry classes over a three-month period. Using interpretive research methods for triangulating qualitative data, we conducted observations of my chemistry classes, directed students to write journals about their learning of chemistry, audiotaped several classes, interviewed five students, and recorded field notes of discussions with five science faculty (Gibson & Humerick, 1997).

Interviews were transcribed, returned to participants for checks on accuracy checks, and divided into "text units." A text unit is a section of text, which may be as short as one word or as long as a whole paragraph or speech. Text units are numbered sequentially, starting at number 1. In our study each text unit, on average, was representative of one, sometimes two and, at the most three, ideas. It was important that the text unit was long enough to allow us not to have to go back to the original document when looking at it, but, that it did not contain too many ideas. The data were analyzed using the NUD*IST (Non-numerical Unstructured Data Indexing Searching and Theorizing) qualitative analysis software program that allows one to organize and explore ideas about qualitative data from many sources (Q. S. R. NUD*IST, 1994). For example, by assigning text units to interview data from faculty it was possible to see that a discussion on the teaching of science centered on the philosophy of a college education, what teachers value about their teaching, techniques used, influences on teaching style, and how a person's teaching has changed over time.

In two of my chemistry classes, I invited students to write responses to the following questions: "Describe a typical chemistry class in terms of what I do and in terms of what you do," and "How do you study chemistry?" The backgrounds of the two classes were quite different. The first was a one-semester class designed for nursing students, most of whom had never before taken a chemistry class, and the second was the first course in a typical college-level general chemistry sequence. All students in the second class had previously taken some chemistry, either at the high school level or in an introductory college course. The classes ranged in size from fifteen to eighteen students, and the response rate to the questionnaire was about 55% for the two classes. Those who responded ranged from the struggling student to the very successful student.

Thus, analysis of student data relating to the teaching and learning of chemistry centered around: (1) my activities in the classroom, (2) student activities in the classroom, (3) student activities outside the classroom, and (4) the issue of where learning takes place.

Teaching Tools for Active Learning

The majority of students in both classes (89%) referred to me as a teacher who explains concepts by using everyday examples and as one who teaches problem solving. They commented on how methodical I am, how I show every step in problem solving, and how I try to teach them a way to attack a problem by questioning themselves and mapping out a path to solve a problem before jumping in. Students commented on how I "break chemistry down into the simplest form possible," and how I "use lots of examples to demonstrate the concepts" I am teaching.

My main teaching tool is the blackboard, and this fact was mentioned by 56% of students. My classroom does not have a computer; the only technology I use is an overhead projector and, on occasion, programs from the World of Chemistry series (Adler & Ben-Zvi, 1990). My overhead projector is scrounged from the English department and, until recently, I was using a blue brick wall as my projection screen. I feel lost without a piece of chalk in my hand, as I need to write and draw diagrams to explain concepts. One student wrote concerning my blackboard use:

> You write everything on the board for your students to better understand. This is most helpful to me. Remember when you were sick and tried to lecture without using the blackboard! You observed our facial reaction of confusion and went to writing on the board so that we could follow and understand.

In reviewing my past chemistry teachers, I can see the influence they have had on my methods of teaching: Their only tool was the blackboard, they used no overhead projectors or television.

One-third of the students commented on interactions between us in class, how they were able to ask questions and not feel intimidated by me or by other students. They noted that I moved around the class to help them with problems individually and had them solve and think through problems in class. Upon analyzing one audiotape of a lecture on chemical formulas and writing and balancing chemical equations, I noted that I initiated forty-nine questions and the students initiated eight questions during the fifty-minute period.

In seeing what the students had mentioned about my teaching, I examined the transcripts of my audiotaped lectures. Analyzing your own lectures can be an enlightening experience. I was amazed at how unconsciously I do many things. I noted that I tend to start a lecture by reminding the class of our overall goal and by practicing what we had just learned in the previous session. I tend also to use a "coaching" method when teaching new skills; in other words, I question, students answer, and then I summarize. I often refer to prior learning as a means of reinforcement and frequently use analogies and personal anecdotes, and I make reference to laboratory experiences. My use of the textbook appears to be limited to giving textbook references for the current topic and using it as a source of problems. When working numerical problems, I give students time to think about and work out the problem, and I ask for the answer before I put a solution up on the blackboard.

The following is one student's response to describing our mutual roles in a typical general chemistry class:

You: You explain the overall concept.

Me: I try to comprehend.

You: You explain what we will need to utilize from prior learning.

Me: I try to remember.

You: You explain, using step-by-step methods, describing something in simple terms. You pause, make eye contact, and allow students time to catch up.

Me: I try to catch up.

You: Then you work a problem from the end of the chapter using the path just taught. You stop before the final answer and walk around the room to see if students are on the right track.

Me: I try to copy from the board the path and information needed. I try to apply the path to the problem and mark in my notes where I lost understanding and just went to copying. I try to show you where I lost understanding and where I need help to comprehend. I try to explain in my notes so I can follow it later.

You: You usually work a second more difficult problem the same way.

Me: I try again to apply the path to the problem and move past the last roadblock.

You:	You move onto the next concept, teaching it the same way as before: path, problem, assistance, problem, assistance.
Me:	I try to comprehend the next concept, note the path and attempt the problem, accept assistance, answer another problem.
You:	At the end of class you are always around to clear up "gray" areas and answer questions.
Me:	I try to ask for assistance if I feel like I have a question in my notes that may make doing the homework hard.

It does appear that this student is engaged in active learning, and for most of the class period is not just copying notes but thinking about the subject. As a student, I often spent lectures copying notes and did not really think about the material until I went home. I have never had a chemistry teacher who moved around the room. This is something that I started during my first year in teaching and have found that it works particularly well for me, especially in teaching numerical problems.

Other tools that I use to encourage active learning are having students work in groups and explain to others how they solved a problem, and students working as a group on the blackboard. I also have students research a topic of interest to them, write a short paper, and produce a poster illustrating the results. Each student may choose his/her own topic (usually an extension of something we have covered in class), and we spend about three class periods presenting and discussing their results. I have found that just having the posters on display in the classroom stimulates discussions in my other classes.

Promoting Learning in Class

We were stimulated by Peter Taylor's visit to St. Johns to ask students and faculty where they believed learning takes place. All of the faculty interviewed professed that the majority of learning takes place outside the classroom and that often time in the classroom is spent organizing material in order to get ready to learn. However, the majority of students told us a different story. In chemistry, 75% of students believed that 50% or more of their learning takes place inside the classroom, with many of them placing it as high as 85%. This response was not confined to chemistry; 62% of biology students also believe that the majority of their learning takes place in class. For those who believed more learning goes on at home, most stated that their professors got them started in the right direction and that they should go on from there.

If students believe that so much learning goes on in class, what types of learning do they actually engage in during class? What one student does in my general chemistry class is described above. Other answers were categorized using NUD*IST, and six major activities were identified: note taking (72% of all students), problem solving (33%), asking questions (33%), thinking about prior learning (11%), following in the text (11%), and audiotaping class (6%).

My nursing students like to create a set of notes and ask questions if they are not following the argument:

> I as the student will write down everything that you write on the board and will frequently ask questions if I am unclear on whatever subject…as a student in her class, I feel I am able to stop her at any point and ask questions about the lecture.

In my general chemistry course, problem solving was cited as frequently as note taking:

> I usually work the problems out with you and concentrate on what the basic concept is in the problem…the class takes notes and is also given time to work problems.

Learning in the Laboratory

All students believe that most learning in the laboratory takes place in class. I was particularly interested in the responses of students who had taken both the nursing chemistry laboratory (a traditional cookbook approach) and our general chemistry laboratory (directed inquiry approach). Our nursing course uses laboratory separates (Chemical Education Resources, 1997). This is a modular laboratory program in which we order the individual self-contained experiments, or "separates," that we wish our students to perform. Students read about the basic theory behind the laboratory in the booklet or "separate," fill out a pre-laboratory sheet, perform the experiment, calculate their results using equations given in the text, and then answer post-laboratory questions. Two students described this experience as follows:

> I told my friends when I was in the Principles of Chemistry Lab that it was sort of like when you buy something in a box and have to take it home and put it together, you have to figure out what you are doing.

> I felt like "a tornado had gone through my head."

This is not to say that all students experience this frustration. Another student told how she was able to look at the overall picture and this increased her learning experience:

> In fact, in chemistry my study partner and I do not rely on just our information, we go and gather other information, we say, "What is happening with your experiment"—it helps because then we get a total overview of everything that is going on.

As a teacher, I have become increasingly dissatisfied by a cookbook laboratory approach, and eventually plan to use a directed inquiry approach in this course as well.

However, I have found that change can be difficult, particularly when we first started using the inquiry approach in laboratories. It took much more effort from me during the laboratory period. At the beginning of the semester, it is vital that I try to stimulate critical thinking among the students and do not answer simply "yes" or "no" when they ask me a question, but try and get them to reason out an answer. I tell them that as long as their conclusions are logical and based on their data, they will get full credit (99 times out of 100 they will also be chemically correct). I have found that it is imperative that they work through the laboratory one step at a time and answer all questions as they come. This proved to be a problem at the start, especially in the graphical analysis area. Students wanted first to do all the experimental work and figure out later what it meant. I now tell them that even if they do their graphical analysis later, they should at least look at the data right away and come to some qualitative conclusions.

Four years ago, our only tool for analysis was graph paper. The second year, I bought a simple graphing program (EZ-STAT), which I placed in the computer laboratory, and spent the first session teaching students about graphical analysis and how to use the program. Two years ago, I acquired two IBM PCs, which we have in the laboratory so that students can look at their data right away. It was also during this period that our mathematics department started requiring graphing calculators for calculus, and I noticed that maybe half of my students owned these. I borrowed a calculator and wrote some simple instructions to show them how to plot their data and perform a linear regression analysis. Now I have a situation where as soon as students collect data in the laboratory they can analyze it using one of the two computers or their own graphing calculators. I have found that this has led to increased understanding and reduced experimental error, with students wanting to repeat measurements if they don't like their initial results.

The advantages of this type of laboratory can be seen in observing a typical laboratory session. I hear students discussing chemistry rather than what they did last weekend. I encourage them to discuss their ideas with other students and with me. The only ground rules are that when they come to writing their explanations down it must be their own work. I have found it useful to require students to write a short statement after completing the laboratory on what concepts they think they learned, what experimental techniques they used, and where their main experimental errors occurred. One student who has taken both laboratories explained that, in the second laboratory, you might do more work but you are using the same skills as in solving word problems, you have to incorporate past knowledge, and you are not told exactly how to do that. I feel that students I teach may not use the latest in equipment or technology, but they are learning skills that will help them in their future scientific endeavors, or, in the very least, will give them a glimpse of what it is that scientists do.

A bonus of laboratory-based data analysis is the opportunity that it provides for me to say that a certain algebraic equation represents the empirical results, and that now we must come up with a hypothesis to explain these results. Thus, I can expect students to be familiar with the inquiry process and to engage in discussion about hypotheses.

It is possible to teach chemistry to an academically diverse student population with a limited amount of equipment and supplies if one is really focused on presenting to students a teacher who is not only caring but can explain difficult topics in an interesting, comprehensible way and is not afraid to change.

Closing Reflections

As an action researcher, it is I who can have the greatest insights into how successful I have been in reaching goals that I have set for myself as a teacher. I have received student evaluations throughout my career and have found some of their comments useful and have made changes in subsequent courses. One of the beauties of an action research study into your own teaching is that you can evaluate your effectiveness during the course of the semester and change almost symbiotically depending on the needs of the class.

Prior to this research, my main sources of inspiration had been reading of professional journals, conversing with others in the field, talking informally to students, and employing trial and error. I often was not aware of what I did that students found helpful until I stopped doing it and they asked me why!

Through action research I have become more aware of what I do in the classroom and laboratory, and how that influences student learning. I have become increasingly familiar with what students believe will help them learn and where that learning takes place.

Rosalind Humerick
Depatment of Chemistry
St. Johns Community College, Palatka, Florida, USA

EDITORS

METALOGUE

Formal Teacher Preparation for Science Professors

KT: Rosalind begins her chapter with a claim that "Many college instructors, including myself, have had no formal teacher preparation and have learned to sink or swim on the job." This draws attention to an important issue. If doctoral students in the sciences are to become the next generation of university and college science teachers, and while obtaining their graduate education, teach undergraduate science courses, then it makes sense to ask about their preparation to teach college science. Although few universities have formal programs to prepare Ph.D. students to teach science, there are strong informal programs in place in all organizations where teaching and learning occur. I believe that the more powerful factors that prepare us to teach are our former teachers (in and out of school) and our own preferences for learning. I think this is one reason why it is so important to have a required course on teaching college science as a part of Ph.D. education in science.

PT: I like to think of Janus, the Roman god who guards gateways to important community buildings, as a metaphor for science professors. Janus has two faces looking keenly in opposite directions. As gatekeepers of the community of science, science professors too need to be keenly aware of two very different (adjacent, rather than opposing) domains. One is the domain of the natural sciences; the other is the domain of education that draws on the human sciences. The problem, however, is that many professors are like a one-eyed Janus, aware only of the domain of science. Unless science professors are well prepared in both domains, their gatekeeping roles will remain impoverished and so too will the future science professors who emulate them.

Formal teacher preparation programs operating in parallel with science programs can stimulate pedagogical vision among graduate students of science. Given that graduate students have well-established learning practices, it makes good sense for them to start reflecting on the process of their own learning as the first step toward developing *learning-centered* pedagogical skills in the context of their formative teaching experiences as laboratory demonstrators and tutors. The domain of Education can provide the tools for this reflective inquiry—*self-study* (or *action research*). Because these tools are transportable, graduate science students will be able to maintain reflective inquiry into their teaching practices as they take up the reigns of professional teaching.

Self-Study of Teaching and Learning

KT: Rosalind shows how college science teachers can do research on their own practices and, in the process, identify issues about which they can reflect, and subsequently identify ways in which they can improve the learning environments for their students. If college science faculty understood about educational research and its similarities and differences with respect to different genres of research in the sciences, they would not only be able to undertake research on their own teaching practices but also they could understand what their colleagues in education were attempting to do in their research projects.

PG: Rosalind's chapter is powerful for scientists, because they can develop a vision of what a scientist can do to change her/his way of teaching. Ros took the same course in *Fourth Generation Evaluation* (Guba & Lincoln, 1989) that I took from Ken Tobin (although we were in different sections). She implemented what she learned in that course to evaluate her own teaching. She took a giant step forward, and started to think about her own teaching and her interactions with her students. As she sorted her qualitative student data while looking for patterns in their responses, she learned much more than if she had just read student comments on end-of-course evaluation forms given by the university. Imagine if many more college and university science teachers did this simultaneously, what a difference it could make in the sociocultural milieu in which we work and in the learning of our students.

PT: Penny, you have raised an important issue of professional identity here. Any professor of science who teaches others about science is, by definition, a science educator. With that realization comes an obligation to be an effective teacher. And to achieve that requires the addition of a *learning-centered* focus to one's repertoire of professional practice. Now one of the best places to start developing an appreciation of how to facilitate learning is with one's own learning. One of the great advantages of undertaking self-study of one's own teaching practice is learning that any claim to be an effective science teacher needs to be supported by evidence and by suitable criteria for judging the adequacy of the evidence. If this process is undertaken with rigor and care, then the science teacher can be assured that her subsequent knowledge claims about the efficacy of her teaching are well founded and professionally justifiable.

KT : What is it that makes an empirical inquiry of this type so compelling?

PT: An empirical inquiry can be both compelling and foreign; and its foreignness can be a great stimulus for critical reflective thinking. What makes such an inquiry foreign is that the standards of judgment are drawn from education and differ from the well-known standards of objectivity that seem to rule the enterprise of the natural sciences. A firsthand experience of the fruitfulness of alternative standards (e.g., credibility, dependability, authenticity, plausibility) can often be enough to promote the realization that effective science teaching (including formal assessment of learning) needs to be regulated by standards other than (or in addition to) validity and reliability. Many of the chapters in this book deal with the issue of alternative standards.

KT: Examining your own *biography* as a teacher and a learner of science appears to be a promising way to begin a self-study of teaching and learning. Who are the most influential teachers in your life? Are they all from schools? Also, how do you prefer to teach in a variety of settings that include classrooms, the field, while working with peers, and in social contexts? Rosalind's chapter commences with an account of the formative experiences in her life that shaped her subsequent vision of university science teaching.

PT: Let me return to the issue of formal teacher preparation for graduate science students, many of whom have been enculturated (unknowingly) into the myth (or worldview) of objectivism during their undergraduate years. This dominant culture is illustrated poignantly in Craig Bowen's chapter (Chapter 2) where the professor displays a predominantly analytical mode of thinking and an educative relationship with students of unremitting remoteness. From a feminist perspective, graduate science students need to learn how to counterbalance this dominant *separate way of knowing* that leads to a view of learning (and thus of teaching) as a solitary intellectual activity fuelled by a predilection for propositional logic and a tendency to reason against others (Belenky, Clinchy, Goldberger, & Tarule, 1986).

Learner-centered teaching requires science teachers to have an empathic orientation to understanding students' struggles to make sense of their learning experiences in forums as diverse as the large lecture theater, the science lab, and the tutorial room. A *connected way of knowing* is essential if teacher–student educative relationships are to promote both the personal agency of the individual learner and his or her students might learn, firstly, how to reconnect with themselves; to develop a *voice* that can give expression to their (suppressed) subjectivities.

Learning to reconnect might be fueled by reflecting on thoughtful questions in the context of their own biographies. Has my learning science always been thus? What are my best, perhaps earliest, experiences of learning science? When did I last feel truly inspired by a teacher of science? Was there a time when my appetite to learn science was sharpened by a particular learning experience? What caused me to take the path of science? Questions such as these can (and should, I believe) be explored by graduate science students. Educational research provides methods of inquiry such as autobiography and narrative inquiry for finding answers to these questions, answers that may enrich a prospective science teachers' self-understanding and thus remind them of the importance of evoking imaginative inspiration in their own science students.

EDITORS

METALOGUE II: Effective Strategies for Active Learning

Diversity in Learning-Centered Teaching

PG: I've long felt that scientists who survived to be college science faculty are the ones who learned well on their own. I was this way myself. I used to study all by myself, falling in love with proofs in plane geometry, making sense of waves in physics, and learning the periodic table by inference. Since college science faculty learn this way, it is easy to assume that all people learn this way. I think that we tend to teach the way we learn, and that since most science faculty learn this way, we only reproduce other scientists who learn the same way as we do. To truly make science for all, we must recognize that others can learn in different ways than we do. We must learn from new findings from cognitive psychology and science education. As we do research in our own classrooms, we can decide what we want to do to improve the learning of our students.

KT: Examining the self as a teacher and learner reminds me of Miller in the study reported by Michael Roth and myself (Chapter 6). Miller always would discuss his own experience as a teacher (as sports coach) and as a learner (physics student who became a nuclear physicist) and then project these preferences onto his students. This tendency links well with Mark's chapter (Chapter 16), in which he warns us about tendencies to project in this way. For example, once we learn about our own preferences to learn we should not then use this as a template to infer that all students prefer to learn in this way. On the contrary, when we consider that the authors of the chapters in this book are accomplished scholars, there would be dangers in assuming that what worked for them would automatically work for their students.

PT: A hallmark of learning-centered teaching is sensitivity to the diverse needs and predilections of one's students and the development of a rich repertoire of professional teaching strategies; a diversity apparent in Rosalind's chapter as well as in those by Abdullah Abbas et al. (Chapter 7), Hal White (Chapter 9), and Penny J. Gilmer (Chapter 17). However, it is as well to be aware of a bitter irony of learning-centered teaching. Some students who have been conditioned into the role of passive receivers of seemingly objective knowledge are likely to react negatively to the prospect of the professor appearing to abrogate her responsibility for teacher-centered teaching. The development of learning-centered teaching is a challenging task that involves re-educating one's students (and colleagues) toward a new set of teaching-learning goals and a diverse range of methods for achieving them.

But what about existing institutional and programmatic constraints that require the efficiencies of teacher-centered teaching? In the short term, these constraints cannot be ignored. Nevertheless, the creatively energetic and inspired professor of science will find a niche for learning-centered teaching strategies. The viability and growth of the niche will be sustained by a powerful pedagogical vision of what constitutes high quality science learning. Other authors in this book provide compelling cases of how their pedagogical visions enable them to maintain the momentum to transform not only their own teaching but the culture of teaching and learning within their departments.

BIBLIOGRAPHY

Abraham, M. R., & Pavelich, M. J. (1991). *Inquiries into chemistry.*Prospect Heights, IL: Waveland Press.

Adler, I., & Ben-Zvi, N. (1990). *The world of chemistry.* 26 thirty-minute video programs. S. Burlington, VT: Annenburg/CPB collection.

Belenky, M. F., Clinchy, B. M., Goldberger, N. R., & Tarule, J. M. (1986). *Women's ways of knowing: The development of self, voice and mind.* New York: Basic Books, Inc.

Chemical Education Resources (1997). *Catalog for modular laboratory program in chemistry.* Palmyra, PA.

Collins, A., & Spiegel, S. A. (1995). So you want to do action research? In S. Siegel, A. Collins, and J. Lappert (Eds.), *Action research: Perspectives from teachers' classrooms* (pp. 117–128). Tallahassee, FL: SouthEastern Regional Vision for Education.

Elmesky, R., Muire, C., Griffiths, N., Taylor, P., & Tobin, K. (1996). *A statewide evaluation of mathematics and science courses in Florida community colleges.* The Florida State University program in science education in collaboration with the Florida Department of Education Division of Community Colleges, Tallahassee, FL.

Gibson, E., & Humerick, R. (1997, April). A community college evaluation study. Region II, St. Johns River Community College. A paper presented at the annual meeting of the American Educational Research Association, Chicago, IL.

Guba, E. C., & Lincoln, Y. S. (1989). *Fourth generation evaluation.* Newbury Park, CA: Sage Publications.

Lewin, K. (1946). Action research and minority problems. *Journal of Social Issues 2,* 34–46.

Q. S. R. NUD*IST. (1994). *Qualitative data analysis software for research professionals: Qualitative solutions and research.* Melbourne, Australia: Qualitative Solutions and Research Pty. Ltd.

Tobin, K., & Muire, C. (1997). A multi-level analysis of mathematics and science education in Florida Community Colleges. A paper presented at the annual meeting of the American Educational Research Association, Chicago, IL.

CHAPTER 9

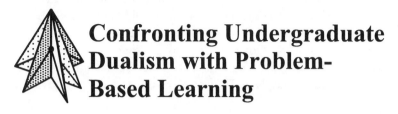

Confronting Undergraduate Dualism with Problem-Based Learning

Harold B. White III

If college teachers define themselves only as content or skill experts within some narrowly restricted domain, they effectively cut themselves off from broader identity as change agents in helping students shape the world they inhabit.
—Brookfield, *Becoming a Critically Reflective Teacher*

The extent to which I was a good teacher for the first two decades of my professorial career I owe to my good teachers, whom I respected and tried to model. Nevertheless, it has only been in the past few years that I have begun to understand more fully the dimensions of what it means to be a good teacher and have developed a personal vision that has redefined my relation to students and their learning. Paradoxically, that vision prompted me to create an environment of managed student frustration in which there is sympathy and support for that frustration. I deliberately challenge the ways students are accustomed to dealing with knowledge and work to have them realize their ability to think and learn for themselves.

Actually, it was my own frustration with the view of knowledge held by many intelligent students that made me question traditional teaching methods and become receptive to alternative teaching approaches (White, 1996a, 1996b). Because I taught courses where I could define the content and because I had tenure, I had the freedom to change course format in an environment that muted the risk associated with change (Groh et al., 1997). I remember well my decided skepticism to the idea that students might learn better if professors lectured less and let students work together on projects. Thus I am quite tolerant of colleagues who reject out of hand some of the things I now do in my classes. My thinking has changed on a number of other issues. Among those I have confronted are, What should be the goal of a college education? How can that goal best be achieved? What should be the role of a college teacher? What do students mean when they say they understand? How do examinations send messages about one's teaching philosophy? These are not new issues. Many before me have pondered them and written about them. However, the issues need to be revisited.

Dualism in Undergraduate Science: The Challenge

Position 1: The student sees the world in polar terms of we-right-good vs. other-wrong-bad. Right answers for everything exist in the Absolute, known to Authority whose role is to mediate (teach) them. Knowledge and goodness are perceived as quantitative accretions of discrete rightness to be collected by hard work and obedience (paradigm: a spelling test).
    —Perry, *Forms of Intellectual and Ethical Development*
                                        in the College Years: A scheme

Most college freshmen view knowledge, learning, and the role of faculty in what William Perry (1968) classifies as *dualistic*: Knowledge comes from the professor, learning and grades should be directly proportional to effort, and problems have a single correct answer. Textbooks and courses that represent science as facts and unquestioned quantitative models reinforce his view. Perhaps because this approach pervades science education, the public perception of science is dualistic. Moving beyond a dualistic view of knowledge, to one that includes uncertainties and using one's conceptual understanding to make decisions, represents a change that a student must make in becoming a scientist (and others should make in becoming responsible citizens in a democracy).

I have always been perplexed by students who do well on objective tests but are unable to answer questions that require analysis, synthesis, and evaluation, the higher-level thinking skills defined by Bloom, Englehart, Furst, Hill, and Krathwohl (1956). Somehow it seemed that there should be a way to teach that would stimulate students to improve their critical thinking skills and conceptual understanding while increasing their content knowledge. Many think they understand when, in fact, they have misconceptions and only a superficial understanding. As an educator, I must assume that among the 70% of adults who seem unable to think abstractly (Adey & Shayer, 1994), many could have been helped by instruction that challenged dualistic thinking.

Since 1993, when I decided to abandon the lecture format in my biochemistry courses, I have used a problem-based learning approach to teaching as a way to address this problem. By demanding higher-level thinking, I have attempted to move students more quickly beyond the dualistic to the multiplistic and relativistic views of knowledge (Perry, 1968), where they can evaluate competing ideas in some context. My objective is to empower students to think critically and to learn on their own. As a consequence, I expect my students to learn how to use the library and practice those skills. I do not intend their searches to be "random" because I expect them to identify what they don't know and need to learn before they go to the library. I expect them to take the initiative and to pursue answers to specific questions they pose. I don't expect to be the source of all the things students learn in my class. In a student-centered environment, the instructor is not in total control of inquiry and learning, so that many times unplanned issues have to be addressed. I want my students to reason with evidence, to know when they don't know, to realize that they have the ability to analyze problems and figure things out, and to learn what powerful resources are available to them and how to use them. How else are students going to gain confidence in these areas unless they are challenged to do so and get practice doing it?

These expectations make some students uncomfortable. Even on evaluations at the end of the course, some long for "objective examinations," "more lectures," and "assignments that are less open-ended." They want to think that "chemistry is not a discretionary science" and that "there is a right answer and a wrong answer" to every question. It may be too much to expect that a single course with a different format embedded in a curriculum of traditional courses will have a profound effect on dualism. More realistically, it provides a different perspective on knowledge and learning early in a student's career so that the transition to multiplicity and relativism may be easier and come sooner. For me, a major objective of a college education is to produce independent thinkers who can work well with others. Problem-based learning embodies the essence of these objectives.

What is Problem-Based Learning Specifically?

Once you have learned to ask questions, relevant and appropriate and substantial questions, you have learned how to learn and no one can keep you from learning whatever you want or need to know.
　　　　—Postman & Weingartner, *Teaching as a Subversive Activity*

My concerns about the limited critical thinking skills of college students mirror the concerns of medical educators about thirty years ago. In an attempt to educate physicians who could use their content knowledge in the context of real patients, several medical schools, initially Case Western Reserve University and notably McMaster University, developed what is now called problem-based learning (Barrows & Tamblyn, 1980). It contrasts strikingly with the stereotypical medical education in which students in large lecture theaters try to assimilate massive amounts of information delivered by a succession of highly specialized physicians. In problem-based learning, medical students work in small collaborative groups with a dedicated faculty tutor. They learn basic science facts and concepts as they encounter them in the context of medical case studies rather than in the order of chapters in a textbook. In other words, a complex, real-world problem initiates learning that occurs in an open-ended, interactive, problem-solving environment guided by a knowledgeable facilitator.

Several features distinguish problem-based learning from traditional lecture-based instruction. Firstly, the problems are qualitatively different from most end-of-the-chapter problems. They often are based on interesting true stories, integrate multiple concepts, and may involve current controversies. They require students to define and gather relevant information and to make assumptions where ambiguity exists. Ideally, the problems initiate discussion that involves analysis, synthesis, and evaluation before coming to some decision after a week or so of study. Second, the instructor functions as a facilitator rather than as the primary source of information. The instructor defines the course content by the problems he or she selects (White, 1996a). Third, students work cooperatively in small groups during class time, often with a tutor present. And finally, though not always, group performance contributes to each group member's grade.

Despite its well-documented use in medical schools, the adoption of problem-based learning strategies in undergraduate education has been limited, particularly in the sciences. It is a different way of instruction, and there are risks. The major deterrents seem to be reservations about compromising content, a lack of suitable problems, uncertainty about the time involved, and loss of control in a student-centered environment. Indeed, one must confront these issues when adopting problem-based learning as an educational strategy. However, these obstacles can be surmounted by instructors who see the potential benefits to the intellectual development of their students. Frequently, those who accept the challenge of adapting problem-based learning to their courses see the change in their students and find the process stimulating in a way that revitalizes their own interest in education.

Creating an Environment of Managed Frustration

Human beings are all right for as long as they are ignorant of ignorance; this is our normal condition. But when we know that we do not know, we can't stand it.

—Thomas, *The Fragile Species*

Is it reasonable to expect freshmen in a biology class for non-science majors to learn about biology by reading highly technical research articles? Herman Epstein thought so about three decades ago when he and several of his distinguished colleagues at Brandeis University took their research interests into the freshman classroom (Epstein, 1970). That radically different approach proved to be exciting for both faculty and students despite the frustration it created initially. By changing his educational objectives, Epstein transformed what he taught, how he taught, and what his students learned. He used a historical series of research articles in his own research field to convey the excitement of scientific discovery as a human activity. In the process, students learned many fundamental concepts in biology within the context of a limited research area. Students who were not destined for scientific careers gained important perspectives on what scientists do, how they think, and the uncertainty absent in textbooks. Epstein thought these were more important things for non-science majors to understand than the traditional curriculum defined by most introductory textbooks.

The issues raised by Epstein for non-science majors are equally important for science majors but are rarely addressed formally. In 1989 when the University of Delaware established its undergraduate major in biochemistry, a course based on Epstein's model became a requirement that may be unique in the country. Initially, Introduction to Biochemistry served about ten sophomores per year, who read research articles on hemoglobin and sickle cell anemia (White, 1992). It used a lecture-discussion format. As the enrollment increased, however, the quality of class discussion decreased. This change coincided with my frustration with the outcomes of traditional instructional methods, my reading of *The Harvard Assessment Seminars* (Light, 1990, 1992), and my exposure to the concept of problem-based learning (Barrows & Tamblyn, 1980; Boud & Feletti, 1991). Since 1993, the course has used the problem-based group-learning format where each research article serves as a problem that needs to be read, reread, and discussed by small groups of students to be understood (White, 1996a). While I still lead occasional whole-class discussions to wrap up a problem, I function primarily as a roving tutor who visits each group for a short time each period. Recently (in 1996), each group acquired a dedicated tutor recruited from third- and fourth-year biochemistry majors. These changes in format represented a shift from teacher-centered to student-centered instruction. They also reflected changes in student frustration and its management.

Undergraduate science majors differ little from non-science majors when they first encounter a research article. They do not understand it. The vocabulary, the methods, the results, and the conclusions are all obscure. This frustrates many students. Why, they think, should they have to struggle with the article when the professor, who understands it, could simply summarize the important points? Although all students know how to use a dictionary, some do not possess one, and few are accustomed to using one. Similarly, most students are not accustomed to or do not know how to pursue information in the library. In a world of instant gratification, it "takes too much time." In contrast to most courses, where books and instructors or standardized examinations define content, students in a problem-based course must recognize and define what they do not know, which becomes the course content. For students to become independent learners, they need to learn how to manage their ignorance. The problem-based classroom must provide support while students deal with these frustrations. It must be safe to admit, "I don't know," and talk openly about where or how to find out what one needs to know in order to understand.

While it is frustrating not to know, it is reassuring to know that everyone else in one's group has the same frustration and that the tutors had the same experience when they took the course. The discomfort is temporary. Confidence that an article can be understood with some work gradually replaces the anxiety associated with each new article. Although the use of peer tutors in this course started only two years ago, the effects are impressive. The class has gotten involved with the course material much faster than in past years because the tutors provide firsthand testimony to the effectiveness of the method and how it works. Based on a self-assessment survey, the students spend six hours per week working outside of class on the course, 25% more time than did the previous class without tutors. Furthermore, the tutors were able to spot and deal with many problems in group dynamics before they festered. At the end of the semester, 80% of the students expressed interest in returning as a tutor in their junior or senior year.

Experiencing Multiplicity and Relativism

The development we trace takes place in the forms in which a person perceives his world rather than in the particulars or context of his attitudes and concerns. The advantage in mapping development in the forms of seeing, knowing, and caring lies precisely in their transcendence over content.

—Perry, *Forms of Intellectual and Ethical Development in the College Years: A Scheme*

As students' views of knowledge change, they progress from a "black and white" dualistic world to a uniformly "gray" multiplistic world where there are many different alternatives of equal value. The multiplistic world gives way to a relativistic world where black and white represent ends of a spectrum in which "grays" are distinguished according to internalized criteria (Perry, 1968). The students who take Introduction to Biochemistry have been conditioned in earlier courses to examinations in which multiple choice, short answer, and quantitative problems have single correct answers. While such questions have value, I deliberately avoid them because they reinforce dualistic thinking. My examinations and assignments typically require higher-level thinking on the Bloom scale and generate a variety of objectively "correct" answers that receive a range of grades reflecting their subjective "quality" (White, 1993; Tobias & Raphael, 1996). Some students express themselves more clearly, provide more or better examples, or display creative approaches that warrant better grades. Specific examples of assignments associated with the first article my students read illustrates the point.

In 1864, G. G. Stokes, future president of the Royal Society of London and prominent Cambridge physicist and mathematician, published an article on the oxidation and reduction of the coloring matter of blood (Stokes, 1864). This enormously rich article describes in archaic language a series of qualitative experiments that Stokes conducted to define chemically the color difference between venous and arterial blood. Unlike most recent publications, Stokes's article lacks an abstract and contains a single hand-drawn figure representing the visible band spectrum of hemoglobin and several of its derivatives. Among the assignments and activities designed to encourage deeper understanding of this article are:

1. In 200 words or less, write an abstract that would be appropriate for Stokes's article.
2. Transform one of Stokes's experiments into an experiment suitable for a contemporary undergraduate laboratory course.
3. Draw a model that conceptualizes Stokes's description of the reactions of hemoglobin.
4. After seeing the classroom demonstration of one of Stokes's experiments, describe in chemical terms what the observations mean.

Clearly none of the above assignments has a single correct answer. Such assignments always involve group discussion followed by an individual or group response. A strategy associated with assignment 3 above has been particularly successful in encouraging multiplistic and relativistic thinking. Answers to the assignment from each of the groups are copied and distributed at the next class period to "scrambled groups." These are new groups in which students from the previous groups are redistributed. The new groups are asked to evaluate and defend, if necessary, all of the models and then create a new better model that incorporates the best features of the previous models. Students especially like this assignment because they get to see the responses of their colleagues in the other groups and become aware of the differences in the way people can perceive the same thing.

In addition to assignments associated with each article, every student must write a term paper that develops some general theme that integrates and illuminates the lives of two prominent scientists, preferably biochemist members of the National Academy or Nobel Prize winners. Interviews are encouraged. This demanding assignment generates a lot of anxiety that must be managed. However, the products are often spectacular as measured by the number of university-wide undergraduate writing awards won by students in the class.

I wish I could report that students who take this course and others that use problem-based learning progress to higher "Perry levels" more rapidly than do students who take only traditional courses. Unfortunately, my student numbers are too small and there are too many confounding variables, a situation very frustrating to someone accustomed to well-controlled laboratory experiments. A single semester is too short a time to see significant changes in attitudes for groups of students. Certainly anyone who teaches this way has many anecdotal stories about particular students who transformed their worldview. I can say that many students reflect on the educational process and their educational goals as a result of problem-based learning, and it is evident that the faculty who use the method do so, too.

In traditional science instruction there is almost an obsession with content that encourages dualistic perspectives. There is the embedded presumption that presentation of the material is sufficient to ensure learning and understanding. Fear that content is lost with a problem-based approach obscures the possibility that more students may understand better and retain more of what they learn. A proper evaluation should not focus on content alone but should also include how well different approaches give students confidence in their ability to think and continue to learn on their own. Content is important and the lecture method is an efficient, cost-effective way to present it; however, I find problem-based learning addresses important aspects of intellectual development that are usually neglected in traditional science pedagogy and that I think must be addressed. Dualistic thinking must be challenged early so that students can make the most of their college education.

> *Students are capable of a great deal more creativity than anyone thinks they are; the only reason they do not show it is that no one asks them to.*
> —Brent & Felder, *Writing Assignments: Pathways to Connections*

Harold B. White III
Department of Chemistry and Biochemistry
University of Delaware, Newark, Delaware, USA

EDITORS

METALOGUE

Beyond Dualism: Teaching Authentic Science

PT: Hal identifies an issue that is crucial for all endeavors to educate from a social constructivist perspective. It is the restrictive way in which students (have learned to) reason dualistically about the world; *dualistic reasoning* masks the creative thinking of scientists, hides the uncertainty of scientific knowledge, and robs students of their agency as self-regulating and collaborative learners. Why is it perpetuated by scientists in their teaching? Probably because it is so much easier to adopt a narrow focus on transmitting objective facts and measuring their reception via multiple-choice examinations or the reproduction of model solutions. Not only does this deception fit neatly into the cultural fabric of many science departments, but its efficiency frees the scientist to expend his creative energies on his true passion—research—which, tragically, is obscured from the students' view. And what is the outcome? Teaching that presents a grossly distorted image of the creative thinking of research scientists and the epistemic status of the knowledge they produce.

Hal White's teaching is designed to counter the tradition of dualistic reasoning by engaging students in making sense of experiences in ways that are much more authentically scientific. Hal aims for students to develop the habits of mind of research scientists who manage fruitfully their own problem-based learning frustrations. He recognizes the value to creative thinking of maintaining a degree of emotional and intellectual dissonance (or disequilibration). These habits of mind enable scientists to pursue collaboratively multiple (rather than single) lines of inquiry and to express the epistemic status of the outcomes in shades of gray (rather than black and white). Scientists such as Hal White have a passion for richly educating their students. Those among them who are prospective science teachers are very fortunate to experience the learning of authentic science.

KT: I wonder about the efficacy of managed frustration. I too have been thinking about *problem-centered learning and doing* science, but at the high school level. Here we have a problem of needing to connect to the interests and passions of the students. If we do not do that as teachers, then the students simply go someplace else—usually onto the streets. So the last thing I want to advocate is frustration. I want to get heads up from the desks and bodies into seats. Success breeds success, so the cliché goes. Accordingly, we are gearing the curriculum toward the interests of the students and managing it so that what they do and learn is connected closely to what they know and can do when they come to class. Being aware of what they do not know and what they would next like to know is an important part of being a learner, and so we are spending some energy to get students to be more aware of their knowledge and their learning goals. As long as the motivation to learn is high, then the frustration of not knowing what to do in a problem context can be offset. Frustration also can be offset by the presence of peers with whom to collaborate and the presence of a teacher to mediate learning. In any problem-centered context I see a valued teacher who can facilitate learning by being a co-learner. I advocate strongly for co-participation as the hallmark of an effective learning environment, be it in a research laboratory or an elementary school classroom.

Beyond Heroics: Collaborating to Transform Culture

PT: Hal warns that, because it is counterintuitive for most college students, becoming a problem-centered learner takes time. I am left to wonder how he has managed to win the approval of his chemistry department for such a radical departure from traditional teaching and learning? To push the envelope like this must surely involve taking great professional risk!

KT: To take risks and do what you believe is right does require an individual to deal with the frustrations of students and the attitudes of colleagues who find what is being attempted to be off the wall and perhaps unacceptable. The social dimensions of reform can never be underestimated. From our research with high school teachers, we know that reform can rarely be sustained unless there is a critical mass of support from within the institution and administrative support as well. We have seen in other chapters that the critical mass is often missing in universities, and disincentives to change the approach to teaching can also be felt from administrators. It seems to me that one source of support can come from the students. If they can understand their roles as learners and perceive that what you are doing is with their interests in mind and directed toward improving their learning, then their support can make all the difference if colleagues raise questions. There is power in student evaluations.

PG: I agree with you, Ken, that it can be lonely for a faculty member such as Hal, who works in a College of Arts and Sciences, where he is trying to change the culture of teaching science while residing in a community that resists change. At some point each of us must decide whether to do what is right or not. The consequence of deciding to be involved in education reform may be that you do not get as large a pay raise this year. In the end, what you choose to do matters most to you and how you feel about your own actions.

KT: Problem-centered learning can be a powerful way of getting students to think and attain the types of goals identified by Hal. He makes a good point, though, when he wonders how much can be attained in just one course. This can be the case especially if other faculty do not support what is being attempted.

PG: We could do so much more if we amplified what each other did, and connected to students' prior understandings and knowledge. Also problem-based learning strategies in one course such as Hal's could be strengthened with other similar strategies in other courses. We need to encourage a critical mass of faculty to become involved.

BIBLIOGRAPHY

Adey, P., & Shayer, M. (1994). *Really raising standards: Cognitive intervention and academic achievement.* London: Routledge.

Barrows, H. S., & Tamblyn, R. M. (1980). *Problem-based learning: An approach to medical education.* New York: Springer.

Bloom, B. S., Englehart, M. D., Furst, E. J., Hill, W. H., & Krathwohl, D. R. (1956). *Taxonomy of educational objectives: The classification of educational goals, Handbook 1, Cognitive domain.* New York: Mackay.

Boud, D., & Feletti, G. (Eds.). (1991). *The challenge of problem based learning*. New York: St. Martin's Press.

Brent, R., & Felder, R. M. (1992). Writing assignments: Pathways to connections, clarity, creativity. *Journal of College Science Teaching 40*, 43–47.

Brookfield, S. D. (1995). *Becoming a critically reflective teacher*. San Francisco: Jossey-Bass.

Epstein, H. T. (1970). *A strategy for education*. New York: Oxford University Press.

Groh, S. E., Williams, B. A., Allen, D. E., Duch, B. J., Mierson, S., & White, H. B., III. (1997). Institutional change in science education: A case study. In A. McNeal and C. D'Avanzo (Eds.), *Student-active science: Models of innovation in college science teaching* (pp. 83–93). Orlando, FL: Saunders College Publishing.

Light, R. J. (1990). *The Harvard assessment seminars: First report*. Cambridge: Harvard University Press.

———. (1992). *The Harvard assessment seminars: Second report*. Cambridge: Harvard University Press.

Perry, W. G., Jr. (1968). *Forms of intellectual and ethical development in the college years: A scheme*. New York: Holt, Reinhart and Winston.

Postman, N., & Weingartner, C. (1969). *Teaching as a subversive activity*. New York: Delacorte Press.

Stokes, G. G. (1864). On the reduction and oxidation of the colouring matter of the blood. *Proceedings of the Royal Society of London 13*, 355–364.

Thomas, L. (1992). *The fragile species*. New York: Charles Scribner's Sons.

Tobias, S., & Raphael, J. (1996). *The hidden curriculum: Faculty-made tests in science*. New York: Plenum Press.

White, H. B., III. (1992). Introduction to biochemistry: A different approach. *Biochemical Education 20*, 22–23.

———. (1993). Research literature as a source of problems. *Biochemical Education 21*, 205–207.

———. (1996a). Addressing content in problem-based courses: The learning issue matrix. *Biochemical Education 24*, 41–45.

———. (1996b). Dan tries problem-based learning: A case study. *To Improve the Academy 15*, 75–91.

CHAPTER 10

 What It Means to Achieve: Negotiating Assessment in a Biology Course

Susan Mattson

Core Team Meeting

Present: Christy, Greg, Keith, Laura, Jill, Sam, Tina

The sound of pages flipping back and forth was all that could be heard at the start of the meeting. They were looking at the proposal for student assessment Sam had given them earlier. The first part listed potential components of the grading system and how they should be weighted on a 100% scale; the second involved a rather lengthy explanation of each component. Most of them hadn't had time to read the whole thing, and parts of it were confusing. They hadn't even heard of some of the things he was proposing.

Christy started. "As a matter of philosophy, I think a grading system should reflect achievement. Some measure of mastery of the material. Everything else is really inconsequential to us." She frowned as she surveyed the outline.

"Well, I know that this isn't typical for biology," said Sam, "but given that it's an experimental course I thought it might be appropriate to try different strategies for assessing learning." He knew it wasn't going to be easy to make a case for what he'd proposed. Most of it reflected ideas that were just gaining popularity among members of the science education community. These ideas were foreign to those in science departments.

Greg nodded at Christy and said, "I agree with you. I think the percentage of the grade for mastery-type items should be much higher. Twenty percent for exams and quizzes is way too low. And 50% for the portfolio seems way too high."

"I disagree," said Keith. "In the unit Jill and I are doing, we really don't have any exams at all. A lot of what they're going to do, like home assignments and write-ups of activities, is supposed to end up in the portfolio. In our unit, the sorts of things we want them to learn don't really conform to tests. "

"Exactly. What if a good bit of their knowledge ends up not being the sort that is best shown on exams and quizzes?" said Sam. "The portfolio and some of these other things are there so we can detect that."

"Well, I don't really understand this portfolio concept," said Christy. "But 50% is way too high. And I'm not clear on what these 'performance assessments' are, but you've given them 10%. Using the library to do a literature search, designing and conducting an experiment, leading a discussion—these sound more like basic participation, which you've also given 10%. That's 20% for them just being there and doing what they're supposed to do!" Students should be expected to participate, she thought, and not be graded on it. That's how students learn what they're supposed to. The way this proposal looked, everything was going to be graded. How could they assess what was really a measure of learning if they were going to spend all their time assessing participation? Ridiculous.

"Performance assessments, as I understand it, are things that we want students to be able to show that they can do," said Keith, "but are the sorts of things that can't be tested effectively on paper or somehow put into a portfolio. And they aren't really tied to specific content but sort of extend through all of the units."

"I think I agree," said Laura, who had worked for several years with the same sorts of students expected to enroll in the course. "Sometimes you think someone's performance is low because they don't understand the material or aren't trying hard enough, and it turns out to be because they can't use a piece of equipment or run a computer program and they get frustrated and give up."

"Things like using a microscope or setting up an experiment shouldn't be assessed," said Christy. "They don't really measure anything, they're practice. Things that measure achievement should be given priority as far as determining the final grade. Not practice, or participation, or whether you like them or not, or subjective things like that." You can't spend all your time checking to see if they know how to do what they should learn to do with practice, she thought. And you can't give people grades based largely on opinion.

"We do need to be careful how we do this," said Tina, "or we could be liable from the university's point of view." Tina was thinking about the problems that could result if grades couldn't be accounted for by referring to specific scores. Subjective grading was always risky.

"Maybe, Sam, if you increase the amount of credit being given for exams and quizzes, it would be more agreeable," said Laura. She didn't really understand some of these ideas either, but she'd worked hard on her unit and wouldn't want to see her own efforts go to waste. Sam had been asked to develop this, she thought. At least they could give it a chance.

"I'd be happier if exams and quizzes were weighted at least 75%. Otherwise, students have no real incentive to do well on them," said Greg.

"I have no problem with adjusting some of this," said Sam. "But I just think that we should try to take into account learning that isn't reflected in answers on exams and quizzes. I've described some ways that these more "subjective" components can be quantified if it would make you feel more comfortable about it. If you could look at this more carefully when you get a chance, then maybe there can be a compromise."

"Well, I'll be glad to look at it, but I think grades should be based on achievement. That's just my philosophy," said Christy. "OK, what's next?"

In the early 1990s, a large research university received a grant for the purpose of planning, implementing, and revising courses in science and science education for prospective teachers. Part of this comprehensive grant dealt with new science content courses for prospective elementary school teachers. One of these courses was a biology course, or BIO 101. Although the development of BIO 101 first engaged diverse stakeholders interested in improving elementary school teacher education in science, a smaller "core team" of course developers, consisting mostly of people situated in the biology department or Science Education Program, took on the task of meeting regularly to develop the course. Eventually, four smaller "unit teams," each of which would ultimately teach one course unit, began meeting concurrently. BIO 101 was a topic of frequent informal conversations among course developers as well.

In the final weeks of course development, the core team turned to the issue of assessment, modifying considerably the original proposal (Table 1) to the plan used (Tab;e 2). Ultimately, 75% of the final grade came from quizzes and examinations; although homework, oral reports, and other forms of assessment were occasionally used. Of the remaining 25% of the total grade, 15% went to portfolios and 10% to class participation. The emphasis thus shifted from what were referred to as more "subjective" forms of assessment in the original proposal, such as portfolios and participation, to those that were more objective—quizzes and examinations.

The cause for this shift became evident during meetings as core team members expressed contrasting ideas about the assessment components that should be included and how they should be weighted. While there was willingness to experiment with other techniques on a limited basis, there was strong objection to departing too much from using examinations and quizzes to assess knowledge and, ultimately, to assign students a grade. When discussing assessment, "objectivity" and "achievement" became referents for making decisions. Contrasting beliefs about these referents emerged in these discussions.

To: Core Team
From: Sam
Re: Assessment

Please refer to previous memos for background information. What I am proposing here is a way of determining students' grades according to a percentage scale.

Table 1: Original Proposal of Assessment Strategy for BIO 101 Students

Component	% of Grade
Student Data	10%
Entry/Exit Assessment	5%
Ongoing Assessment	5%
Portfolio	50%
Unit I	10%
Unit II	10%
Unit III	10%

Continued on next page

Table 1 continued

Journal	10%
Extensions	10%
Performance-based	10%
Quizzes and Exams	20%
Participation	10%

Table 2: Actual Assessment Strategy for BIO 101 Students

Unit	% of Grade
Unit I	10%
Unit II	25%
Unit III	20%
Unit IV	20%
Portfolio	15%
Selected assignments from Units I–IV	
Journal	
Enrichment projects	
Participation	10%
Attendance	
Entry/Exit questionnaires	
Course evaluation	
Ungraded assignments	
Overall participation	

Objectivity Versus Subjectivity

Beliefs about objectivity and subjectivity influenced the positions core team members took on appropriate assessment practices. The position that Christy took when she rejected the idea of assigning grades "based largely on opinion" is the same position taken by several other team members. To these members, "being objective" meant removing one's personal biases to the greatest extent possible when making judgments about the adequacy of students' understandings of science. By deferring to the established conventions and collective knowledge of a "higher authority"—in this case, the appropriate scientific discipline(s)—the belief that assessment could approach objectivity was widespread among core team members.

"Being subjective," on the other hand, meant having a more personal, and hence, idiosyncratic means of making judgments. This was of significant concern to some course developers because the assessor could no longer defend a judgment by deferring to the authority of science. Rather, the authority shifted to a subjective assessor. This idea raised concerns about accountability among some course developers who thought that objectivity protected students from assessment practices that could be considered "unfair" and instructors from accusations by disgruntled students. Tina's opinion that "subjective grading was always risky," and thus undesirable, was shared by several other team members for precisely this reason. Others believed that forms of assessment not based on appealing to the authority of science for judgment were simply inappropriate for a science course.

Achievement Versus Practice, Participation, and Portfolios

When Keith says that the "things we want them to learn don't really conform to tests," he articulates his concern that a good deal of knowledge vital to building a strong foundation in teaching elementary biology is *not* best represented in traditional modes of assessment—quizzes and examinations. This idea was expressed by some core team members at various times during course development, particularly with regard to building knowledge through "practice," "participation," and "portfolios."

"Practice" came to be understood as the act of building and applying knowledge of science by using specific cognitive processes, methodological techniques, or physical skills. This knowledge was expected to grow in class or home activities that provided opportunities for repeated experience. In the original proposal for assessment, the type of knowledge built through practice was thought to be something that should be assessed by application in a realistic situation, hence "performance-based assessment."

"Participation," as a category for assessment, was originally established to provide a means for judging the quality of students' engagement in course work. Those supporting the original proposal envisioned that instructors would periodically give students feedback regarding their participation—especially in the event of perceived problems—where individualized solutions would be negotiated between instructor and student.

Another alternative proposed for the assessment of students' knowledge was the portfolio. The portfolio, as originally envisioned, was meant to contain evidence of student learning during the four course units and to be a place where students stored items that might be of use in the future. The portfolio was also to contain journal entries; entries could act as a "diary," a place to express feelings, a notebook for questions of interest, and a way for the instructor and student to communicate on an individual basis. Another part of the portfolio was to include evidence of "extensions"—activities selected by students to strengthen knowledge in areas of their own interest. Portfolios were thus meant to be individualized forms of assessment, with no two portfolios to be the same.

Team members did not deny that the above forms of assessment could serve as useful indicators of what students understood about biology. However, Christy's opinion that "a grading system should reflect...some mastery of the material" and Greg's comment that "the percentage of the grade for mastery-type items should be much higher" reflected a predominant belief in the group that the knowledge represented in practice, participation, and portfolios *should not* be the focus of assessment—this knowledge did not represent the type that should "count" in a significant way. From this perspective, what *should* be the focus of assessment was achievement, and, to most of the group, quizzes and examinations best measured achievement.

The course developers modified the original proposal for assessment for the first implementation of BIO 101. The predominant view about practice, which was regarded by most team members as a required form of participation, resulted in the elimination of performance-based assessment from the assessment strategy for the course. Participation *was* included as a category for assessment, but was based mainly on class attendance. Portfolios were also included; however, their relative weight within the final assessment strategy was considerably lower than envisioned.

Objectivity and Achievement

Highly valued in traditional views about the scientific enterprise, the desire to be objective is tightly linked to the belief that objectivity is possible; that is, when making a judgment, one *can* remove or at least significantly reduce the influence of one's personal biases. When deferring to a body of scientific knowledge as the authority for assessment purposes, one narrows down the list of "right answers," whether they be facts, algorithms, techniques, concepts, habits of mind, or values, to those that have been agreed on by the scientific community. Essentially, these right answers form an "objective" standard against which we judge all answers. It is this standard—the accumulated body of knowledge that is a product of consensus in a scientific discipline—that the majority of core team members thought evidence of achievement should be based on. Beliefs about objectivity thus closely aligned with beliefs about achievement: By orienting assessment toward achievement of a consensual standard, one can "be objective." The desire to be objective in this way when assessing student knowledge was pervasive among group members.

A contrasting perspective, while not predominant, was linked to concerns about more subjective forms of assessment. Like Sam's comment that "we should try to take into account learning that isn't reflected...on exams and quizzes," some course developers shared beliefs about the nature of knowledge that are having increasing influence on educational theory and practice. These ideas assert that objectivity is impossible and that forms of assessment based on the desire to "be objective" are inadequate ways of determining what students have learned. One suggested direction for reform thus involves a movement away from "objective" forms of assessment.

Unit Team Meeting

Present: Greg, Laura, Vicky

"So we're just supposed to ignore that there are proven scientific facts and theories? Anything goes because nothing's for certain? Then why even bother teaching science?" said Greg.

Here we go again, thought Vicky. "I'm not saying that facts and theories don't exist. I'm saying that they don't exist outside of our minds."

"Even if they can be tested, time after time? Regardless of who's doing the testing?" said Greg. "That's the power of scientific knowledge. If I drop an apple from a tower, I can prove it will fall every time!"

"Under all conditions?" said Vicky. "And who comes up with the criteria for deciding what counts as proof? Who decides what methods are appropriate? Who makes the observations that lead you to do an experiment in the first place? A human being. And our senses and minds are limited to knowing things only from a human perspective. There is no objective knowledge; we can't know anything in an absolute sense."

"Well, of course," said Greg. "Scientists don't claim to be able to prove that anything is true in an absolute sense. The idea is to prove what is false so we can at least get close to what is true."

"Vicky," said Laura, "isn't this more a philosophical argument than anything else? Does it really matter whether you believe there's an absolute truth or not? I mean, if you asked a question like, How do plants get energy? wouldn't you want students to know that the answer was photosynthesis? Even if everyone agrees that we can never know everything about photosynthesis or that people were the ones deciding what photosynthesis is?"

Good point, thought Vicky. This always came up when they talked about what knowledge was. It was as if her beliefs about the nonexistence of objectivity and objective knowledge implied that facts and theories were something that scientists just "made up" and could be changed according to human whims. And what did this really have to do with how students were assessed?

"Laura's right," said Greg. "Why does it matter whether or not objectivity or objective knowledge exist? There are still things that we want students to know."

"It doesn't matter whether they exist, but it does matter how your beliefs about knowledge translate into assessment," said Vicky. "It just seems like the knowledge that is represented on tests—this so-called 'objective knowledge'—has come to symbolize the only knowledge of value. And objective knowledge equals right answers, and right answers take the form of definitions, descriptions, and maybe a few applications on paper."

"What's so wrong with wanting to find out if students know the right answers, whatever form they take?" said Greg.

Vicky was getting frustrated. "Nothing is wrong with wanting to find out!" she said. "But when judging the extent of students' knowledge is based only on a search for whether or not they have 'the right answers' in that form—the objective form—we've missed the forest for the trees. It's like saying, Well, if you know the answer is photosynthesis, you know enough."

"Well, we can't assess everything," said Greg, "and there are facts and theories that are fundamental to understanding biology! The way I see it, it's my responsibility to make sure that I teach these things. And tests are the most efficient way I know of assessing what they know," said Greg. "And the fairest way of assigning a grade," he added.

Tests might be the easiest way, thought Vicky. But not the best. And certainly not fair in all cases. "Well, I hate to throw another wrench in the works, but there's one more thing. I really question the purpose of assessment in this course. It sounds like it's just to provide us with something to base grades on."

We need to move on, thought Laura. They would never see eye to eye on this.

Traditionally, science has been portrayed as a "search for truths," with the "scientific method" being the most reliable pathway to knowledge of these truths. It is these truths—the "products" of science—that have become the "right answers." And yet, when the idea of right answers emerged in the context of assessment, reactions of course developers were highly dissonant. These reactions could be linked to ideas about the nature of scientific knowledge and what constituted evidence of "having" scientific knowledge. At first, it seemed as if disagreement revolved around conflicting beliefs about the nature of scientific knowledge—that there "is" objective knowledge, objectivity, and "certain" scientific ideas that are simply indisputable. But just as Vicky and Greg "sounded like they both agreed," group members also seemed to agree on several points.

On the Nature of Scientific Knowledge

It was clear that all group members believed that the phenomena of the natural world, as described through the senses of scientists, are *really there*. In other words, members believed there is a universe that operates by "rules" that are independent of any observer: Absolute truths exist. The course developers also shared other beliefs. When Vicky implies that no one can know absolute truths because human perceptive and cognitive capabilities are limited, she rejects the existence of objective knowledge. When course developers took objective knowledge to mean knowledge that exists without mediation by the human mind, there was agreement that objective knowledge, like objectivity, *does not* exist.

The reactions of course developers to the idea of "right answers," in relationship to assessment, thus didn't seem to be a product of disparate beliefs about the nature of scientific knowledge. Rather, it seemed to be a product of the extent to which core team members believed that more traditional forms of assessment provided adequate evidence of students' knowledge of science for the purposes of the course.

The Limitations of Quizzes and Examinations

In early discussions about assessment, it at first seemed easy to conclude that the emphasis on using quizzes and examinations reflected beliefs that, as Vicky says, "the knowledge represented on tests...[was] the only knowledge of value." As discussion progressed, however, it became clear that this was not the case. Core team members, most of whom had advanced degrees in biology and/or many years of experience in biology or biology education, shared an understanding of scientific knowledge that included many more dimensions than those that could be represented on traditional tests—dimensions that were more fundamental to the advancement of scientific knowledge. The position taken by Greg when he says, "we can't assess everything," however, was echoed by course developers who admitted that tests were limited but were, nonetheless, the appropriate form of assessment for the purposes of the course.

Course developers in favor of alternative forms of assessment objected to tests because they believed them to be inadequate indicators of what students know in a variety of ways. Several group members believed that tests indicated little about the extent of student knowledge with regard to a particular topic. Knowledge about photosynthesis, for example, could be limited to a definition, a formula, or a description. Alternatively, students might be able to show how knowledge of photosynthesis can be applied to revive a dying house plant or hypothetically to solve the problem of the "greenhouse effect." Because tests are usually created in such a manner that responses are expected to converge on a limited set of right answers, however, opportunities to detect what a student knows *instead of* or *in addition to* these answers are missed.

For the same reason, several course developers believed that tests can rarely be used to diagnose learning barriers because, as one core team member said:

> [Tests] aren't designed to tell you what students *don't* know, just whether or not their knowledge looks like it matches [the instructor's]. If students don't give the right answer, especially when the question involves higher-level cognitive processes, there could be numerous gaps in their knowledge that can't be located. Tests simply don't probe deep enough into what students do or do not know.

An additional shortcoming of tests cited by another core team member was that "tests were inappropriate ways to assess the extent to which students could 'do something.'" Those supporting performance assessment, which allows students to demonstrate what they've learned in the context of solving an authentic, complex problem, were particularly concerned that, for example, "the fact that students could write about doing an experiment would be accepted as evidence that students could *do* an experiment."

A related concern was that some students had difficulties representing what they knew in the form required by tests:

> Some students just don't have the writing skills to show that they know something. Or, they might not perform well under time constraints. But that doesn't mean they don't have the knowledge. There need to be other ways that students can show what they know.

The limitations of quizzes and tests articulated by course developers who supported the use of alternative forms of assessment can perhaps best be summarized by the idea Vicky expresses when she suggests that by emphasizing quizzes and examinations for assessment purposes, "we've missed the forest for the trees."

Other concerns about the shortcomings of more traditional forms of assessment emerged in discussions about the purpose of assessment.

Informal Conversation

Present: Sam, Vicky

"I don't know, Vicky," said Sam, shaking his head. "It just seems like they're missing the whole point." They had been talking about how

no one seemed to understand what portfolios were all about.

"Another problem is that portfolios didn't work in the other class," said Vicky. "So now the general opinion is that they'll never work." Portfolios had been tried in a similar chemistry course that had been piloted a couple of semesters earlier. The instructor had never really known what portfolios were supposed to assess, and the students had also been confused. Their portfolios had reflected this confusion.

Sam shrugged. "So much for withholding your conclusions until you have replicates." He'd run into this problem before. When a new strategy for improving teaching and learning in science didn't show immediate success, there was a tendency for people to dismiss it after the first try. He'd found this to be particularly true of people situated in science departments, especially when no empirical evidence existed to support claims that a new strategy represented an improvement over a more traditional one.

"You know, the main problem people seem to have with portfolios is that they don't lend themselves to an 'achievement-oriented' grading system. 'It's too subjective,' they say. 'How can we compare what students know if everybody has something different in their portfolio?'" said Vicky, referring to some of the comments others had made.

"Another thing," said Sam, "is that this overriding interest in achievement focuses exclusively on what students know rather than what they've learned. I mean, how can you really tell if any learning occurred if all you're interested in is seeing if students have 'mastered' the material? For all we know, this is one big review of biology for some students." Tests are like snapshots, he thought. They capture knowledge at a specific point in time. Portfolios, on the other hand, could actually capture learning by allowing students to revisit original understandings and somehow add to them.

"Well, I can understand how they might be concerned about having to evaluate twenty-four portfolios containing very different things. And I guess it would be hard to give the same grade to someone who had, say, fifty tries versus someone who had twenty-five," said Vicky.

"Maybe. But you've got to look at these portfolios more like artists' portfolios," said Sam. "No one expects two artists to have the same portfolio. In fact, that's the beauty of them. Students can become their own assessors with portfolios."

"I'm not sure what you mean," said Vicky.

"Well, learning about some scientific concepts would be like learning a new sport," said Sam. "Take tennis, for example. You get out there for the first lesson and whack the ball around without much of a clue. You know what you should look like because you've seen Wimbledon, but you're not there yet as a player. So you practice, and the next time, you do better. You know where you were, you know where you are, and you know where you want to be. You assess yourself, and it's progress that gets assessed."

"That's an idea. Portfolio as a way of participating in assessment rather than spectating!" said Vicky. "Now, how do we sell this?"

When the course was implemented, portfolios continuously came up as a source of confusion for both the instructors and the students. Ultimately, they became a "depository" for almost all of the written assignments. Essentially, most of the core team thought that there was no purpose to the portfolios because they contained things that had already been graded or that shouldn't be graded. Vicky's comment that portfolios presented problems for people because they "don't lend themselves to an achievement-oriented grading system" reflects beliefs held by the majority of the group regarding the purpose of assessment. More specifically, it was clear that most course developers associated assessment primarily, if not exclusively, with determining what students had "achieved" in the form of knowledge represented on tests.

One concern about this view, as expressed by Sam when he compares tests to "snapshots" of knowledge, parallels the beliefs of course developers who thought that assessment should be a measure of how much students learned *beyond* what they knew when they entered the course and that tests were not an adequate indicator of this learning. Arguably, certain forms of tests or the administration of pretests and post-tests might be one way of addressing this problem. But the predominant belief among course developers—that quizzes and examinations were the most appropriate forms of assessment in BIO 101—implies assessment that allows only a "one-shot" comparison of what students know to what instructors think the students *should* know, according to instructors' expectations rather than what students might have actually started out knowing.

Another perspective that dominated how the instructors ultimately assessed the students was that the primary purpose of assessment was evaluation—the assignment of grades. A contrasting belief expressed by some course developers, which parallels Sam's position that the "beauty" of portfolios is that they allow insight into each student's perspective, was that at least some component of assessment should represent the unique understandings of the individual. A reason for this belief was expressed by one group member:

> When [students] create a portfolio, they are essentially revealing themselves for who they are. This makes it risky, but also...a potential source of great pride....It feels good to look at something that you alone have done.

Related to this view were beliefs that the function of assessment be not so much to inform the instructor for grading purposes, but to inform the student so that s/he may monitor his or her progress toward a learning goal that s/he had a role in determining. That is, rather than viewing assessment as an endpoint, it is viewed as a turning point for continued progress toward learning goals. This perspective, however, was not predominant during core team meetings that focused on assessment.

The Call for Reform of Assessment in Science Education

Changing the Nature of Assessment

Like other aspects of the educational system, the ways in which students' knowledge is assessed are being targeted for reform (American Association for the Advancement of Science, 1990; National Research Council, 1996, 1997; National Science Foundation, 1993). The problems associated with traditional forms of assessment have been recognized for many years. With regard to these problems, Malcom (1991) makes the following observations concerning the nature of assessment:

> The difficult task in reform is not only figuring out what is wrong with the current system, but also identifying (or inventing) replacement components. The limitations of traditional modes of assessment are regularly outlined by reformers both in and outside of school. These include the failure of most standardized testing to assess what is most important for students to learn or to inform practice; the disconnection of testing from real world problems; and the inequity that testing reflects and supports. (p. 187)

Rather than detecting the breadth, depth, and/or connectedness of a student's understanding, standardized tests tend to emphasize scientific knowledge represented as isolated facts, algorithms, and skills requiring mainly rote memorization. While BIO 101 did not involve standardized testing, its counterparts—"objective" quizzes and examinations requiring responses that converged on "right answers"—were the primary means of formally assessing student knowledge. Because of the types of knowledge and cognitive processes involved, both standardized tests and their classroom counterparts are thought to assess only a superficial understanding of science.

Another shortcoming of these more traditional forms of assessment is that they emphasize the products rather than the processes of science, whether in thought or action:

> Traditional tests...yield no direct evidence about the thought processes that underlie that competence. In this sense, they provide at best, a partial, and at worst, a misleading, picture of student capabilities. (Campione, 1991, p. 303)

Because they are poor indicators of the extent of student knowledge, traditional tests fail to assess the degree to which students actually exhibit scientific habits of mind and action under circumstances in which scientific knowledge is relevant. Mainly, students are not given an opportunity to demonstrate—or even realize—that knowledge of science can be, and *is*, used to solve problems in their everyday lives.

A more subtle problem indicated by Malcom involves the role that more traditional forms of assessment have in creating and perpetuating inequity in the opportunities that students have to demonstrate and *benefit*—academically or personally—from what they know. While she is referring primarily to the sociocultural bias she believes implicit in standardized testing, her demand that assessment "give all groups a fair chance to demonstrate what they know" (Malcom, 1991, p. 324) can be extended to apply to the inequity that might be created when these so-called "objective" tests are the primary means of assessing student knowledge.

Her suggestions imply, among other things, that assessment should provide students with opportunities to represent knowledge in ways that suit them best, and to base at least a portion of assessment on knowledge that is thought to be of value for the student in his or her own sociocultural setting.

Changing the Purposes of Assessment

In discussing the features of portfolios, Collins (1991) reveals an alternative view of the purposes of assessment:

> The processes of conceptualizing and designing a portfolio present opportunities to teachers and students alike to identify clearly the instructional goals for the science class, to articulate the criteria for success, to negotiate publicly what will count as evidence, to participate in both the design and development process, to express individual strengths and become self-reflective, and to become co-learners. (p. 299)

This view of portfolios represents a significant departure from traditional perspectives on who should control assessment and what the ultimate goals of assessment should be. Until recently, the instructor has been considered to be the appropriate authority for determining instructional goals, criteria for success, and acceptable forms of evidence. With regard to BIO 101, one member of the core team expressed concerns about shifting some of the authority for assessment from the instructor to the student:

> [How can] students be the ones to determine what they should know? They have no idea of the universe of possibilities, nor what they really need to know. And they have no way of really being sure if they've achieved [a goal]. That's what we're here for. To determine what's in their best interest. If you let [students] decide how they should be graded, their standards will almost always be lower than the teacher's. They just aren't good judges of what excellent work is. And most of them want to get A's anyway, so they're going to set standards so this is possible.

These beliefs promote an assessment system that ignores the student as a legitimate participant in determining what s/he is expected to know and how this knowledge is to be expressed.

While few would disagree that the instructor, as an expert, is *one* appropriate resource for setting instructional goals and evaluating progress toward them, there is currently disagreement about the extent to which students should also be resources. In addition to the inequity associated with standardized testing cited earlier, Malcom highlights the inequity perpetuated by traditional views on who should control assessment—namely, students have little or no voice in determining what it is they should know and how this should be assessed, even though the case is continuously made that this knowledge is "in their best interest."

Other concerns about traditional assessment are evident in one course developer's criticism that "this whole system [of giving tests] seems ironic because it's designed more for the teacher's convenience than the student's learning. And yet it's the students who we're supposed to be serving." This same course developer went on to say: "What's even more ironic is that it's the student who usually gets blamed for getting bad marks, not the teacher. How many teachers really change anything [based on test grades]? And how many tests really tell you what you need to do differently?"

Apparent here is an interest in assessment that is instructional in nature, for both the student and the teacher. Baron (1991) echoes this interest when considering the merits of performance assessment:

> By situating...performance assessments...in the classroom, both teachers and students will be in a position to talk about the quality of students' thinking and the understanding that is being demonstrated. Opportunities to gain access to students' thinking and to discuss its quality using shared standards have dramatic potential for teaching and learning. (p. 250)

One common theme among proponents for changing the purpose of assessment is that students should be included in *at least some* decisions about instructional goals and how progress toward these goals is to be assessed. Mainly, it is thought that sharing authority for assessment has the potential to enhance learning because students have a greater sense of "ownership" in their learning *and* in demonstrating what they know. Traditional assessment is considered to fall short in promoting meaningful learning because students become disenfranchised—that is, students are apt to merely "go through the motions" as directed by a higher authority rather than actively participating in building and assessing knowledge that is of value to them.

Toward Improved Assessment

As part of more comprehensive reform in science education, improving how we assess students' knowledge of science demands development of new types of instruments as well as a reconceptualization of the purposes of assessment. Haertel (1991) describes the type of shift he considers necessary:

> In all kinds [of assessment]...teachers must maintain a proper relationship between testing and instruction....Systematic classroom observation and performance assessment should be the mainstays in determining the results of "hands-on" science education. (pp. 243–244)

What is clear in Haertel's suggestions is a movement toward alternative forms of assessment that allow students to represent the varied types of understandings representative of a strong foundation in science. He does not suggest that more traditional forms of assessment be abandoned altogether; he does suggest that they be limited to circumstances where preceding instruction targets the particular types of knowledge that they generally assess. Kulm and Stuessey (1991) suggest that

> student learning and attitudes in science must be viewed, and therefore assessed, from multiple perspectives....Above all, assessment in science should promote and encourage learning and focus on what [students] can do rather than what they can't do or only on what they have memorized. (p. 83)

The message of the above authors, and many others, is clear. Namely, assessment should be designed to enhance the whole of science teaching and learning rather than merely measure a limited spectrum of scientific knowledge.

Making changes will not come quickly or easily (Moscovici & Gilmer, 1996). If progress is to be made toward reform in science education, however, it will be important to reconceptualize not only the nature and purpose of assessment, but also the relationship of both traditional and progressive forms of assessment to grading: Progressive forms not only need to be used more, they need to *count* more as indicators of knowledge. Wiggins (1989) perhaps says it best:

> If tests determine what teachers actually teach and what students will study for—and they do—then the road to reform is a straight but steep one: Test those capacities and habits we think are essential, and test them in context. Make them replicate, within reason, the challenges at the heart of each academic discipline. Let them be authentic. (p. 13)

Methodology

As one of the seeds of Transforming Undergraduate Science Teaching, this story grew from my dissertation entitled *When World Views Collide: A Study of Interdepartmental Collaboration to Develop a Biology Course for Prospective Elementary School Teachers* (Mattson, 1997). The purpose of this research was to learn more about interdepartmental collaboration between those situated in science departments and those in science education. At the time, I was a staff member in the biology department assigned to the core team and a doctoral student in the Science Education Program engaging, for the first time, in interpretive research. I have also been a student and teacher of biology for many years and, more recently, a science-teacher educator. Informed by those perspectives, I thus lived the development of BIO 101.

My goals as a writer were to find ways to voice these different perspectives and to create a text that actively engaged readers in thinking about and, hopefully, making progress toward goals for reform in science education. I therefore structured the text to contain elements that are descriptive, interpretive, critical, and propositional in nature. The preceding text represents an excerpt from this effort.

In considering my writing goals, I found a compelling case for including narrative, particularly as a form of description. Narrative is receiving increasing attention as a means of reporting interpretive research (Bruner, 1986; Clandinin & Connelly, 1994; Josselson & Lieblich, 1993; Richardson, 1994; Van Maanen, 1988). Specifically, I became convinced that narrative could "do things" that other forms alone could not do.

For example, narrative allows an opportunity for what I call "condensed description" (as opposed to the "thick description" referred to by Geertz, 1973) where events that are thematically related but separated in time and space (e.g., discussions related to assessment) can be "crystallized" into a single literary entity. Another case for using narrative is that it allows key data to be communicated in a storylike manner that is familiar—and perhaps more comprehensible—to a general audience by virtue of being "lifelike."

Yet another reason for using narrative pertains to the possibility of engaging readers on an affective level with the intent of provoking a quality of interaction with the research text that might otherwise not occur using more traditional forms of writing. One final reason for incorporating narrative into a public research text is the possibility for such a text to act more effectively as a "springboard" for active reflection than can those that are strictly expository. This is based on the assumption that readers who can "position themselves" in a text are more likely to make meaning of the text by calling on their own prior knowledge—their own stories—as a way of establishing the text's viability. While this is certainly true of all text forms, traditional expository forms are more apt to position readers at the "receiving end" of a text's message, thus diminishing the potential for readers to include themselves as parts of the field to which its message applies.

The narratives are meant to portray events that occurred during the collaborative course development process. Associated with each narrative are underlying themes and variations that relate to key issues that arose during this phase. Each narrative begins by identifying the type of meeting described and those who were present.

All narratives are "fictionalized"; that is, they make use of fictional characters and situations. They are, however, based on actual participants and events. There were several reasons for fictionalizing real data for the purposes of description and interpretation. First, it was important that I protect the anonymity of the participants involved in course development. Second, I wished to emphasize that underlying themes and variations are representative of similar collaborative course development efforts rather than isolated to the specific people involved in this particular case. Finally, using fictionalized narratives allowed me a certain "poetic license" that could be used to extend my descriptive and interpretive capacity for persuasive purposes.

Another feature of the narratives involves point of view, or "the central consciousness that narrates the tale" (Stern, 1991, pp. 184–199). I chose third person dialogic narrative form with a third person narrator, where characters "speak" to one another and the narrator acts to give readers access to their thoughts and feelings as well as providing additional commentary. This form provides the flexibility of representing and expanding on the numerous points of view belonging to the various characters.

Each narrative is then followed by interpretive commentary, which is intended to direct the reader to questions of meaning—to expose the themes and variations underlying each passage. To substantiate my interpretations, I include other types of evidence, such as verbatim statements from (unidentified) course developers, parts of relevant documents, and synopses of actual events. Frequently, I link this evidence back to its fictional counterpart in the narratives—such as referring to a character's quote—as a way of helping the reader make connections between the fictionalized and real data.

Because one of the primary purposes in undertaking and reporting this research was my interest in continued progress toward goals for reform in science education, it seemed particularly important that a critical component be present in the text. Eisner's (1991) ideas about "educational criticism" as a literary form were useful in this regard. In the present study, educational themes—and more specifically, the variations of themes and their consequences—provided an opportunity for criticism. This criticism is made in relationship to goals for educational reform as detailed in seminal reform documents. The critical component is then followed by a section proposing how continued progress toward goals for reform might be made.

Susan Mattson
Leon High School
Tallahassee, Florida, USA

EDITORS

METALOGUE

Fictionalized Research: From Credibility to Believability?

KT: Sue used a novel methodology in this study. She did not want to be critical of people in any way. But at the same time she did want to learn from the experience of participating in the planning of a biology content course for prospective elementary school teachers. The decision to identify the main themes from the many hours of transcribed oral texts and then to build constructed cases and scenarios enabled her to make significant points about planning to teach college biology, especially for courses taken by prospective elementary school teachers. The university committee on Human Subjects in Research approved Sue's project and consent forms, and she provided participants in her study with copies of the materials she wrote. Because of her methodology, nobody could point to any character in the chapter and proclaim with certainty, "That's me." Sue managed to retain *credibility* and present what she learned from her research in ways that are likely to be appealing to college science teachers and science educators.

PT: Further to my comments on researching nonexemplary teachers (see the metalogue for Chapter 6), Sue developed a narrative research methodology and produced fictionalized texts based on interpretive research data (rather than her imagination). Thus, the characters are not identifiable as actual and specific professors of science. These dialogical texts were designed to engage the reader in pedagogically thoughtful responses about key issues. Now, the interpretive research standard of credibility (Guba & Lincoln, 1989) was evoked, not in respect of portraying accurately aspects of participants' viewpoints, but in identifying key issues associated with their collaborative activity as they negotiated an assessment policy for a new science curriculum. I wonder about the appropriateness of the standard of credibility in this research; especially inasmuch as it would require the researcher to check her interpretations with the subjects of her inquiry. How could this be achieved when none of the participants were able to identify themselves in the text? Might it not be sufficient for the text to be rich in *believability* as part of an essential dialogical quality (Denzin & Lincoln, 1994; Van Manen, 1990)? After all, the purpose of the text is to provoke thoughtfulness about educational issues rather than to paint realistic portraits.

KT: The credibility of a study can be based on a variety of different warrants. In this particular study member checks and peer debriefing were obtained from a variety of sources prior to the concatenation of data. When the drafts of manuscripts were prepared, these were submitted to the participants in the study and their suggestions for change were elicited. These acts by the researcher all address issues of credibility. But I do agree with you that the bottom line is one of believability. The research reported in this and other chapters will or will not have an impact on the practices of others, depending on the extent to which they find what they read believable, personally relevant, and compelling.

Building Relationships for Authentic Collaboration

PG: Sue started out her doctoral defense with an audiotape of an enacted play of a fictionalized committee meeting, similar to what she had witnessed during the development of the biology course. It focused her audience's attention and encouraged me to visualize what she had written and described to us at the defense. Sue highlighted her theoretical frameworks of constructivism, worldview theory, and semiotics. The issues of power and culture came up repeatedly, as in many of the other chapters in this book. These issues come up powerfully also in Hedy Moscovici's chapter (Chapter 4). Hedy studied a university biology classroom as well, and the power inequality was between the teacher and the students. In Sue's chapter, the power inequality was between the scientists and the science educators in the committee meetings.

KT: As in other chapters in the book, the conflict between objectivism and constructivism as ways of making sense of education and science are very evident in Sue's chapter. The oppositional nature of people who make sense from objectivist perspectives seems to preclude them from accepting alternatives. Implications for the nature of science, teacher and learner roles in a college science classroom, and the manner in which students are assessed all come to the surface in this chapter. As is evident in other chapters in this book it is possible for negotiation to lead to consensus in epistemology. However, for negotiation to occur there has to be a willingness to negotiate and learn. Sue's chapter does not present evidence to suggest that such collaboration would be possible. On the contrary, the objectivist perspectives are regarded as right and constructivist perspectives are objects for scorn.

PG: One big question I still have after reading Sue's chapter is, How can we learn from Sue's study to make collaboration more productive, so people can learn from each other, so people can trust each other? It is as if we need to have a common theoretical ground on which to collaborate. How do you get there? There is considerable discussion now about collaboration (Bruffee, 1993), as much research both in science and science education demands collaboration. To solve the tougher problems, we need to work together and learn from each other's disciplines and ways of thinking. For instance, at Florida State University hundreds of scientists here and elsewhere from around the globe work collaboratively with each other at the National High Field Magnetic Laboratory. Also in science education, the National Science Foundation funds grants called *Collaboratives for Excellence in Teacher Preparation, Urban Systemic Initiatives,* and *Rural Systemic Initiatives,* which highlight collaboration among various individuals and groups from different disciplines and different types of institutions, including K-12 schools, community colleges, colleges, universities, science centers, and industries.

PT: Perhaps the answer lies partly in the conflicting nature of the rationalities that scientists and educators bring with them into the incipient discourses of collaboration. As is evident in Hal White's chapter (Chapter 9), one of the hallmarks of objectivism is a *dualistic rationality* based on propositional deductive logic, which (like the myth of Nature abhorring a vacuum) abhors contradiction. The objectivist standard of validity requires arguments to be free of internal inconsistency, particularly contradiction. If two opposing ideas occur, one must go! Now, if a dualistic thinker also happens to be a separate knower—which, according to feminist theory, is a way of knowing that is analytical, oppositional, and impersonal—then we are likely to have a person who is predisposed to engage others in unempathic critical discourse (see the metalogue for Chapter 9). Here, one's expressed ideas are regarded as objects in the public domain and, therefore, as a legitimate target of strong contestation. Such a combination of attributes is likely to constitute a person's worldview and thus be highly resistant to change. Could it be that college science professors are imbued with such a strongly objectivist worldview? Judging by their pedagogies, it would seem that Professors Stern (Chapter 1), Mendelson (Chapter 3) and Miller (Chapter 6) are likely candidates.

If these science professors were to agree to collaborate with science education colleagues, one would imagine a compelling need for opportunities to (re)negotiate the rules of discourse for educational decision-making. If an ethos of authentic negotiation is important to maintaining collaboration, then so too is the need to establish respectful and trusting relationships among the scientists and educators; a way of knowing that constitutes relationship building or *connectedness* (see metalogue for Chapter 8). It seems to me that the rules of discourse should include, amongst other things: (1) legitimation of *dialectical reasoning*, where opposing ideas, such as divergent metaphors of teaching and learning, can coexist in tension; (2) acknowledgment of the importance of empathic concern for others' ideas; and (3) time-out for *metadiscourse* for settling disputes about the (often hidden) rules of discourse.

Assessment of and by Participation

KT: Something that seems to arise in a number of chapters is the notion of what classes are for. Are classes places for practice and participation or are they sites for absorption of knowledge? It seems as if some teachers are reluctant to acknowledge that learning is a process that is built around *participation in a discourse*. In Sue's chapter, the necessary roles of students that were implicit in the arguments over the grading distribution appear to be listening attentively, copying notes, and learning through home study and out-of-class activities like reading in the library. Some professors were reluctant to give credit for participation. Hard quantitative data were required as evidence of learning. Hard data in this case are to be equated to performance on tests and examinations.

PT: Assessment strategies indicate to students which types of learning are valued by the teacher. If learning to co-participate in the development of a community of inquirers is a valued teaching goal, then assessment must assign value to that type of activity.

KT: What is the meaning of assessment? When one thinks of assessing what students know from a constructivist perspective, it makes sense to see it as a process in which activities are designed to allow students to show what they know. Use of a verb such as *show* is to emphasize that students might have more opportunities to decide the what, when, and where of assessment. However, I do not think that students should have sole responsibility for showing what they know or for that matter what they do not know. I believe the responsibility should be shared with the teacher. In most cases it is usually the prerogative of the teacher to decide what is to be assessed, how it is to be assessed, and when the assessment will take place. If a constructivist perspective is used to make sense of learning, then it also is an imperative for teachers to have ongoing insights into what all students know and can do. In these circumstances the development of *portfolio cultures* in classrooms becomes an increasing priority.

PT: In a portfolio culture, *dialogue journals* can be a promising means of monitoring student learning, providing important feedback to students, and obtaining evidence of the efficacy of teaching. They can be as simple as a weekly half-page of reflective thoughts by each student about the process of teaching and learning. The professor (perhaps in a tutorial context) reads, reflects, and responds in writing to the student. Dialogue journals can bring teaching, learning, and assessment into a much more interconnected relationship. They can be used not only to monitor students' active participation but also to foster co-participation among teacher and students.

PG: I think that assessment is a critical issue, because how the teacher decides to assess the students drives what students try to learn. If you ask students for answers that they can memorize, they will memorize. If you ask them to synthesize various commonalities in their understandings, the students will try to do that. Getting students involved in *self- and peer-assessment* via dialogue journals encourages them to formalize the criteria for evaluating student performance. Students learn from what is possible, especially when they assess another student's writing. It gives them ideas of what they can do to improve their own work. As students critically assess each other's work, they learn what makes good writing and good science. Likewise, students in a learning community give formal feedback to their peers when their peers' writing needs improvement. Students become aware of what components are important in excellent writing, and they learn to incorporate those elements into their own writing. If a student's ideas are flawed, the peer needs to support his/her criticisms with appropriate references. Therefore, students receive feedback not only from the teacher but from their peers. The students self-assess in writing how they did in each assignment; that is, they address what they did well and where they need improvement, using the feedback from the teacher and peers. The goal is to improve the learning of the entire community.

I learned to use this type of assessment via a website, first as a student (in my studies toward a second doctorate) and second as the teacher. This can be done on the same electronic website that Ken Tobin describes in Chapter 13. The teacher evaluates how well the student has done the peer-assessments as well as the self-assessment. Finally, the teacher gives each student a grade and written feedback on the *assessment of assessments* for each weekly assignment. All this takes time, but the time is worth it, because the students learn through the verbal interaction and become more confident in themselves.

PT: In thinking about assessment we also should consider the notion of *accountability*. What I have in mind is professors being accountable to their students for the quality of their teaching. For some years a major focus on quality teaching has been gathering momentum in Australian universities. One of the means of achieving this quality teaching is to require university professors to administer *student assessment of teaching* surveys to their classes at the end of each semester. When portfolios for promotion are being prepared, the professor is required to include evidence of student appraisal of his or her teaching and a commentary on his or her responsiveness to this important feedback. In this way, professors are being compelled to acknowledge the importance of their teaching and of their professional identities as science educators. The past two years have seen university administrations also developing policy for the training and certification of college professors as professional teachers.

BIBLIOGRAPHY

American Association for the Advancement of Science (1990). *Science for all Americans.* New York: Oxford University Press.

Baron, V. (1991). Performance assessment: Blurring the edges of assessment, curriculum, and instruction. In G. Kulm & S. M. Malcom (Eds.), *Science assessment in the service of reform* (pp. 247–265). Washington, DC: American Association for the Advancement of Science.

Bruffee, K. A. (1993). *Collaborative learning: Higher education, interdependence, and the authority of knowledge.* Baltimore, MD: The Johns Hopkins University Press.

Bruner, J. (1986). *Actual minds, possible worlds.* Cambridge, MA: Harvard University Press.

Campione, J. (1991). Dynamic assessment: Potential for change as a metric of individual readiness. In G. Kulm & S. M. Malcom (Eds.), *Science assessment in the service of reform* (pp. 310–312). Washington, DC: American Association for the Advancement of Science.

Clandinin, D. J., & Connelly, M. F. (1994). Personal experience methods. In N. K. Denzin & Y. S. Lincoln (Eds.), *Handbook of qualitative research* (pp. 413–427). Thousand Oaks, CA: Sage Publications.

Collins, A. (1991). Portfolios for assessing student learning in science: A new name for a familiar idea. In G. Kulm & S. M. Malcom (Eds.), *Science assessment in the service of reform* (pp. 291–300). Washington, DC: American Association for the Advancement of Science.

Denzin, N. K., & Lincoln, Y. S. (Eds.). (1994). *The handbook of qualitative research in education.* Thousand Oaks, CA: Sage Publications.

Eisner, E. W. (1991). *The enlightened eye: Qualitative inquiry and the enhancement of educational practice.* New York: Macmillan.

Geertz, C. (1973). Thick description: Toward an interpretive theory of culture. In C. Geertz, *The interpretation of cultures* (pp. 3–30). New York: Basic Books.

Guba, E. G., & Lincoln, Y. S. (1989). *Fourth generation evaluation.* Thousand Oaks, CA: Sage Publications.

Haertel, E. (1991). Form and function in assessing science education. In G. Kulm & S. M. Malcom (Eds.), *Science assessment in the service of reform* (pp. 233–246). Washington, DC: American Association for the Advancement of Science.

Kulm, G., & Stuessey, C. (1991). Assessment in science and mathematics education reform. In G. Kulm & S. M. Malcom (Eds.), *Science assessment in the service of reform* (pp. 71–88). Washington, DC: American Association for the Advancement of Science.

Josselson, R. & Lieblich, A. (1993). *The narrative study of lives.* Thousand Oaks, CA: Sage Publications.

Malcom, S. (1991). Equity and excellence through authentic science assessment. In G. Kulm & S. M. Malcom (Eds.), *Science assessment in the service of reform* (pp. 313–328). Washington, DC: American Association for the Advancement of Science.

Mattson, S. A. (1997). *When world views collide: A study of interdepartmental collaboration to develop a biology course for prospective elementary school teachers.* Unpublished doctoral dissertation, Florida State University, Tallahassee, FL.

Moscovici, H., & Gilmer, P. J. (1996). Science faculty try alternative assessment strategies: The ups and downs. *Journal of College Science Teaching* March/April, 319–323.

National Research Council. (1996). *National science education standards.* Washington, DC: National Academy Press.

National Research Council. (1997). *Science teacher preparation in an era of standards-based reform.* Washington, DC: National Academy Press.

National Science Foundation. (1993). *Proceedings of the National Science Foundation on the role of faculty from the scientific disciplines in the undergraduate education of future science and mathematics teachers.* Washington, DC.

Richardson, L. (1994). Writing: A Method of Inquiry. In N. K. Denzin & Y. S. Lincoln (Eds.), *Handbook of qualitative research* (pp. 516–529). Thousand Oaks, CA: Sage Publications.

Stern, J. (1991). *Making shapely fiction.* New York: Bantam Doubleday Dell.

Van Maanen, J. (1988). *Tales of the field: On writing ethnography.* Chicago: University of Chicago Press.

Van Manen, M. (1990). Researching lived experience: Human science for an action sensitive pedagogy. London, Ontario: SUNY Press.

Wiggins, G. (1989). A true test: Toward more authentic and equitable assessment. Phi Delta Kappan, *70*(9), 703–713.

CHAPTER 11

Team Teaching in a Restructured Physical Science Course

Sabitra Brush

In recent years there has been much talk about a decline in American competitiveness in the areas of commerce, industry, science, and technology (National Commission on Excellence in Education, 1983). The decline of American achievement in science has been linked to problems that arise early in the educational system, specifically at the elementary and middle school levels (Krieger, 1990). In 1993, a large southeastern university designed a reformed physical science course on a National Science Foundation (NSF) grant that aimed at restructuring science and science education courses for prospective elementary school teachers.

The reformed physical science course integrated the laboratory with the lecture, in a small group setting that allowed increased student–teacher interaction, and provided prospective teachers with the confidence and understanding needed to teach science effectively in their own classrooms. This course was unique in that it involved team teaching by two faculty from physics and one from chemistry. After two weeks of introductory physical science taught by the chemist, the two physics faculty each taught about four weeks, followed by the chemist returning for the four final weeks. Even though the content areas of physics and chemistry were presented separately, one of the major goals was to show interconnections between these two content areas.

In order to develop a better understanding about curriculum change, this chapter focuses on the pedagogical approach that was used by the three college instructors of the restructured physical science course for prospective elementary school teachers. In particular, the chapter examines the extent to which the beliefs of the instructors, coupled with their various philosophies about teaching and learning, affected the way in which team teaching in this class actually evolved. A comprehensive account of this study can be found in Brush (1993).

This team of professors claimed that their major goal was to stimulate the students in a challenging way, so that they could have a hands-on, minds-on approach to science. They also desired to enhance student understanding by showing underlying themes whenever it was possible so that students would view chemistry and physics in an interrelated manner, and not as two entirely distinct subject areas. They sought to present students with the view that it was all science, just different aspects of science, in an attempt to broaden students' perspectives.

Research Design

An interpretive research approach guided my involvement in the study (Guba & Lincoln, 1989). In my role as a nonparticipant observer in the restructured physical science classroom, I generated data from extensive interviews of each of the three professors, from observations of their teaching actions, and from student interviews and students' daily reflective journals. Data analysis involved generating interpretive assertions, based on both confirming and disconfirming evidence, and in checking with the professors my inferences of their intentions and meaning perspectives. All names used in this report are pseudonyms.

Participants

There were twenty-five students in the class; most of them college freshmen who aspired to enter the College of Education majoring in elementary education. There were two white male students and twenty-three females (two of whom were African Americans). I selected two "focus" students: Latoya, a black female medium-high achiever, who sat in the front of the class and seemed to be quite conscientious; and Mary, a white female medium-high achiever, who sat in the middle of the classroom and was always asking questions, or volunteering solutions to questions that had been asked. Both Mary and Latoya were college freshmen who had studied physics, chemistry, and biology in high school.

Mary had also studied anatomy in high school, and she took two science honors seminars at the college level. She considered herself to be a strong student in both science and mathematics. When asked why she decided to go into early childhood education she said, "I want to teach because I feel that I will be able to shape lives and therefore make a difference."

Latoya had taken a college level introductory biology course in which there were usually about 1,400 students. She did not view mathematics and science as her strong point. Latoya was undecided about the level at which she wanted to teach, even though she was interested in teaching. She had attended the university developmental school from the seventh to the twelfth grade, so she claimed to be at ease in the type of atmosphere that existed in the restructured physical science classroom.

The Restructured Physical Science Course

The course combined physics and chemistry, with the laboratory integral to the course. The topics included the following: matter; energy; waves and light; electricity and magnetism; astrophysics; atomic structure; gases; stoichiometry; and solution chemistry. The course was divided into three sections, with each section being taught by a different professor and lasting about four weeks.

The faculty designed the schedule to be flexible in order to empower the students. If students wanted to spend more time on a particular topic because they were more interested in it, that was possible. The actual content of the schedule was typically on the same topics that the regular chemistry and physics courses offered for non-science majors. However, it had been proposed that the method of presentation, the emphasis on certain topics more than other topics, as well as the integration of a laboratory component, would serve to distinguish this course from the regular courses. In addition, an attempt would be made to present the material using a more hands-on, minds-on approach. The goal was to stimulate the students to learn science, and to arouse their curiosity by what they experienced, so that they would start feeling more comfortable with science, a feeling that would carry over into their own classrooms when they eventually became teachers.

Although the concept of team teaching was embraced in the course, it materialized mainly in the weekly meetings and "behind the scenes" planning, rather than in the classroom. Most of the time a single instructor taught his own section and tried to demonstrate the links to the other sections whenever possible, a technique that I have termed "turn teaching."

The Team Members

The two senior physics professors involved in the study, Miller and Smith, had known each other for a long period of time. They were very much at ease dealing with each other, and with giving criticism to each other, and this candor oftentimes erupted into open arguments between them. They both insisted that there was no genuine hostility between them. The junior member of the team was Adams, who had been a chemistry professor at that university for almost eight years. He was a very enthusiastic faculty member and one whom students found very approachable. Adams took on the role of non-instructor and sat in on the other professors' sections in an attempt to further integrate the sections of the course.

When all three men were present at a course development meeting, it was easy to sense Adams's uneasiness and to detect that a strong brotherhood existed between the two senior faculty members, Miller and Smith. As a result, Adams tried to be very diplomatic in his dealings with them. Adams felt that it would be important for him to participate as a non-instructor in the physics component of the course, since he believed that it would give him some idea of what the students had learned before they came to his section of the course. He felt that this experience would be tremendously helpful in helping him plan this final section.

Adams was very youthful in his appearance and attitude. Students were able to trust him instantly, and it was rare to hear a negative comment ever made about him, except with regard to his "open book" tests, which always appeared to have no relationship to the course content or the textbook. That was an intriguing aspect of this instructor. His examinations were always very challenging, and students endured the humiliation, gracefully thinking that it must be some shortcoming that was inherent within them. When Adams was in the class he made an effort to appear as "one of the guys" and his frequent use of humor kept students relaxed and at ease.

Miller was a tall, lean man in his late fifties who came across as very tough and businesslike. There was no time for dwelling on many questions or straying too far from his agenda for the day. He had a goal to accomplish. He would cover the material within the time frame he had selected. Having taught this material for about twenty years, he was definitely coming across to students as the expert. In addition, his experiences as a coach for many years also strengthened his role as an authority figure in the classroom. To survive in his class students needed to be good "players" and struggle silently with whatever had been meted out to them.

Smith, on the other hand, appeared to students as a very happy-go-lucky type of guy who bore a striking resemblance to Santa Claus, with his white beard and rounded figure. However, in spite of his appearance, he was not a very approachable professor because he appeared so knowledgeable about everything that it was very easy for people, including his students, to feel quite intimidated in his presence. His classes were rich experiences in which he reveled in historical and comical anecdotes about almost every topic in science. Smith was a dynamic expert who could easily fascinate while instilling awe.

This is what the team looked like. They were three professors as unlike each other as possible, with differing attitudes and philosophies about education. Here was a team in which there existed a wealth of talent but little else in common. Could the three successfully team teach a course that was going to try to revolutionize the way prospective elementary school teachers were being taught science at the university level?

A Peek into the Team-Teaching Classroom

Today is February 10, and Miller is getting ready to begin class. The students are coming in and getting settled down. It is now 11:15 A.M., and Miller goes to the podium. In the last two weeks Miller has been teaching about motion and acceleration. The following is an excerpt from the class.

Miller: I am passing out a list of books that you can select from to do your book report. You don't have to select from that list. But if you decide on something else then you should run it by Dr. Smith. If we think of any more we will add to the list. The books are available in the library; if not, come and see one of us. (He walks over to the front desk and opens his book.)

Adams:	I have a question. Did you see on NBC news about the spark that was planted on the truck to see if it exploded on impact or not? Let me explain what happened. In GM trucks, due to the position of the gas tank, there is a danger of exploding. So NBC wanted to explain this using electronic spark devices in rocket engines under the tank, and they had two cars slam into each other to see if they got an explosion. Sure enough they got an explosion. What I was curious about is was that fair or unfair? Were they just trying to trick the public to think the trucks exploded, or was it a situation where sparks or something can set off an explosion?
Student:	I thought it was very unfair.
Adams:	Was that the general consensus here? A lot of people die because of these trucks exploding. The reason I want to give you this to think about, and use your journals to comment on this, is because I think it was unfair to put a rocket engine under a truck. Also to repeat the experiment a number of times to get an explosion was wrong. But I'll tell you what they are doing is to model a situation, from their point of view, that could be real.
Student:	But they needed more people.
Adams:	And the people watching it need to understand how science works. It is the same thing when someone reads that they pumped a rat full of saccharin and it caused cancer. Does that mean you'll never touch saccharin again? Of course not. But they are trying to model what happens over extended exposure. Is that fair to do or not?
Miller:	I think there are two numbers you have to state: the number of occurrences or the number of times you tried, and under what conditions, and then you have to say the exact number of explosions you got. This helps to relate your model to the real-life situation. It turns out that the second number is the hardest one to get when you make these models. The first number is easier to get, because you know how many times you've tried it. It still does not mean your car is going to blow up, because the chance of you getting one of these sparks is very, very

small. And you do not design any engine which has combustible fuel, that is why that other number which is the one that nobody ever talks about is the important one.

With respect to Adams's questions, I have a very good friend, a physicist, who has actually computed, based on various experiments, what your cancer risks are every time you eat a steak that's been prepared on a grill, your cancer risks when you smoke a cigarette, and the cancer risks of the sex organs with respect to "package wear" clothes. For example, the higher the temperature of the sex organ, the higher the risk of cancer. So if you wanted to minimize the risk of cancer, you'd in fact have to go nude. It's only if you look at all these risks in the broad picture that you get the true feeling of what the real risk should be.

Even though Adams was the non-instructor, he would ask questions during the class, and he stimulated discussion such as was described above in order to get the students thinking. However, Miller was fidgeting with his lecture notes the entire time this was going on, impatient to start on the day's topics. On most occasions, though, the questions were directly related to what was being discussed by the instructor.

Most of the students did not ask many questions, and most of the time the only interactions occurring were those between the instructor and the non-instructor. However, by the time they had gotten to the second instructor, Smith, students were asking a lot more questions in class. These types of interactions between the two professors could have had some delayed impact on students' abilities to ask questions about the things that they did not understand in order to experience meaningful learning with the content being presented.

Interactions in the Team-Taught Classroom

The interactions that occurred in the team-teaching classroom can be grouped into several categories: teacher A–student, student–student, but also teacher A–teacher B, as well as teacher B–student. In the small group tasks, the interactions were mainly student–student, with some student-teacher interactions. In the whole class activities, several teaching genres were used by the three instructors: *questioning, lecture–demonstration,* and *whole-class discussions.*

The instructors utilized questioning extensively throughout the course. In the questioning technique, called "volunteer turns," the teacher asked the class a question, and the students took the initiative to either call out the answer or raise their hands (Anderson, Evertson, & Brophy, 1979). In the *question-answer-evaluation* teaching style, the instructor modeled questions and answers for the students by seeking specific answers to the questions asked. This proceeded in a systematic step-by-step manner. In the absence of questioning, *exposition* consisted of a monologue in which the instructor advanced the thematic pattern of his section through talk that consisted of narratives, historical and personal anecdotes, and formal science. This technique involved mainly the chalkboard or the overhead projector.

In the lecture-demonstration teaching genre, the whole class was involved, and the discourse was built around a science demonstration. The teacher related the tasks directly to the demonstration.

Whole-class discussions took various forms. In the *conversation* teaching style, the instructor interacted with the whole class, and the discourse was made up of conversations between the instructor and a select group of students, usually called "target students" (Tobin & Gallagher, 1987). These students were intensively involved in the classroom discussions, and they were the ones who provided answers to the teacher's questions. In the *student-led discussion* approach the lead question came from a student, and the student momentarily took control of the classroom learning environment.

In the small group tasks, many students asked questions and raised issues. The main teaching style here was student-led discussion. The students asked many questions, to which the teaching assistant, the instructor, or the non-instructor responded. This technique was used by all three instructors.

Predominant Patterns of Interaction

The predominant style in Miller's classroom was question-answer-evaluation. His approach was to ask the students a question from the homework or one of the exercise sheets, and then to evaluate the answers that were presented in order to model concept-building. Sometimes Adams (as teacher B) would ask a question, or make a comment, and this would result in teacher A–teacher B interactions, or teacher A–student, or teacher B–student interactions. Miller did not really approve of this approach involving teacher B, and found that it was distracting to concept-building and the thinking strategies he was trying to develop in the students. He emphasized thinking about abstract concepts, graphing, and problem solving in lecture as well as laboratory sessions. These class discussions were mainly teacher-student type, with the instructor adopting a very authoritarian role. He considered it his job to develop these concepts in the students' heads, and he proceeded to do that without much regard for how the students felt about it.

In Smith's section, the predominant interaction style was exposition, with him more in a lecture mode as the brilliant storyteller, weaving history around the scientific information that he presented. There was no doubt that he was in command in the room. There were also many occasions of teacher A–teacher B interactions that blended right into the class, and Smith just continued moving along with the lesson. He found the teacher–teacher interactions to be stimulating to his class. He gave many examples from everyday life as he lectured. In his classes, he used many interesting illustrations of science that caught the students' attention. One of these was the demonstration of how an electric current was generated by cutting the lines of force in a powerful magnet. At times, Smith reverted to a lecture style, but he still maintained the students' attention most of the time, because what he was talking about seemed so interesting. However, during the first part of his section he tended to do most of the talking, and the students did not seem to have been sufficiently engaged in constructive thinking.

In Adams's section, there were no teacher–teacher interactions because neither of the other instructors were present. The main approach was a modified form of lecture-demonstration, more like demonstration-questioning. Usually he would do a demonstration and then, once the students' curiosity was aroused, he would make them think about what was happening and would introduce concept-building. He also had more student-led discussions in his section than did the other two instructors. This seemed to result in: (1) the students enjoying the chemistry sessions more, and relating to them better than the physics sessions; and (2) the students being more comfortable with Adams, probably because he had been sitting with them in class for the entire semester.

Students appeared most interested in what was going on in Adams's section, and they appeared more comfortable with their learning. There was also a great deal of interaction among the students, and also between students and Adams. Most of the chemistry content appeared relevant to everyday life. For example, around Easter time the students did experiments that involved dyeing eggs, an activity related to the concept of acids/bases. Other experiments dealt with common indicators and lightsticks. Students were able to relate these concepts to what they thought was necessary to teach elementary school students.

Although some methods of classroom interaction, such as small group work, questioning, and lecture-discussion, were utilized by all three team members, there were certain characteristic patterns that dominated each member's section. In Miller's classroom it was "question-answer-evaluation," in Smith's it was "lecture-discussion," and in Adams's it was "demonstration-questioning." The classroom interactions in this course were very varied, and they were heavily dependent on each of the instructors and their philosophies of teaching.

The Case for (and Against) Team Teaching

Six interpretive assertions about team teaching were formulated from an analysis of the data obtained during the study (Brush, 1993).

Assertion 1: Team teaching can improve student participation

The two focus students, Mary and Latoya, were interviewed about their views on team teaching. Both reported only positive experiences. Latoya claimed that "there was always someone to go to if you had questions":

> It is less stressful on the teachers too, because they do not have to do everything all the time. I think this is a good idea for education in the '90s because the classrooms are getting larger, and one teacher cannot control all the students all the time. This was also good in that it taught you how to relate to different instructors.

Mary also saw advantages to team teaching.

> There were advantages in that if you did not like to hear one of the professors it was not for the whole time. It was also interesting to see Adams learning with us in the class during physics, because you'd think that he is a professor, but he does not know this either, and that's pretty neat. Sometimes I wondered about his questions, though. "Does he really not know this, or is he just asking questions to kind of spur us along?"

Team teaching exposed the students to different ways of teaching science, since each member of the team employed different patterns of interaction with the class. According to Davis (1976), one of the main advantages of team teaching is that it improves the possibility of students receiving individual attention. These students had a team of three faculty members to choose from whenever they had questions. Kirwan and Willis (1976) found that when instructors alternated in their teaching, the change of pace lengthened the attention span of the students. Using different instructors with distinct personalities also presented students with a greater choice for seeking help. Students often would not seek help from a particular instructor due to a personality difference. Using team teaching enabled various team members to be approached preferentially by various students. Also, if a student did not particularly favor one of the approaches that was being used, it lasted for only one-third of the semester, and so it was more tolerable than being in the traditional class.

Personal interviews with several students in the class revealed that the questions Adams asked in the classroom had a big impact on the participants, and many of the students really felt that his presence had a positive impact on their learning outcomes. It was evident from the interviews that many of the students did not feel sufficiently comfortable to ask questions, and several did not know what to ask in the beginning since it was a class of predominantly freshman students. However, by the time Smith, the second professor, started his section of the course the students were asking more questions. The increase in the number of questions by participants could have been due to a combination of factors, including the presence of the nonlecturing faculty member (or non-instructor; see below) asking questions, and the students feeling increasingly relaxed as the course progressed.

Assertion 2: In team teaching perceived power inequalities among instructors can hinder collaboration

According to Armstrong (1977), if students perceive the teachers involved in team teaching as sharing equal status, there will be no problem with students ignoring or paying less attention to the subordinate teacher. This will create a better classroom atmosphere for discussion, questioning, and any type of clarification. Therefore, when organizing the course, each team member should take the primary responsibility for certain sections of the course. The difference in the expertise available on the team should also be emphasized, to make the students aware that scientists are not experts in all fields (Belts & Walton, 1970).

In this restructured physical science course no attempt was made to make instructors appear equal in the eyes of the students. Neither was there any attempt to emphasize the talents of each instructor. Miller said:

> We needed a couple weeks of introduction. But Adams established a certain style in the classroom during the first two weeks, and then there is a sudden abrupt change in style when I start. Then there is another abrupt change when Smith takes over. The difference between Adams and Smith is not so great in terms of personal style. We just did not do this very well this time, and we suffered from some shock value. Since we all did just a 'teeny' bit in the first two weeks, and Adams did most of it, many of the students also got the impression that Adams was in charge of the course. It is bad when people think that someone is in charge of the course when they are not.

The three instructors in this course were not even comfortable with the idea of themselves as equals on the team, as was evident in the following comment by Adams:

> Since I did not know Smith and Miller as well as they knew each other, I was much less inclined to come out and say "This is not working" or "That's not a good idea." Instead, I tried to bring it up in a very diplomatic way, such as "What about this? Maybe we can try this," not "This is not working" or "This is wrong." Because Smith has been teaching twenty years longer than I, and he is a very good teacher, I'd feel a little bit funny to tell him how to do his job, even if I felt my way was better.

The attempt to work as a team in presenting the content to students had inherent flaws because members had mentally already placed themselves in certain power positions, and they had difficulty voicing opinions to each other. This was easily apparent to students, who began to view the process as three separate courses, with Adams the only common factor throughout.

Assertion 3: In team teaching, differences in instructors' teaching styles can be an advantage or disadvantage

I interviewed each of the instructors about the impact of the various teaching styles on the learning in the classroom. Miller felt that it was a huge mistake to have instructors with such different teaching styles on the same team. But Smith felt differently:

> I think that one of the advantages of a team taught course is that students are exposed to different teaching styles. But I think that advantage is, of course, usually greatly overrated. Team teaching can also be cooperation versus competition. I don't want this team teaching thing to be a competition to see who is the best teacher, even though I think I'd win. I don't think that's appropriate. I think the teaching styles were different enough from one person to another, and I also think there are a lot of correct ways to teach.

When Adams was asked to comment on the various teaching styles presented to the students, he responded:

> I did not think it should be taught as three separate courses, and I think to a large extent we did that. I am aware that we are talking about three people who, as it turned out, had some pretty different ideas about what teaching involves, and what works, and what we should do in the classroom. Sometimes we got frustrated or irritated by the discussions of what to do in class. When I think of the three of us, I am much closer to Miller in terms of my philosophy of the instructor's role than I am to Smith. The difference is I am more of a salesman in the sense that I work much harder to bring the students on board, and to get them psyched, and to try to get them caught up than Miller would. Whereas Miller, in spite of the strength of his conviction, would just say, "This is how you learn, and if you don't appreciate that now, trust me, you'll appreciate it later." But when it comes to this idea of what learning is, and how it works, Miller and I were right together on it.
> It was the way we interacted with the students that was basically different. Smith was more of a traditional lecturer. He would spin tales and weave stories of the science, and he had much less of a constructivist classroom. But he'd have these rich collections of ideas, and ways of thinking about things that he could lay out before the students.

Adams and Smith were more effective instructors in the eyes of the students than was Miller. Both Adams and Smith were very enthusiastic presenters but with widely differing philosophies about what teaching should involve. Adams felt that his philosophy was closer to Miller's, but it was easily apparent from interviews with Miller that he resented Adams's teaching style. As he phrased it, Adams viewed himself as one who was "down in the trenches with the students." Adams, on the other hand, appeared to be a bit intimidated by Smith and was very impressed by this man who was "simply a rich collection of anecdotes and history" that he could connect to science at will.

Assertion 4: The different roles of the instructor and non-instructor can be complementary or conflictual

Prescott and Anger (1972) described the concept of the non-instructor as one of the advantages of team teaching. With one instructor sitting in the class during the presentation there can be a better feel of whether students are understanding what is going on or not. In cases where what is being said by the speaker is not clear to the listener, the nonlecturing team member can try to correct it. Any errors on the board or during a lecture can also be picked up by this team member. Errors of this type in a liberal studies class could be detrimental, since the students are not scientifically sophisticated enough to pick them up on their own. Dialogue between team members and questions by the non-instructor were found to stimulate classroom discussions (Kirwan & Willis, 1976; Prescott & Anger, 1972). Usually, when teaching non-science majors, it is a problem to maintain the students' interest and attention.

Steiner and Lesiecki (1980) studied team teaching in which one member of the team was always in the role of non-instructor. Initially, there was some difficulty in creating a free-flowing exchange between the instructors in the classroom. This was attributed to two main reasons: (1) the nonlecturing team member did not feel sufficiently comfortable to interrupt the speaker in order to modify an explanation, or to disagree with him on an issue; and (2) their planned dialogues failed to materialize very smoothly in the classroom. As a result, they decided to abandon that second tactic and to just let the dialogues happen spontaneously as they went along.

In the restructured physical science course in this study the second team member to teach was Smith. He had a favorable response to Adams's presence in his section of the course:

> The students were afraid of me and Miller. I think the students were less afraid of Adams because he made a point to sit among the students. The main reason I didn't sit back there among the students was because I need

to get up and get out sometimes. The reason that I didn't want to sit in Miller's class, and it is not meant as any insult to Miller, or that I don't like either his class, or the way he teaches his class, but it is just the material. I know that material....I have been through it so many times myself. It has also been particularly helpful having Adams sitting among the students, because he was pretty much into it, and he asked all sorts of things. He and I sometimes got dialogue going, and I think that was a particularly valuable thing for the students. The dialogue was between teachers from two different disciplines, and students can learn from that. Students must have found that interesting.

It was a very different situation between Adams and Miller, however. While being interviewed, Miller reacted irately:

I will never allow him [Adams] in the room at the same time that I'm in the room ever again. When I'm the coach I'm the only coach. When people coach my children they are the only coach. I don't intervene even if I don't agree with what the person is doing, or if I believe I have a better method. The first rule of this learning exercise is that there can be only one director of the process. Many directors will affect you during the course of your life, but in any given activity there can be only one director. What happened is that we ended up splitting the class during the group events because I'd be talking to half the class, and Adams would be talking to the other half of the class. In other words, I did not want them to have his version of what I'm saying, I wanted them to have only my version when they are doing my exercise. Likewise, I want them to have only his version in his exercise. There is more than one way to get an idea across, but it is ineffective to get too many across at the same time. This leads to cross-purposing.

In order to successfully utilize a non-instructor in the classroom, the instructor and the non-instructor have to work out the details about what each expects before the course starts. If Adams's actions were disruptive to Miller, and prevented Miller from accomplishing his goals with the class, then they should have discussed the problem and tried to figure out a way in which Adams was working with Miller and not against him. It was a tough situation for Miller, because not only was this his first time teaching non-science majors at this level, but his style of teaching was not perceived as very "student-friendly." Also, the material that he was teaching was very abstract to students with very little experience in science. Therefore, in trying to get them to develop these concepts he had a very difficult time. When this was coupled with having Adams in the room, it just added to the problems because Adams's approach to teaching was very student-friendly. When students felt that they were facing an obstacle while trying to comprehend the material that Miller was presenting, they resorted to Adams. The situation, very unintentionally, turned into one of the non-instructor and the students versus the instructor. For this technique to be successful, the teaching styles of the instructors should be complementary or else, as happened here, students will get confused, and turn to what they think is the easier alternative, the non-instructor.

Assertion 5: Team teaching does not necessarily enhance teacher morale

Steiner and Lesiecki (1980) conducted a study of a teaching team of two chemists (one organic, one physical) who aimed to alternate lecturing on the concepts and the applications of chemistry and to become engaged in dialogue on important points. This group thought that their team teaching experiment was a success. The teaching styles of the team members showed dramatic improvements as the term progressed. They attributed this positive effect to the natural desire to impress one's colleagues with one's own teaching abilities, as well as to after-class in-depth discussions of their presentations. Each instructor ended up with a small following of students, and that allowed them to reach more students than each instructor could have done if they were teaching individually. In addition, they reported that over 50% of the students indicated on the university evaluation that their interest in the subject matter had increased. The class consisted of thirty-seven students, the majority of whom had indicated a low interest in science when they started the course. Steiner and Lesiecki evaluated the students' perceptions using an evaluation questionnaire specifically aimed at team teaching issues. The results of the survey indicated that the students liked the idea of getting different viewpoints; more questions got answered because team members could answer questions on their specific disciplines readily (both lecturing and non-lecturing members); and the discussions caused the class to be livelier.

In the physical science course, team teaching failed to boost teacher morale because the instructors seemed to have placed themselves in some sort of hierarchy where it was difficult for them to really work and interact together as a team. Instead, each team member had confirmed mental pictures of what was right and wrong for the students and believed his ideas were the correct ones.

Assertion 6: Team teaching can be very time-consuming

Although the teaching responsibility is divided up in a team teaching approach, each instructor has to be aware of all the material and of what is going on in the other sections of the course (Kirwan & Willis, 1976; Prescott & Anger, 1972). In order to be aware of what was happening in each other's classes, the three professors met every week to discuss their plans for the next week. In addition, they met with the members of the Course Development team every two weeks to give the team feedback on what was happening. The Course Development team included the three instructors (Smith, Adams, and Miller), the two Principal Investigators (one from the College of Education, and the other from Arts & Sciences), a teaching assistant, one elementary school teacher, and one student (who had previously taken a restructured physics course and excelled in it).

Smith summed up the general feeling of the team members in the following words:

> Team teaching is not my idea of a good time. I feel less strongly about that than I used to, because of the research that I am now involved in. I used to be part of the High Energy Experimental Unit here, and that went on for many years. I was actually hired to do that. In about twenty years of doing that kind of research, only once did I publish a paper with just my name alone on it. During that period of time I had published enough papers to get from where I was to full professor. Now, I did not like that very much. I like to feel that I could be something of an individual. I had to be part of a team in research. So for years I rebelled about being part of a team. If I have to submerge my personality when doing research, then at least let me have a little individuality in teaching. This course was also too time-consuming. I could not make any decisions by myself. I always had to consult with Adams and Miller. It was more time-consuming, not more difficult. You always had to explain to everyone what you were doing, or sometimes spend time discussing the most suitable way to do something.

Adams and Miller had similar views on the issue. At least there was one aspect of team teaching that all the members were in agreement with: "It is time-consuming when you attempt to teach this way."

Conclusion

The approach that was used in team teaching at the large southeastern university was very interesting, and some major issues surfaced from this experience. The meetings every week and the "behind-the-scenes" planning made the class a team effort. However, the class was presented to the students as three separate sections. This may have been because of the very different teaching styles of the three instructors, since teaching is a function of one's personality. Also, each person seemed to be undergoing a struggle for individualism, because it appeared that the level of cooperation that was finally obtained was one that was brought about through struggle. And even so, team members did not take criticism easily from each other. Added to that was the "invisible hierarchy" that Adams felt placed him on the lowest rung of the ladder and that created the biggest obstacle to having a true team effort.

Instructors involved in team teaching probably should be perceived by the class as equals, and in order for the class to have that perception the professors involved need to be selected carefully. If there is a dramatic mismatch of teaching approaches or philosophies, then instructors might not feel comfortable working with each other, thus creating a possible source of conflict. Therefore, even though the concept of team teaching has great potential, it calls for active teamwork, not only behind the scenes, but also in the eyes of the students. What seemed to have actually happened in this particular physical science class could more accurately have been described as "turn teaching" rather than "team teaching."

Sabitra Brush
Department of Chemistry
Armstrong Atlantic State University, Savannah, Georgia, USA

EDITORS

METALOGUE

When Teaching Metaphors Collide

KT: The role of Adams and his questions in this study really reminded me of the very different beliefs teachers have about teaching and learning. Miller, for example, did not believe that students had to make sense of science in class. He believed that at some later time students would figure out physics for themselves. In contrast, Adams taught in a very student-oriented way. His teaching was quite consistent with constructivism, and he wanted learners to accept responsibility for their own learning. Thus, when he attended Miller's class as an observer he wanted to model for students how to be a good learner. However, his questions were a source of frustration for Miller, who wanted to present the physics content in his own way. Miller wanted to be in control of the content dissemination and made sense of his role as teacher through a metaphor of the teacher as coach. His perspective was that there can be only one coach on a team and that Adams should remain silent when Miller was the coach.

PG: Here again we see the importance of a teacher's construction of a metaphor for the teacher's role. Here the physics professor Miller saw himself as the coach, and he didn't want to share the power with anyone, especially his junior colleague, Adams, from chemistry. This metaphor of coach constrained Miller's approach to teaching and learning. Miller's metaphor contrasts with the metaphor of tour guide used by Mark (see Chapter 7) to go where the students wanted. Mark had firm ideas about where to go in studying chemistry, and he took his students there, but he would follow his students' ideas, interests, and questions.

PT: So it appears that learning to transform one's way of teaching is not necessarily an easy task, even when rich resources are at one's disposal. In this case, the resources of a team included: *critical friends*, who might be trusted to provide both supportive and critical commentary; *peer* *teaching*, which might provide exposure to different metaphors of teaching and learning; and science education researchers, who might provide *a means of evaluating* the impact of teaching innovations on student learning. But these resources did not seem to be well utilized. Why not? Sabitra's chapter points out the strong sense of *individualism* among the professors that resulted in "turn teaching" rather than coteaching. There is also evidence that hierarchical power relationships amongst faculty obstructed their collaborative teaching. Perhaps what was missing from the team was a feminist perspective on human relationships, which emphasizes the importance of connecting with one's colleagues on the basis of mutual respect and reciprocal trust (see metalogues of Chapters 8 and 10). Without such a perspective it is difficult to see how co-learning to team teach might occur. Kate Scantlebury helps us to understand the importance of a feminist perspective on teaching and learning (Chapter 5).

Why Model Collaborative Teaching?

KT: There were a variety of approaches to team teaching displayed in the study. As Sabitra points out, there were advantages in having three different teachers involved in the course. However, because the teachers had very different beliefs about the roles of learner and teacher there was a possibility that students could become frustrated by the different expectations. Since most of the course was taught with only one teacher at the front of the classroom at a time, the potential problems of the different expectations being manifest in the same lesson only occurred on occasions when Adams's questions were a source of frustration to Miller. However, I am thinking about the desirability of prospective K-12 teachers being taught in the ways in which they are expected to teach. Of the three university teachers described in this chapter, only Adams used strategies consistent with those encouraged by the National Research Council (1995) in the *National Science Education Standards* that the prospective K-12 teachers should use when they teach in their own classes. The National Research Council (1997) addressed how university science teachers can improve teaching and learning in university science classes in its *Science Teaching Reconsidered: A Handbook.*

PT: Just think of the enormous potential of team teaching for rich learning to be modeled for students who themselves will one day be responsible for facilitating students' learning of science. In the middle of a class, two professors engage in an expert dialogue, puzzling over an issue, questioning one another, proposing alternative points of view. Planned or unplanned, such a dialogue could, if conducted with due sensitivity to students' linguistic and conceptual abilities, model science as a powerful and dynamic means of inquiry that involves not only specialist knowledge but the ability to collaborate dialogically. Of course, this type of team teaching requires professors to be communicatively competent scientists!

PG: I keep thinking of more research that needs to be done. Would it not be interesting to continue the study by visiting and interviewing the teachers who were in this class, now that they have their own classrooms, to see what sort of metaphors they utilize now to define their role(s) in the classroom? Do they use a metaphor of a guide like Abdullah described for Mark (same person as Adams in Sabitra's study) or a metaphor like a coach as used by Miller (Chapter 6)?

Transforming Cultural Forces via Collaboration

KT: The differences in power between teachers in an institution need to be considered when collaboration occurs. In this study two senior full professors from one department collaborated with an untenured assistant professor (who became tenured and was promoted during the study) from a different department. It seems as if the power differences in this study may have led to a situation in which negotiation was difficult. Each of the teachers believed in his own approach to teaching, and only Adams was interested in undertaking research on his own teaching. Miller and Smith were agreeable to doctoral students and others undertaking research in their own classrooms, but when it came to making decisions about teaching, learning, and curriculum they listened to their own voices to a much greater extent than to what was being learned from the research.

PG: I think that Ken Tobin has raised an important point about unequal power relationships between participants in a collaboration. When I was an assistant professor I was told to conduct my biochemistry research only within my own research group and not with other faculty, or else at promotion and tenure time, people might just figure I had ridden on the coattails of someone with more knowledge and power. The problem is that if we are ever to change the way we teach in universities, it seems a shame to have to wait to collaborate on teaching or research until the faculty member is tenured. That's waiting five to seven years of a 25- or 30- year career. Even when science or mathematics faculty wait until promoted to associate professor, they may be taking a chance that they might not get promoted to full professor if they get involved in teaching and learning issues. One way to avoid that problem, as I mentioned before, is if the candidate gets it in writing that published scholarly work in teaching will contribute to his/her research effort.

PT: So the culture of the college science department seems to be a powerful influence on the role-determining beliefs that professors enact in their teaching. Can the reverse also be true? Can a professor expect to obtain legitimation from her department for her transformative teaching practices on the evidence of good-quality student learning? How far beyond the confines of her own classroom or laboratory can the individual professor facilitate the process of pedagogical reform? In science departments where the prevailing professional ethos is governed by a strong sense of competitive individualism, where impersonal and hierarchical collegial relationships prevail, and where a hostile attitude to *unscientific* ways of knowing is built in, it seems unlikely that the conditions are ripe for legitimating constructivism as a referent for science teaching. Even the most senior professors might be expected to experience obstructive barriers. The significance of Sabitra's chapter is the way that it points to the need for skilled and committed teamwork among professors in order to begin to transform the powerful cultural forces of the department, which render as natural the barriers to reform.

BIBLIOGRAPHY

Anderson, T., Evertson, C., & Brophy, J. (1979). An experimental study of effective teaching in first-grade reading groups. *Elementary School Journal 79*, 193–223.

Armstrong, D. G. (1977, Winter). Team teaching and academic achievement. *Review of Educational Research 47*, 65–86.

Belts, D. S., & Walton, A. J. (1970, November). A lecture, or "Anything you can do I can do better." *Physics Education 5*, 321–5.

Brush, S. (1993). *A case study of learning chemistry in a college physical science course developed for prospective elementary teachers.* Unpublished doctoral dissertation, Florida State University, Tallahassee.

Davis, E. J. (1976). *School Science and Mathematics 76*, 466.

Guba, E. G., & Lincoln, Y. S. (1989). *Fourth generation evaluation.* Newbury Park, CA: Sage Publications.

Kirwan, D. F., & Willis, J. (1976). Co-teaching experiment at the University of Rhode Island. *American J. Physics 44*, 651–4.

Krieger, J. (1990). Winds of revolution sweep through science education. *Chemical and Engineering News 68*, 27–43.

National Commission on Excellence in Education. (1983). *A nation at risk: The imperative for educational reform.* Washington, DC: U.S. Department of Education.

National Research Council. (1995). *National science education standards.* Washington, DC: National Academy Press.

———. (1997). *Science teaching reconsidered: A handbook.* Committee on Undergraduate Science Teaching, Washington, DC: National Academy Press.

Prescott, J. R., & Anger, C. D. (1972). Team-teaching freshman physics. *American Journal of Physics 40*, 311–4.

Steiner, R., & Lesiecki, M. (1980). Team teaching chemistry. *Journal of Chemical Education 57*, 353–4.

Tobin, K. G., & Gallagher, J. J. (1987). What happens in high school science classrooms. *Journal of Curriculum Studies 19*, 549–560.

CHAPTER 12

Evolution of a University Biology Teacher's Classroom Interactions

Carol Briscoe

Calls for reform within the last fifteen years (Carnegie Forum on Education and the Economy, 1986; National Commission on Excellence in Education, 1983) have led to investigation of questions regarding change in science teaching in the K-12 classroom. Various studies have demonstrated that change is influenced by teachers' understandings of their instructional roles (Briscoe, 1991; Duffee & Aikenhead, 1992) and curriculum (Crawley & Salyer, 1995), their beliefs about students and how they learn (Appleton & Asoko, 1996; Tobin, Tippins, & Gallard, 1994), the culture of the school and classroom in which learning takes place (Eisner, 1992), and social and political factors that influence learning (Sweeney, 1996).

With the recent publication of the *National Science Education Standards* (National Research Council, 1996), research has focused on investigating change in university level teaching, particularly in classes designed for prospective teachers (Fedock, Zambo & Cobern, 1996; Gilmer, Barrow, & Tobin, 1993). Based on recommendations in the *Standards*, university science classrooms should be places where teachers continually create opportunities that challenge students and promote inquiry through their questioning, and where students practice science and participate in discourse about scientific ideas. However, research indicates that teaching as described in the *Standards* is rare in university classrooms (Barnes, 1983; Dillon, 1990; Fassinger, 1995; Roth & Tobin, 1996).

Implementing educational reform that is consistent with the recommendations in the *Standards* will not be easy for those involved. Long-held traditions related to a transmission model of learning represent significant social forces that have constrained instructional practices (Ellner & Barnes, 1983; Tobin & McRobbie, 1996). If university classrooms are to provide models for student inquiry and become communities where there is a shared discourse of science, university teachers will need to make substantial changes in how they interact with students.

The *Standards* suggest that, in order to effect change, teachers "should have opportunities for structured reflection on their teaching practice with colleagues, for collaborative curriculum planning, and for active participation in professional teaching and scientific networks" (p. 58). Yet at the university level, teaching contexts are not conducive to such collaborative activities. With the demands placed on university teachers to participate in research and service in addition to their teaching, there may be a perception that time invested in improving teaching is not worth the cost in meeting other requirements for tenure or promotion. As Fullan (1991) notes, teachers "must have the opportunity to work through the [change] experience in a way in which the rewards at least equal the cost" (p. 127).

Perspectives

Change is an evolutionary and adaptive learning process. When a teacher attempts to implement innovations in curriculum and materials, well-understood contexts change as interactions among persons, things, and processes associated with the innovation take place. Through reflection teachers make sense of their experiences and the contexts that they construct through these interactions. In this way teachers construct knowledge regarding their work and their roles (Briscoe, 1997; Clandinin, 1986; Tobin & Ulerick, 1989). This knowledge is adaptive in the sense that it enables a teacher to cope with experiences, to communicate, and to function socially within the educational setting (Glasersfeld, 1989).

The change experience also is dynamic. That is, it has a historical foundation, as well as a present and future. As participants make sense of where the context and interactions are leading, the past shapes the construction of new knowledge that influences the teaching. In turn, what the teacher learns influences future adaptations in the curriculum (Connelly & Clandinin, 1988).

Thus, if we are to understand how individual change occurs as university faculties create new science courses to meet the needs of prospective teachers, we must come to understand how participants construct the meaning of curriculum and how classroom interactions affect the learning experiences of the individuals involved. We also must understand how traditional views of teaching and learning affect instructional practices. Accordingly, I began this study with the understanding that the communicative interactions experienced by the students and the teacher had a social as well as an individual history that would influence the manner in which participants interpreted these interactions.

A Study of Change in Classroom Communication

In this study, I focused on the kinds of communicative interactions fostered by Chris (a pseudonym), a university biology teacher, in the context of teaching an innovative introductory biology course for prospective elementary education majors. Chris is a teacher with a number of years of experience at institutions of higher education, including both community colleges and universities. However, unlike many other science teachers at the university level, Chris is not a scientist. Rather, the perspectives that he brought to this study were those of one who has spent a lifetime learning science associated with teaching science.

The course differed from the traditional introductory biology course taught at the same university. It differed in terms of laboratory and lecture experiences planned for the students. A major feature of the course design was that enrollment was limited to less than twenty-five students in order to facilitate increased opportunities for students to participate in laboratory work. Although the curriculum development team did not set out to make increased communication a goal of the course, it seemed that a lower student–teacher ratio would naturally facilitate multiple opportunities for students and teachers to interact with one another. Accordingly, the manner in which Chris adapted communicative interactions in the portion of the curriculum that had traditionally been taught in large group lecture settings became a focus of the study. Furthermore, I investigated how Chris's experiences in this alternative context for teaching biology led to evolution in the curriculum and enhanced interactions with students over the two years of the study.

An Innovative Curriculum

A team of teachers, including Chris, who were to share in the teaching planned the curriculum for the introductory biology course. They met with various stakeholders, including classroom teachers, scientists, philosophers, science educators, and a district science supervisor, all of whom helped plan and evaluate the course. The team took particular notice of what the practicing teachers identified as science concepts that were important to understand in order to teach children. As an outcome of the meetings with concerned stakeholders, it was agreed that the course would provide a "broad overview of biology" based on major themes. The team decided also that laboratory activities and field trips should be important components of the curriculum. Chris was responsible for teaching in six class sessions the portion of the curriculum concerning topics related to biochemistry, the cell, cell division, human biology, and genetics.

Interpretive Research Approach

I employed an interpretive research design (Erickson, 1986) that focused on constructing an account of how Chris made sense of teaching and learning in this special course.

Videotapes served as a primary data source for examining Chris's teaching practices. I viewed several times the tapes of ten lectures, totaling nearly nine hours of teaching, and transcribed all sections that depicted interactions between Chris and the students. I viewed each tape several times using a holistic sociolinguistic frame for analysis, and constructed a general categorization to represent the various contexts of classroom interactions. The analysis was conducted by recording the number and purpose of utterances that initiated "science talk" between Chris and the students or between students. The categories of teacher questions that I established—*motivating, focusing, presenting*—were based on the context in which interactions took place (Carlsen, 1991; Cazden, 1986). Tables 1 and 2 show summaries of the categorization and analyses of questions that Chris asked during each year of the study.

As well, I conducted similar analyses of students' utterances, based on segments of transcribed tapes, and I constructed inferences that led to categorizations of students' communicative interactions—clarification questions, information-seeking questions, informative comment (Tables 3 and 4).

My primary role in this study was as a nonparticipant observer. My understanding of what was happening in the classroom was based on the videotaped data and communication with Chris throughout the data analysis process, primarily through e-mail interviews and telephone conversations. As I identified various interaction patterns between Chris and the students, I asked Chris to respond to assertions that I had constructed, to explain how he interpreted what was happening and his role in the class. Although our interactions provided Chris with opportunities for focused reflection on his practices, the interviews were not specifically designed to facilitate change in those practices.

Analysis of the data took place on a continuing basis throughout the study. As Chris responded to the assertions, additional data were gathered and we negotiated and clarified the meaning of the assertions. Through this negotiation process and continued reviews of the data, I constructed final assertions that encompassed all the data. The credibility (Guba & Lincoln, 1989) of the assertions is supported by the use of multiple data sources, the long term of data collection, and Chris's involvement as a stakeholder in the process of analysis.

Details of the interpretive data analysis of the study have been reported elsewhere (Briscoe, 1997). In this chapter, I present a synopsis of the findings and illustrate them with selected vignettes.

Table 1: Number of Teacher-Initiated Questions by Category
Year 1

Category of question	Content presented		
	Motivating	Focusing	Presenting
Photosynthesis (48 minutes)	3	7	13
Cell structure & function (37 minutes)	5	10	6
Human body structure & function (48 minutes)	3	1	5
Mitosis (34 minutes)	4	0	8
Genetics (70 minutes)	2	0	10

Continued on next page

			Table 1 (Continued)
Category of question	Content presented		
	Motivating	Focusing	Presenting
Total	17	18	42

Total time: 237 minutes

Average participation: one question initiated by teacher every 3.1 minutes

Table 2: Number of Teacher-Initiated Questions by Category
Year 2

Category of question	Content presented		
	Motivating	Focusing	Presenting
Mitosis and meiosis I (51 minutes)	6	3	1
Mitosis and meiosis II, Genetics I (73 minutes)	8	4	2
Genetics II (48 minutes)	8	0	9
Human body–Reproduction (78 minutes)	1	2	4
Sexually transmitted diseases (40 minutes)	1	0	0
Total	24	9	35

Total time: 290 minutes

Average participation: one question initiated by teacher each 4.3 minutes

Table 3: Forms of Student Communicative Interaction
Year 1

Category of Interaction	Content presented		
	Clarification Question	Information-Seeking Question	Informative Comment
Photosynthesis (48 minutes)	7	2	3
Cell structure and function (37 minutes)	0	0	1
Human body structure and function (48 minutes)	3	4	0
Mitosis (34 minutes)	0	3	0
Genetics (70 minutes)	11	9	2
Total	21	18	6

Total time: 237 minutes
Average participation: one student communicative interaction every 5.3 minutes

Table 4: Forms of Student Communicative Interaction
Year 2

Category of interaction	Content presented		
	Clarification Question	Information-Seeking Question	Informative Comment
Mitosis and meiosis I (51 minutes)	3	18	1
Mitosis and meiosis II, Genetics I (73 minutes)	6	5	4

Continued on next page

Table 4 (Continued)

Category of interaction	Content presented		
	Clarification Question	Information-Seeking Question	Informative Comment
Genetics II, Cell structure and function (48 minutes)	6	2	2
Reproduction (78 minutes)	8	13	6
Sexually transmitted Diseases (40 minutes)	1	7	5
Total	24	5	18

Total time: 290 minutes
Average participation: One student communicative interaction every 3.3 minutes

Classroom Interactions: Year 1

Analysis of videotape recordings of lectures during the first year of the study indicated that Chris used questioning in three different contexts: (1) to motivate students' learning, (2) to focus students' attention on their own prior knowledge so that they could form connections between what they already understood and new material, and (3) as an alternative to lecturing when Chris presented new material. In the third context, questions served to assist students to organize previously learned information in new ways and to develop an understanding of new relationships. It was in this context that Chris most often initiated interactions with students. As shown in Table 1, over half the questions asked by Chris in the first year were in the "presenting" category.

A primary factor that influenced Chris's use of questioning to promote interactions was his understanding of students' previous experiences with the concepts to be addressed. However, when introducing new material with which Chris believed students had had little experience, his use of questioning as an interactive strategy diminished substantially. For example, during a presentation on concepts related to human body structure and function, he spoke without interruption for fifteen minutes.

Although videotape analysis indicated that Chris almost always used questions in some context during a lesson, the format of the questions did not provide for prolonged dialogue. Most often, his questions resulted in interactions taking the form of *triadic dialogue* (i.e., teacher question, student response, teacher evaluation; Lemke, 1990), which did not require students to respond with more than a word or two. Nevertheless, most questions were an open bid for any member of the class to answer, thus reducing the intellectual risk for individual students.

There was one exception to the triadic pattern of dialogue in the genetics lecture. In the context of teaching genetics, Chris involved students in using the Punett Square to present crosses on the white board. The other members of the class chose the crosses to be presented. As students worked at the board, others completed the crosses at their seats. Students checked one another's answers and discussed phenotypic and genotypic ratios. Because Chris asked fewer questions, students had many opportunities to ask questions and make comments as the class progressed. Students clearly engaged in a more active role in constructing the interactions.

However, as shown in Table 3, this session in the first year was an exception. In the four class sessions, not including genetics, a total of only nineteen questions were asked by students. Although students commented occasionally on some point, relating it to personal experience or linking it with previous knowledge, during most class sessions students rarely participated verbally other than to answer questions posed by Chris.

Constraints

Chris believed that questioning was an important strategy for promoting classroom interaction and that students asking questions was an important way to learn how they made sense of biology concepts. However, I did not observe him use questioning in the lecture setting that had potential to engage students in open discussion of scientific ideas. From interview data, I identified two major factors that constrained the nature of Chris's communicative interactions with his students: (1) his curriculum decisions regarding what students needed to learn; and (2) his beliefs and metaphors associated with various instructional roles.

Constraints Related to Curriculum Decisions

A major factor that influenced communicative interaction during class was curriculum decisions that team members made in planning the course. Based on the comments of elementary school teachers who were part of the planning group, the planned curriculum emphasized breadth of content. In five lectures in the first year Chris dealt with concepts of photosynthesis light and dark reactions; plant and animal cell structures and functions; the blood, circulation, respiration, excretion, glandular control of homeostatic mechanisms; mitosis and cancer; and genetics, including monohybrid and dihybrid crosses of dominant, co-dominant, and sex-linked traits. Chris justified the need for teaching such an extensive array of content in the following way:

> These people teach our children, and all of the students that come to [the university] later become biology and chemistry and whatever majors. Young people in elementary school get no science and that's because they are taught no science. The elementary teachers in our planning group said that one teacher in each school teaches science and sometimes no teacher. They told us it was because they never had science and never understood science! We were trying to build a program where the next generation of elementary teachers liked science, were not afraid of it, and saw that it was fun and that they could do some of this.

Because most of each two-hour class session was devoted to laboratory activities, Chris had only forty to fifty minutes to present material. Thus, in order to cover the content, there was pressure to present the material at a pace that did not allow time for extensive classroom interactions. Although Chris believed that it was important that students make connections between the information presented in the lecture and other activities that were part of a unit, the structure of the course did not facilitate making those links. For example, in one activity he observed that several students were unable to connect what was said in the lecture with related laboratory experiences on photosynthesis, and he expressed the following concerns about the structure of the class:

> When we did the photosynthesis laboratory, I was trying not to tell them what was happening and let them discover for themselves. And they were

so frustrated. We had already had the lecture and discussion of photosynthesis and they could not relate the two. Then I didn't have time to go back and run the lab again or talk about it because [I only taught] every other day and it wasn't my day.

The pace of the class also did not provide opportunities for students to reflect on the presentations or what they understood so that they could contribute to classroom interactions. Aware of the differences in prior knowledge that students brought to class, Chris often provided lengthy explanations for those students with poor backgrounds, frequently stopping during the lectures and asking, "Are there any questions?" However, students rarely responded. When time ran short, Chris suggested that students come by his office for further assistance.

Clearly, overplanning the curriculum for the time available was a significant factor that denied students the opportunity to interact with each other and with Chris, and to share their constructions of scientific ideas. Each time the students met, Chris taught a new topic in order to implement all that had been planned in the curriculum.

Constraints Related to Beliefs and Metaphors

The metaphors that teachers construct regarding their teaching roles can be influential guides to practice (Briscoe, 1991; Tobin, Kahle, & Fraser, 1990; Tobin & Ulerick, 1989). In this study, I identified two metaphors that influenced Chris's classroom interactions. As Chris noted:

I want the students involved in their learning and for me to be a guide to that learning and an encourager for that learning. I use lecture and probably always will. When the student does not know the topic or have any clue about the topic, even after reading an assignment, then I will use lecture and questions to introduce this content.

Chris believed that, as a guide to learning, the teacher held primary responsibility for students' learning. Thus, in the role of guide he selected what was important for the students to know and used lecture to ensure that the students would get the information.

Chris believed also that learning biology should be an enjoyable experience for the students, and that it should be a priority to assist students to develop a strong motivation to learn science. However, believing that learning science was not a priority for most of the young men and women enrolled in freshman biology, Chris faced a dilemma that he elaborated:

> The students, as freshmen, are not interested in science nor talking nor thinking about it....They are immature in some of the decisions they make about the importance of academics in a university. They are really into sororities, college life, their personal life, sports! Very few are into university biology topics....I want them to like biology and to learn specific topics, to be able to ask anything they want to know and to feel comfortable doing that. [But] as freshmen their willingness to engage in class is limited and they are shy in front of these people they don't know.

Thus, in order to encourage participation, Chris always responded with positive comments and body language to students' questions, comments, or responses to the questions he posed. He listened attentively to every interaction, accepted, praised and, somehow linked it to the topic of discussion. Even students' incorrect responses to teacher-posed questions were turned into positive contributions to the lecture as Chris asked additional questions to elicit additional responses that reflected what he wanted them to know about the topic at hand.

Throughout the study, Chris's love of teaching and personal love of biology, and his view that knowledge of biology would be very important in the lives of students, were apparent, as represented in the following statement:

> I want all students to have a love of biology, I know it won't be everyone's favorite course, but a love and appreciation for life and biology is important. That's what I try to do. I hope that no student that leaves my class will say "I hate science!" or "I hate biology!" I have fun teaching and I enjoy the students! We learn together! We grow together! Biology is fun! Life is fun!

Chris's metaphors of guide and encourager were grounded in an overarching framework of his enthusiasm for student learning. However, coupled with a belief that the implemented curriculum should expose students to breadth of content, these metaphors constrained Chris to approach teaching with a main focus on lecturing. Consequently, during the first term, classroom interactions were very constrained and students were provided with little opportunity to participate in discussing scientific ideas.

The Change Process: Year 2

Recognizing a Need for Change

Chris was genuinely concerned that students were not benefiting from the course as anticipated. Reflecting on his classroom experiences during the first year, Chris expressed a personal need to change what was happening. He explained:

> A curricular change was imperative to do the class differently and have time for the student reactions and responses. We needed more talk and discussion and questioning time. More time to finish thoughts and whatever....[I tried] to maybe do a little too much methods of what an elementary [school] teacher might use in her classroom....I was handing out diagrams and activities and ideas that they could use later and trying to accomplish the goal of a course where they learned and liked it....I really thought that what I planned was the best that I had ever planned for a section of a course! Until we heard students say that it all seemed like busywork and was too much, and they hated all the handouts and couldn't learn for all of the requirements....I began to weigh the activities and papers and handouts and I could see their frustrations....I thought I was helping them to give them things they could use in teaching. Then I realized that was not my job.

Planning and Implementing Change

In the second year of the course, Chris planned to decrease the number of biological concepts introduced. He justified decisions regarding which concepts were important for students to learn about biology in the following way:

> Through the years with more and more information, I have tried to sift and sort through all of this information and determine what would be important to a citizen of this city, state or the USA...what knowledge in biology is important to a student's health, life, and our environment. Also there are a few topics that I just want them to know that they can learn (no matter how hard other people think they are). I want the students to know that they can learn and be successful in college.

The reduction in number of topics covered during the lecture portion of the course in the second year increased opportunities for interaction between students and Chris (compare Table 4 with Table 3). However, there was not a substantial change in the triadic nature of interactions he initiated. His questions continued to serve the same purposes as in the previous year, but the frequency changed slightly, from one question every 3.1 minutes to one question every 4.3 minutes (see Tables 1 and 2).

On the other hand, given the additional time to think about the ideas presented, students more often took the initiative and asked their own questions related to topics being discussed, as noted in Table 4. In particular, there was an increase in the number of information-seeking questions asked by students and also in the frequency of the comments that they offered voluntarily in relation to the lecture topic. Students were able to ask questions and make comments regarding their own ideas of the ramification of the topics. This kind of interaction was not common when Chris first offered the course.

Clearly, the reduction in number of topics afforded students more opportunities to interact with Chris and encouraged them to do so. Yet the students did not take the opportunity to take discussion a step further and develop a common discourse for their ideas, mainly because Chris continued in the role as "guide" and used lectures to ensure that students learned what was expected.

An important change in the curriculum in the second year was in Chris's use of students' journals. In the first year, Chris had not focused on the journals as an important way for students to communicate their ideas. However, in the second year he asked students to write on a regular basis in their journals their responses to class activities. Through the journals, he opened a dialogue with students who might not otherwise participate in class. He explained the process:

> I really enjoy using the student journals (for the first time in my life). I have the students write every day: what they learned, what they still did not understand from that day, and any question for me.

The use of the journals helped Chris learn more about students and provide better instruction based on their comments as noted:

> Each student comments on topics, questions, concerns for family, etc., that never are mentioned in class. Also if there is a misconception....I can pick that up on these personal individual pages. I noticed by the journal pages that a misconception about cancer was developed in one discussion about inheritance of cancer. We talked about this topic again with the whole class during the next class time, and I also wrote to each student individually who had commented on this.

Chris's use of students' journal questions to introduce topics in class often promoted increased interaction. The journals increased the ways that Chris provided for students to communicate their scientific ideas. Furthermore, his use of journals was consistent with his belief that students who preferred not to communicate their ideas in a public forum could be encouraged to make known in this way their questions and ideas. In turn, student journal entries could assist him to better address students' learning needs.

Conclusions and Implications

The purpose of this study was to investigate how a university biology teacher's practice changed as a result of experiences in teaching an innovative introductory biology course for prospective elementary education majors. In light of reforms called for in the *Standards* regarding the need to provide students with a classroom environment that fosters the development of discourse about scientific ideas, I focused on how the classroom interactions facilitated by the teacher evolved over two years.

The results of the first year indicate that the interactions of the teacher followed similar patterns to those reported by Barnes (1983). Although Chris used questions during the lecture component of the course for several different purposes (Table 1), little dialogue was initiated by the questions. Typically the triadic nature of the dialogue supported by Chris's questions provided few opportunities for students to engage in discussion regarding their scientific ideas.

The results also demonstrate major factors that influenced the nature of the science talk that Chris promoted in lecture sessions during the first year: (1) a curriculum that demanded more time than that allowed by the parameters of the course; and (2) metaphors that supported Chris's understanding of his pertinent teaching roles. However, as Chris reflected on what was happening and learned that students wanted more time to discuss ideas in class, he decided to change the curriculum and his teaching practices in order to facilitate increased classroom interaction.

The change process was not easy because Chris held conflicting beliefs that constrained the change process. His belief that students needed more time in class to participate in discussion of ideas, to address questions that came up in the laboratory activities, and to come to consensus regarding their scientific understanding conflicted with his belief that breadth of content coverage was important in a course for future teachers. Furthermore, Chris felt constrained from changing classroom interaction patterns by his belief that maintaining control of discussion, answering questions, assessing students' learning by asking questions, and focusing learning on predetermined topics were the essence of the role of a guide and encourager. These beliefs held unequal status in influencing decisions Chris made regarding how to engage students in developing discourse about their scientific ideas. Thus, in non-laboratory contexts, Chris tended to use lecture as a primary means of delivering content and tended to maintain the triadic dialogue in interactive situations.

Beyond individual factors that influence change, there also are social influences that can affect teachers' understanding of their work. For example, Fassinger (1995) argues that, in order to increase interaction in university classrooms, not only do teachers need to change, but also students need to be emotionally supported and assisted to build confidence in their own knowledge and ability to participate. The introduction of journals to encourage less confident students to share their ideas was a significant change for Chris. The use of journaling was one way that Chris believed students would be able to build the necessary confidence to discuss their own ideas. However, further research is needed to determine whether communicative confidence can be supported through nonoral means of communication such as journals.

From a constructivist perspective, social interaction is important in the generation of viable knowledge regarding teaching and learning (Tobin & Ulerick, 1989). Not surprisingly, perhaps, change occurred when Chris paid attention to ideas communicated by students and reflected on feedback that indicated students' need to discuss their scientific ideas. It was important also that Chris interacted with peers who shared in the planning and teaching load for the course. These types of interactions may lead eventually to additional changes in the course and in Chris's teaching practice. It is through the interaction of all participants that meaning is constructed for the need and viability of changes that can lead to development of educational settings such as those envisioned in the *Standards*.

We need to engage in further research to understand better the institutional constraints that influence the beliefs that university teachers use to make sense of the way they teach science, and to find out how they can be assisted to reflect on their conceptions of teaching and learning as they attempt to implement innovative practices. If "structured reflection, opportunities for collaborative curriculum planning, and active participation in professional teaching and scientific networks (p. 58)," as described in the *Standards*, are goals for facilitating change in science teaching, then research is necessary to establish means for precipitating and rewarding these kinds of activities in higher education.

Carol Briscoe
Department of Elementary and Middle Level Education
University of West Florida, Pensacola, Florida, USA

EDITORS

METALOGUE

How the Habitus Constrains Reform

KT: The habitus associated with the teaching of college science seems to support coverage of content at the expense of students having adequate opportunities to interact in class with the teacher, with the purpose of using the teacher as a source for building understandings of biology. Carol's study clearly shows that the possible reactions of Chris's peers mediated the manner in which the biology curriculum was enacted. Although there was evidence that the teacher wanted his students to participate in science, enjoy what they were learning, and connect what they learned to their roles as teachers, there was a disinclination to respond to his own beliefs.

PT: In a world of exponential knowledge growth it is becoming exceedingly difficult to justify a science curriculum that emphasizes the learning of a minuscule portion of what is known at the expense of other important outcomes such as *learning how to learn*. This is not an argument for abandoning content, but an argument for making a pedagogically sound decision about the multiple purposes of teaching science, *one* of which is to foster canonical knowledge development.

Another important purpose that we see driving college science curriculum reform in this book is to cater to the future-oriented science teaching goals of many students. By taking account of these students' aspirations, those who are charged with the considerable (moral) responsibility for planning college science courses are less likely to fall into the trap of catering exclusively to the narrow purpose of reproducing the (sometimes socially dysfunctional) culture of their own college science communities. In the case of students who aspire to become science teachers, course planners must provide equal emphasis on the development of disciplinary knowledge and (linguistic/experimental) skills and the development of reflective, collaborative, and communicatively competent learners.

PG: Carol said that the National Research Council (1996), in its *National Science Education Standards*, suggests that, in order to effect change, teachers "should have opportunities for structured reflection on their teaching practice with colleagues, for collaborative curriculum planning, and for active participation in professional teaching and scientific networks" (p. 58). The problem at the university level is that teachers are not rewarded for reflecting on their teaching practice with colleagues. In fact, thinking about your teaching, at least at a major research university, is often discouraged. Ken Tobin mentioned earlier that he knew geology faculty who were discouraged outright from doing research on their own teaching and from collaborating with education faculty.

Deficit Model Versus Co-Participation

KT: An assumption that the students coming to class have no motivation to learn is consistent with a *deficit model* of the learner. Does it make a difference in this case? Presumably the guide and encourager metaphors by which Chris showed his enthusiasm for biology would have provided a context in which the students also could become enthused, whether or not they were initially that way. However, as a general rule the deficit model is very damaging and can lead to a distancing of the teacher from his students. If Chris had assumed that the students shared his enthusiasm for biology, then there may have been more examples of co-participation in which excited students brought forth their interests and carried evidence of their interests from the world outside of the classroom and into the classroom.

PT: That's a good point, Ken. We need only look as far as Craig Bowen's chapter (Chapter 2) to see what can happen to a student who brings unrecognized enthusiasm into the science classroom. In that case, the teacher's distancing epistemology snuffed out Diane's initial eagerness to learn. Students' initial enthusiasm for a subject can serve as a conceptual and emotional resource for the teacher who is looking for means to precipitate student participation early in the semester.

Control Versus Communication

PG: Carol identified two metaphors that Chris used—a guide and an encourager—to identify his role in the classroom. Metaphors are powerful referents for our beliefs and actions. The problem was that Chris's two metaphors constrained the students from becoming more active learners. In addition, Chris felt like he would lose control of the classroom if he assumed any other metaphors for his role. We saw the importance of a teacher's metaphor not only for Chris but also for the role of metaphor with Mark (in Chapter 7) and with Miller (in Chapter 6). It is interesting that Mark and Chris both utilized the metaphor of being a guide, but it worked differently in the two cases. Mark was a "tour guide," while Chris's guide was more that of a mentor. Chris's two metaphors, both guide and encourager, modulated each other.

KT: The tension felt by Chris is very common. There is an expectation within the fabric of many science departments that content coverage is an important goal of teaching at the college level. This goal came into conflict with Chris's growing realization that more interaction with the students would enable them to learn more effectively. I think the key implication is that teachers like Chris need support if they are to practice their teaching of science in innovative ways.

PG: I felt encouraged that Chris utilized written journals from his students to learn of issues and concerns from the students. It informed Chris of where there were misconceptions on the students' parts. There was communication both ways, from students to Chris and from Chris to the students, as he returned journals to the students. This is similar to what I did with electronic mail correspondence on students' understanding of chemistry seminars in my own study (Chapter 17) within a chemistry honors course. As the teacher, you learn not only what science they understand and what they don't, but also you learn their goals and interests.

BIBLIOGRAPHY

Appleton, K., & Asoko, H. (1996). A case study of a teacher's progress toward using a constructivist view of learning to inform teaching in elementary science. *Science Education 80*(2), 165–180.

Barnes, C. P. (1983). Questioning in college classrooms. In C. L. Ellner & C. P. Barnes (Eds.), *Studies of college teaching* (pp. 61–82). Lexington, MA: D. C. Heath and Company.

Briscoe, C. (1991). The dynamic interactions among beliefs, role metaphors, and teaching practices : A case study of teacher change. *Science Education 75*(2), 185–199.

———. (1997, March). Making sense of teaching biology to elementary majors: The evolution of a college biology teacher's classroom interactions. Paper presented at the annual meeting of The National Association for Resarch in Science Teaching, Oak Brook, IL.

Carlsen, W. S. (1991). Questioning in classrooms: A sociolinguistic perspective. *Review of Educational Research 61*(2), 157–178.

Carnegie Forum on Education and the Economy. (1986). *A nation prepared: Teachers for the 21st century.* New York: Carnegie Corporation.

Cazden, C. B. (1986). Classroom discourse. In M. C. Wittrock (Ed.), *Handbook of research on teaching* (3rd ed.) (pp. 432–463). New York: Macmillan.

Clandinin, D. J. (1986). *Classroom practice: Teachers' images in action.* London: Falmer Press.

Connelly, F. M., & Clandinin, D. J. (1988). *Teachers as curriculum planners: Narratives of experience.* New York: Teachers College Press.

Crawley, F. E., & Salyer, B. A. (1995). Origins of life science teachers' beliefs underlying curriculum reform in Texas. *Science Education 79*(6), 611–635.

Dillon, J. T. (1990). *The practice of questioning.* New York: Routledge.

Duffee, L., & Aikenhead, G. (1992). Curriculum change, student evaluation, and teacher practical knowledge. *Science Education 76*(5) 493–506.

Eisner, E. W. (1992). Educational reform and the ecology of schooling. *Teachers College Record 93*(4), 610–627.

Ellner, C. L., & Barnes, C. P. (Eds.). (1983). *Studies of college teaching.* Lexington, MA: D. C. Heath and Company.

Erickson, F. (1986). Qualitative methods in research on teaching. In M. C. Wittrock (Ed.), *Handbook of research on teaching* (3rd ed.) (pp. 119–161). New York: Macmillan.

Fassinger, P. A. (1995). Understanding classroom interaction: Students' and professors' contributions to students' silence. *Journal of Higher Education 66*(1), 82–96.

Fedock, P. M., Zambo, R. & Cobern, W. W. (1996). The professional development of college science professors as science teacher educators. *Science Education 80*(1), 5–19.

Fullan, M. G. (1991). *The new meaning of educational change* (2nd ed.). New York: Teachers College Press.

Gilmer, P. J., Barrow, D., & Tobin, K. (1993, April). Overcoming barriers to the reform of science content courses for prospective elementary teachers. Paper presented at the annual meeting of the National Association for Research in Science Teaching, Atlanta, GA.

Glasersfeld, E. von (1989). Cognition, construction of knowledge and teaching. *Synthese 80*(1), 121–140.

Guba, E. G., & Lincoln, Y. S. (1989). *Fourth generation evaluation.* Newbury Park, CA: Sage Publications.

Lemke, J. L. (1990). *Talking science: Language, learning, and values.* Norwood, NJ: Ablex Publishing.

National Commission on Excellence in Education (1983). *A nation at risk: The imperative for educational reform.* Washington, DC: U.S. Government Printing Office.

National Research Council. (1996). *National science education standards.* Washington, DC: National Academy Press.

Roth, W. M., & Tobin, K. (1996). Staging Aristotle and natural observation against Galileo and (stacked) scientific experiment or physics lectures as rhetorical events. *Journal of Research in Science Teaching 33*(2), 135–157.

Sweeney, A. E. (1996, November). *Teacher beliefs, cultural and linguistic diversity, and the discourse community of science: An emergent theoretical analysis.* Paper presented at the annual meeting of the Southeastern Association for the Education of Teachers of Science, Smyrna, GA.

Tobin, K., Kahle, J. B., & Fraser, B. J. (Eds.). (1990). *Windows into science classrooms: Problems associated with higher-level cognitive learning.* Philadelphia: Falmer Press.

Tobin, K., & McRobbie, C. J. (1996). Significance of limited English proficiency and cultural capital to the performance in science of Chinese-Australians. *Journal of Research in Science Teaching 33*(3) 265–282.

Tobin, K., Tippins, D. J., & Gallard, A. J. (1994). Research on instructional strategies for teaching science: A project of the National Science Teachers Association. In D. L. Gabel (Ed.), *Handbook of research on science teaching and learning* (pp. 45–93). New York: Macmillan.

Tobin K., & Ulerick, S. (1989, March). An interpretation of high school science teaching based on metaphors and beliefs for specific roles. Paper presented at the annual meeting of the American Education Research Association, San Francisco, CA.

SECTION III

 **Potentialities…**

CHAPTER 13

Learning to Teach Science Using the Internet to Connect Communities of Learners

Kenneth Tobin

Teaching is a profession that, by necessity, adheres to a philosophy of lifelong learning. To remain effective facilitators of their students' learning it is essential that teachers themselves are avid learners, learning from professional experience, through their lives outside of the classroom, and also through their efforts to gain formal course credit from universities. Traditionally, science teachers have experienced difficulties in taking courses in the fall and spring semesters of the year and, in many universities, summer offerings are scarce, particularly for studies outside the school of education. Since schools and universities tend to offer their classes at approximately the same times, access has been a problem for teachers wanting to study courses offered by universities. When university courses are offered, teachers frequently are engaged professionally and cannot conveniently access opportunities to further their education. Even when universities seek to compromise by offering courses several hours after the school teaching day is completed, it may not be convenient for teachers to participate. The professional roles of teachers extend beyond the classroom, and it is a challenge for teacher educators to provide convenient access in ways that empower teachers to choose not only what to study but also when and with whom. The recent rapid development of resources on the Internet has done a great deal to change the potential for universities to offer courses in ways that address issues of accessibility, convenience, and professional autonomy.

Teaching often is regarded as an isolated profession. What is meant by the phrase is that teachers spend their time with their students, assisting them to learn, and infrequently have opportunities for professional interaction with adult colleagues. Although it makes sense to describe teachers as belonging to a discourse community, it needs to be recognized that participation in that community usually takes place in the absence of other adults. Other participants in the practice domain of teachers comprise students, parents, and school administrators. A goal of staff development programs for teachers usually is to bring teachers into a proximity such that they can communicate with one another about their professional practices, learn from their interactions, and apply what they have learned to improve their teaching. Even though the population density is high, nowhere are the problems of isolation and inaccessibility of appropriate university staff development options more evident than in the large urban communities of the United States. Heavy workloads often militate against teachers availing themselves of opportunities to participate in staff development activities that are ongoing in nature. For example, in Dade County, Florida, one of the largest school districts in the United States, teachers have to contend with increasing class sizes; students with very different ethnic, social, cultural, and economic backgrounds; decreasing school budgets; and community expectations that teachers can do more with less if they are held accountable. The day-to-day professional lives of teachers in these circumstances are challenging and take their toll to the extent that, at the end of the school day, most are unable (or unwilling) to commit to further study.

Purpose

The purpose of this chapter is to provide an overview of the use of interactive computing to provide practicing teachers with convenient access to graduate level courses and degrees in science education. In the following sections I describe how, over the past five years, I have developed interactive computer courseware to allow teacher education students to communicate with one another and instructors and to complete the requirements of graduate courses.

In the Beginning...

My initial application of technology in science education courses was to use electronic mail (e-mail) in all of my courses at both the undergraduate and graduate levels. At that time, in 1993, the biggest challenges for students related to the availability of computers and accounts. They did not have ready access to high-quality hardware or software that could edit documents and send them conveniently in a way that retained an appropriate format. On the contrary, the access they had, for the most part, was outdated, difficult to use, and permitted only the most elementary forms of text transfer. The asymmetry between the tools I could access from my office or home and what was available to students was striking, and it took several visits to a computer laboratory in the nearby science library to convince me that it actually was very difficult for students to log on and respond to the enticing issues I expected them to discuss. The click-and-drag environment of the Macintosh that made it so convenient for me to edit, copy, and paste into a message area or to attach and send formatted documents was not even a remote possibility for many of my students, who were required to use old technology and obsolete software.

Despite the seemingly insurmountable challenges faced by students in getting access to suitable tools, the potential of e-mail to transform the ways students engaged in science education courses was immediately evident. Two desirable trends emerged. First, students provided thoughtful and extended responses on a wide variety of issues that were pertinent to science education. The points of view expressed in the e-mail messages were not necessarily those that ordinarily came up in class and, as a consequence, the discourse that evolved was qualitatively different than what usually occurred in science education courses. Because we created a class account, all messages were sent to all class members and, as a consequence, the discourse was able to mediate the learning of all readers. Second, the availability of e-mail as a tool for expressing a point of view appeared to be empowering to students who might not regularly communicate in class time. I was surprised by the high quality of the written texts of many students when considered alongside the oral texts of the classroom. E-mail provided students with opportunities to engage in different ways and enabled those with a preference to express their thoughts as written dialogue to do so. I could also see the extent of the empowerment in the way that students began to direct their written comments to additional faculty on campus and then to scholars in the research community. Thus, the tools provided opportunities for students to create bigger and more diverse communities of co-learners.

I began to see different ways to organize science education classes to better exploit the potential of interactive technologies such as e-mail.

However, I realized that interactive technologies would be no panacea. For example, if they were to realize their full potential, e-mail conversations needed to be structured. Accordingly, I created the roles of initiator and synthesizer. For each topic, or electronic conference, an initiator would be identified with the role of getting the discussion going and then staying tuned throughout so that further input could be provided to keep the discussion alive. Typically an initiator would have the role of making a written contribution at the beginning of the week (e.g., Sunday night) and then making another toward the middle of the week. The synthesizer also would follow the discussion throughout the week and provide a synthesis for all to read and respond to by the end of the week. The other participants in the class were expected to contribute to each conference at least once between Sunday and Thursday. All contributions were to be "bite-sized" so that reading from the screen would be feasible.

Distance Learning on the Internet

Over a series of courses, procedures such as those listed above evolved, and glimpses of the potential of using the Internet became more visible among the myriad challenges and problems that presented themselves. These glimpses were of sufficient promise that I received a modest grant to offer a course on research methods in science education on the Internet. The course was an immediate success at the conceptual level, with groups from all over the world requesting permission to participate formally or informally as "lurkers." The popularity and associated large numbers of students revealed some problems that might otherwise not have emerged and led to the development of a different mode of delivering courses to students in off-campus locations. To explicate some of these issues I briefly discuss three distinctive groups of participants from Florida.

One group from Florida consisted of community college instructors from around the entire state. These science and mathematics educators were learning to do research that was focused on their own practices and on those of science and mathematics educators in their own institutions. For the most part they were geographically remote from one another, but the grant provided funds for three meetings during which we were able to discuss how to send and receive e-mail and to clarify conceptually difficult material. A second group was a cohort of doctoral students, each of whom was a practicing teacher. These teachers were from geographically dispersed regions and met twice during the semester to discuss aspects of the course and to plan for their progression through the degree program. The final group was an on-campus class consisting of candidates for the doctoral and master's degrees and one colleague from the department of chemistry. This latter group participated fully in the e-mail part of the course and also met once a week for a period of three hours. Thus, the on-campus group had experiences with the course content that were different from those of the off-campus groups. Those different experiences were frequently the foci for rich e-mail conversations and enabled all participants to elaborate the content of the course. Conversely, the off-campus participants connected the issues of the course to their professional experiences and also were able to focus discussions on a variety of professional practices associated with teaching and learning science in elementary, middle grade, high school, and college levels. The diversity of the group within the domains of science and mathematics education provided a breadth of relevance that enriched the learning opportunities of all. However, the volume of e-mail soon became a limiting factor that began to affect the motivation of some to participate.

"Do you have any idea how it feels to have fifty messages download every time you log on to the computer?" A telephone call from Peter Taylor broke the news that was already becoming a stark reality. The students perceived themselves to be in a deluge of mail and the relentlessness of it was becoming a problem. A solution was obvious: break up the participants into smaller groups. The problem was solved almost immediately, but an unintended and undesirable outcome was that all groups except for the on-campus group became silent and many students who previously had been active participants in the computer-oriented conversations became less engaged.

Had I characterized the problem differently, an alternative solution might have been possible. For example, on reflection, the biggest problem of the e-mail deluge was that users had no choice about when to access e-mail and what to download. For most students it was not possible to quickly preview the pages and then decide whether or not to download or print. In most instances e-mail files were downloaded from a mainframe computer to a desktop computer as soon as students selected to check their mail. Depending on the modem speed and the size of the files the downloading process was slow and frequently expensive. Students had to accept files without knowing whether or not they were relevant to their needs. Any long-term solution to this problem needed to take into account the desirability of participants having the autonomy to identify and select resources to support their learning. As I thought about the direction in which we should proceed, the desirability of maximizing the autonomy of learners became increasingly important.

The main difficulties experienced by the participants in these initial courses centered on the extent to which they could learn how to use the technology while at the same time learning the substantive content of the course. Indeed, many students regard learning how to use the technology as one of their most important accomplishments. Some participants were virtually illiterate when it came to computers and did not know even how to switch one on and get it to do anything useful. Their challenge became one of learning how to use a variety of applications and then how to interface these applications with communications software. In addition, many students had not typed before and were very slow. For some, it was a major effort to reliably type several sentences, whereas others typed several pages in the same period of time. Lack of facility with computer and related interactive technology became an obstacle to the learning of some participants. A second problem, related to the first, was the rapid change in the software and hardware tools available to support interactive learning. During this period there were rapid transitions in terms of the development of better e-mail resources and a growing awareness of the importance of the Internet. Freenet and private servers became available and software that emulated the ethernet (known as Point to Point Protocols or PPP) enabled home computer users to access the World Wide Web and use more sophisticated e-mail packages such as Eudora (1997). The rapid technological advances created a context in which my imagination could run wild as I endeavored to connect technological possibilities with social constructivist perspectives on learning (Tobin, 1993) and a value commitment to providing learners maximum autonomy and power over their learning. By the time we had reached the midpoint in the research methods course it was apparent that the future growth of computer-oriented courses would focus on the use of websites as organizers for interaction and access to resources.

Use of Websites for Distance Learning

In the following semester two Tallahassee-based groups enrolled for credit in a doctoral course titled Teaching and Learning Science. A goal for the development of interactive computer support for this course was to begin with an e-mail system developed in the previous semester and to gradually phase in the use of a website. I focused on the development of a website that would do the following:

1. Provide students with autonomy to access education when they have the time.
2. Provide options to permit synchronous (i.e., real-time interactive) and asynchronous (interactions that occur when participants are not present at same time) learning environments.
3. Allow public and private interactions between faculty and students.
4. Provide convenient access to resources.
5. Limit the inconvenience of account clutter.

Over a two-year period I collaborated with colleagues in computer science and science education to develop for practicing teachers an interactive learning tool, *Connecting Communities of Learners* (CCL) to provide convenient access to learning (Tobin, 1998). I anticipated that participants would learn by engaging in a virtual community using tools created especially for this project and accessible on the World Wide Web (WWW). The idea of a virtual community incorporated the perspective that students would interact using a variety of primary and secondary discourses (Fairclough, 1992; Gee, 1989, 1990) that reflected their social, cultural, and professional histories. I anticipated that a shared discourse would emerge from the interactions between participants, a discourse that would transcend the software used in the delivery of resources to support learning, the classrooms in which participants taught, and the myriad interactions that occurred between participants in the community. The emerging discourse would change continually, and the associated discursive resources would encompass a variety of ways of knowing about teaching, learning, and science education. To the extent possible I wanted to create a community without boundaries, one that was inclusive not only of the teachers enrolled in the program but also their students, colleagues, administrators, parents, and friends. I felt that learning opportunities would be greatest if diverse participants could interact in a co-participatory way (Schön, 1985), each contributing cultural capital (Bourdieu, 1992) from which others could learn. I regarded diversity in the primary and secondary discourses of participants as an essential foundation for an emerging community of professional practitioners.

In the sections that follow I describe the development of the CCL, an interactive learning tool developed to support learning at a distance, and provide insights into its strengths and weaknesses in the context of a project involving teachers from Dade County, Florida. Dade County is an urban setting that includes the city of Miami, a region that has traditionally been a challenging place to get an adequate education. In the fall of 1995 I negotiated with Dade County Public Schools (DCPS) to commence a distance learning degree program using the CCL. The county agreed and helped me to recruit 200 elementary and middle school teachers to commence advanced degrees in science or mathematics education. After the first year of the project, an additional 50 middle school teachers were added to the program. The degree consisted of coursework totaling thirty-three semester hours and was offered over a seven-semester period, including the summers of 1996, 1997, and 1998 and the intervening fall and spring semesters. The following subsections provide an overview of some of the most significant parts of the CCL and an explanation of how participants from the DCPS project used the CCL to interact asynchronously and synchronously. Each subsection that follows examines one of the functions of the CCL and describes how participants engage in the process of learning to teach science. The functions that are described are critical reviews, dialogue journals, conferences, notice boards, library, and portfolio.

Critical Reviews

One of the essential conversations that I believe ought to occur at a graduate level involves the critical review of written texts. To facilitate discussions about common readings, the CCL enables students to post thoughtful and well-edited comments regarding chapters of texts and other resources used to promote learning. All students from a given section of a course prepare a critical review of a given resource and post it in a text window prepared for that purpose.

A file is allocated to each chapter, and participants in the course can write one critical review of 300–500 words for each assigned chapter or article. Unlike other sites, it is not intended that the discourse here be spontaneous or real-time interactive. On the contrary, each contribution to this section is carefully prepared and edited before posting and each participant is able to post only one critical review. Critical reviews focus on an issue identified as salient to the professional practice of science education. As it is posted, the text is date- and time-stamped and identifying information of the author is provided. The reviews are public and can be foci for critical dialogues between students. Only the author is able to edit the file. Initially we did not provide participants with editing privileges; however, after we had used the website for a semester, we decided to place more responsibility on the student and permit editing to occur after the initial post. Interactions among students about the posted material also can occur and are stored on notice boards that are linked to the critical reviews. The interactions differ in form from the original contributions in that they are likely to be unedited, spontaneous, and shorter (approximately 150 words).

The critical reviews for a student are automatically posted to an electronic portfolio where each contribution can be self-assessed and graded by peers and the instructor. The critical reviews section is a public, and hence tangible, sign of the emergence of a critical discourse within a learning community.

The following comment from a participant provides an indication of the value of the critical reviews.

> I have enjoyed the critical reviews because they give me the opportunity to learn from others' points of view. I don't always agree with their perceptions, but at least I know where they are coming from.

An excerpt from a critical review written by a middle school teacher provides an indication of the extent to which she is able to link her review to the posts of others, to her wider reading, and to her classroom practices:

> This article presents an investigation undertaken to study how questioning techniques used by the teacher affect the learning of science concepts. The study is exemplary, as it provides the reader with a sample of sound qualitative research done in a classroom setting. A part that I thought was outstanding is the manner in which the author describes and supports her motives for undertaking this study. The inclusion of the reflective self-

analysis, the description of her teaching philosophy, augmented by the critical examinations of her inherent weaknesses and brief account of teaching experience with Limited English Proficient (LEP) and bilingual students, present a strong and convincing platform on which the investigation is built. The literature review is concise yet effective in supporting her claims and concerns. The methodology is emergent, as described by Guba and Lincoln, and it is in this that we find the major strengths of the study. The following excerpts will focus on some of the salient features of this methodology. The author begins by identifying the full array of stakeholders and follows with obtaining informed signed consent from all the stakeholders involved. This is an essential step in building trust and obtaining credible data, as reported by Guba and Lincoln (1989). The author uses the literature in a postmodern manner to obtain input for her emergent constructions. In other words, the "units of information" are introduced for consideration but without being attributed with the authority of a typical positivist study. Unlike previous studies in this monograph, the criteria for analysis for instance of the videotapes are clearly described and explained with the corresponding rationale. As my colleagues have previously pointed out, the methodology presented has fulfilled the quality criteria proposed by Guba and Lincoln. This is achieved by the use of the hermeneutic dialectic circle in three different occasions and complemented by the informal member check implemented at the end. As a result, the author is able to offer an authentic, credible, dependable and easy-to-confirm report in which the emerging joint shared constructions have important implications both for her and other classrooms.

. . . As a teacher of LEP students it has been my experience that in order for the children to learn science and obtain a high level of achievement it is necessary to integrate the content area as a context for the oral and written language production in the social process of talking and doing science. My students are encouraged to use any language or symbols that enable them to communicate with others. The content is not "watered down" and the emphasis is to solve problems and to investigate new areas that contribute to their understanding of science as producers (scientists) and wise consumers of science and technology.

Dialogue Journals

Based on our experiences with science education graduate students and research on teacher learning and change, we knew that critical reflection on action was an important component of learning. To facilitate critical reflection we set up what we refer to as a dialogue journal (DJ). In the DJs, students select one or two partners with whom they engage in a critical discourse on issues that are germane both to science education and to their interests. The partners collectively edit a common file and take turns initiating discussion and responding as critical friends. The purpose of this part of the website is to enable participants to relate their professional practices to their personal beliefs, associated literature, and research outcomes. It provides an opportunity to reflect on practice and to link those reflections to what is taught in the courses. These reflective remarks are then shared with others in the small group.

Writing in the DJs is restricted to group members and instructors. The DJ is password protected and can only be edited and read by members of the group. This degree of autonomy carries some potential risks because simultaneous editing offers the opportunity for multiple versions of the same file to coexist. However, over a period of time we have adapted the dialogue journal to minimize the problems. At the present time participants can add commentary, view it before posting, and then post it to the journal. The faculty also can read and comment on any part of the DJ. Entries from the DJ are simultaneously posted to an electronic portfolio and also can be read from there.

The following comments from a participant provide insights into the way students have perceived the DJs.

> The DJ helps keep you on track. It allows you to know that you are basically hitting the right chord. Sometimes we need reinforcements to let us know we are not "crazy" and that others share or can understand our points of view. I think the only negative is when you have a dialogue partner who does not participate or uses the DJ to strike down your ideas rather than encourage a different mind-set.

The following excerpt taken from the dialogue journal of three elementary school teachers is typical of the conversations that occurred.

> I agree that Chapter 7 may have helped me to understand Guba & Lincoln a smidgen better, but not enough to make me feel it was worth all the effort to try and understand it. That book is written on a level way over my head. Regarding research, I do have an idea of what I want to do, but I still am not quite sure of exactly how to put it all together. Gena, yes, the freedom to teach can be wonderful, but as you seem to have noticed some people abuse it. I guess that is just human nature—to take the easy way out. I agree that the primary years are of utmost importance. Private schools are much stricter about what is to be taught, which does have some advantages. But I do feel there can be a very happy medium out there. If only all teachers would take their job seriously and treat it as a real profession. For anyone who reads this—I do realize that the number

who abuse the system are few. But the number of students affected by the few can be a lot! It can be really detrimental if some students have the misfortune to run across several of these lackadaisical teachers in their years of education. I also realize that some days we all become stressed out and may not do our very best job. That is very understandable. But those days should be kept to a minimum. We as teachers have an obligation to teach and prepare these students for the future. Anyone who is not willing to take on that responsibility should not be a teacher.

Conferences

Students may also engage in on-line conferences in the conference centers. Each class section can access one or more conference rooms that focus on designated topics. We have tended to use the conference centers to disseminate what was learned from the classroom-based research undertaken by participants. For example, when the participants were undertaking research on language and the learning of science, each of the five class sections engaged in a conference on that topic. The proceedings of the conferences were available to all participants in the course, and it was possible for a reader of the proceedings to append comments and interact with those who had contributed to the conference. Like professional meetings, conferences were scheduled for given rooms to commence at a given time and continue for a specified period of time (e.g., two weeks). After that period, the proceedings from each conference were archived in the library, where they can be retrieved and accessed by users who can append notes that are subsequently available for other participants to read.

Four different roles were created for the conference center: convener, presenter, participant, and synthesizer. The style of discourse that characterizes the conference center is formal. The convener, presenters, and synthesizer are expected to plan their conference in the week preceding the scheduled conference and must be ready to participate fully throughout the duration of the conference.

The convener is required to make an introductory presentation on a given topic. The introduction is thematic, addresses some of the major trends in the research relating to the topic, identifies some of the key scholars in the area, and introduces the subsequent participants. The convener has the added role of focusing the discourse in order to deal adequately with the major issues. So that the text can be conveniently read on-screen, its length is restricted to 300 words. It is anticipated that the text is carefully edited, is not a first-draft effort, and is pasted into the conference center from a word processor.

Presenters contribute to the conference after the convener has introduced it. Each presenter selects a topic that relates to the conference theme and contributes a scholarly presentation of 300–500 words that is grounded in research and theory. The presenters are to enhance the learning of all students and the instructors too. Accordingly, the topics of each presentation are coordinated in such a way as to minimize redundancy and to cover the scope of the topic of the conference.

The participants in a conference are class members who read over the presentations and ask questions, comment on what transpires, seek clarification on certain points, and request elaboration and justification. When the conversation lulls, the convener is responsible for initiating dialogue. To ensure full participation it is required that each student in a class section participate in each conference at least three times a week. Formal participation is required in an assigned conference room; however, participation is also permitted in the other conference rooms as well.

The synthesizer has the task of pulling the different presentations together in a posting of approximately 500 words to conclude the conference. The synthesis may include texts from the convener, presenters, and participants as well as selected material from the published literature.

An example of what is posted in a conference room is shown in the following excerpt from a presenter on the topic of co-participation in science classrooms.

> It is wonderful how co-participation helps the students to be involved in the class activities. When we were talking about Pyramid of Energy and Food Chain, most of the students were confused with the new vocabulary such as chlorophyll, producers, decomposers, consumers, chemical reactions, and others. The students started talking and disrupting the class. I understood that they were not understanding the lecture. I decided to show them a short video about photosynthesis. After that, I organized them in small groups to analyze and discuss the green plants chemical reaction. They drew their favorite green plants, and animals, and they established the relationship among solar energy, plants, animals, and human beings. After that, most of them drew the pyramid of energy representing their favorite food. For example they represent black and red beans. Hispanics labeled them in their own language: "frijoles, caraotas, and habichuelas." Other groups were talking about green color, killers,

and eaters instead of chlorophyll, producers, and consumers. At the end of the class most of the students had a clear idea about the process. On the basis of their understanding, we were able to move to a deeper and higher level of scientific knowledge.

Notice boards

One component of Web-based dialogue is the notice board, where individuals may broach any topic in any way they choose. There is no identification necessary, no censorship, and no permission required to post. Entries may range from informal, unedited, and unfocused to formal, edited, and topic-specific. Like a graffiti wall, postings are removed periodically to make room for other announcements. Occasionally gems spring up that are transferred to the library and catalogued to benefit learners in current and future courses.

The purpose of the notice board is to foster communication among a group of students. We have found it desirable to provide notice boards for sections of twenty to thirty students, and although we do not restrict postings to members of that group only, the identification of a notice board with a group of individuals tends to constrain the extent to which others refer to it and hence the tendency to use it to communicate with persons belonging to the identified group.

The following excerpt from the Hot Topics notice board provides an example of how this function was used during the DCPS project.

> Greg, it would be counterproductive to restrict the construction and reconstruction of knowledge to any set of activities or any one method of instruction. "The process of eliciting, clarification and construction of new ideas takes place internally within the learner's head. This occurs whenever any successful learning takes place and is independent of the form of instruction." (Millar 1989, p. 589). (If you have the time, the entire article is quite interesting.) No, there is no set of a priori criteria that one can apply to a set of activities to determine their constructivism. I think that constructivism has been purposefully defined as a way of teaching and learning so that we (the teachers and learners) can focus on the larger issues of co-participation, equitable appropriation of the discourse, and the subsequent entrance into the larger Discourse community (the cases of the letter d are used as Erickson [1998] uses them in his article). These are not just nice-sounding phrases. If we agree that any meaningful learning implies either the construction or reconstruction of knowledge, then access to the language of the subject matter and the power to validate that language must be given back to the learner. We've always heard that knowledge is power. Just look around at the

communities who have little or no access to quality education and knowledge. As I read the literature, it is becoming more and more evident that power relations are a central issues. Two good examples are Kincheloe (1998) and Erickson (1998), who speak of the power inequalities that exist in our learning communities and the need to redress them. But no piece of paper will guarantee that. No paper has ever guaranteed anything. The people who read and interpret the words on the paper are the "keepers of the faith." So how does one redress power inequities and injustice??? One person/teacher at a time. Lucy Drage's eloquence is a perfect example of the kind of change that is occurring. (Very well written, Lucy!!) I'm trying also. You are too. Who knows, maybe DCPS, in its own inimical and bureaucratic way, is also trying a little reconstruction.

The above excerpt illustrates most of the advantages of having notice boards. First, the author is responding to a classmate and relates what has been said to the literature, to other posts from the website, and to his professional activities in DCPS. Posts such as this one were among the richest aspects of the learning environment created around the various functions of the CCL.

Library

The library provides student access to a variety of papers. In the initial stages, these papers were written by faculty and were prepublication versions (to avoid copyright violations). As time passed, I obtained permission from colleagues to include some of their papers in the library, thereby expanding its utility as a learning resource. The library continues to develop in much the way large libraries develop within an institution. As the number of articles has increased, the need has arisen to classify the materials so that students participating in a diverse array of courses can access them. Therefore, we have consolidated materials into separate collections for the science and mathematics education libraries. Each instructor can decide what resources to include in the library for given courses.

A second function of the library is to provide direct links to existing electronic resources, such as catalogues and databases, at main libraries on campus and within the state. Two other resources are available in the library. Student work from other relevant courses, and also from a variety of sources, is included in the library and can be accessed. In the near future search engines will be available to identify potentially useful material from an archive of student work. Also, with the permission of publishers and authors some copyrighted work has been made available in electronic form in a password-protected part of the library.

Portfolio

An electronic portfolio is automatically created for each student and consists of contributions the student makes in the conference center (sorted by topic), DJ, and critical reviews. The portfolio is accessible to students for reading and to the professors for commenting and grading. Each component of the portfolio can be assessed by an instructor, one or more peers, and by the individual student who has contributed the work to the portfolio (i.e., self-assessment). A five-point rating scale of superb, very good, good, satisfactory, and unsatisfactory is used to assess the performance of each artifact in the portfolio, and provision is provided for private comments to justify the assigned rating. It is also possible for the instructors and students to negotiate which artifacts are to be included in a portfolio and provision is made for students to include, adapt, or delete artifacts from their portfolio. I regard it as a high priority to develop options so that the autonomy for the instructors and students can shape each portfolio to the extent that the participants in the learning community negotiate the contents.

Current Applications

The CCL is used extensively in a number of universities for distance learning and to augment on-campus courses. Since I have assumed a position at the University of Pennsylvania I have continued to develop the functionality of the CCL to meet our visions of what is desirable to promote interactions and learning among participants in the communities that participate in activities using the Internet tool. For example, the assessment center is a relatively new addition in which students can take tests, build and use a portfolio, undertake assessment of their own work and that of peers, evaluate the course or aspects of it, and examine a gradebook in which all of their progress is recorded in relation to the performance of others. Another function that has been added is the lecture hall, which allows instructors to add hyperlinks to a variety of resources on the Web or to add resources to facilitate student learning for any of the components of the course.

For the past two semesters I have used the CCL with a group of prospective teachers enrolled in a master's degree program. The high quality of the interactions has been astonishing, and these students have used the CCL to promote their own learning and to form a community in which each participant not only facilitates the learning of others through cooperative activities but also provides emotional support in times of difficulty. The manner in which the CCL has served to provide a strong link between the academic parts of the course and the student teaching experience is very promising. An example of the reflective potential of conversations in the dialogue journal is evident in the following remarks from Shirley (a pseudonym), who is in her final semester of student teaching prior to graduation.

> I think that Sherry is absolutely right in saying that critically reflecting on our teaching practices and roles is the mark of a developing teacher. I don't think that the process ever ends (or should end) if we want to reach our full potential as teachers. I see too many people in my small learning community whom have become very satisfied with the idea that they have the "right" answers in how kids should learn. Each teacher seems to have his/her own style, and no one criticizes one another. It's a very "live and let live" environment in which no one is pushed to question himself/herself.
>
> When I began teaching my own class, I was shocked at how little critical feedback I received. My co-op and my supervisor seemed very impressed at how well I handled a class of thirty-seven students, but their praise was focused only on how I managed my "clerical" tasks of composing lesson plans, making tests, and keeping track of student grades. If that is all it takes to be a good teacher...
>
> I knew at the time as I know now that something was wrong with the situation. My co-op left after the first week of my teaching and hasn't returned for a full class since. I've had to try and figure out things on my own and with the help of fellow secondary ed students.
>
> It's definitely been a struggle, and the struggle seems to become harder and harder when I see that my students' performances aren't improving. Over half the class has failed the last two tests. And the only input I get from my co-op is "Let the kids take responsibility." That really doesn't get me too far as I worry about how to get the kids to learn.

Posts with as much candor are commonplace in the dialogue journals and, because the instructor is able to read and comment, the opportunity exists for rich conversations among students and also between students and the instructor in a course. However, because of the incidence of such comments we decided to adapt the CCL to allow prospective teachers to communicate with their cooperating teachers and their supervisors. Dissatisfaction with the quality of the oral feedback was universal, and we felt that the CCL might provide a context in which asynchronous communications would occur more often and would be of a higher quality.

The next round of adaptations for the CCL will be to develop additional functions for the professional development of practicing teachers so that communities of teacher learners can use the CCL for reflective conversations on their practices.

Conclusions

The development of a website while learning how to meet the needs of students who are studying at a distance is a work in progress. In leading the development of the CCL I have remained open to the perspectives of others and have endeavored to build a tool to enable participants in discourse communities to co-participate. I have regularly undertaken evaluations of the courses and the CCL to enable me to be responsive to the changing needs of participants while fully exploiting the features of the rapidly changing technological tools. The following excerpt of an evaluation from a middle school science teacher is representative of the perceived advantages of using the CCL as a focus for learning.

> The advantages of using the Internet as an educational format far outweigh the disadvantages. As a wife, mother, teacher, Girl Scout leader, church participant, etc., I enjoy being a member of a learning community that is available 24 hours a day, 7 days a week. I can participate and improve my practice of teaching without inconveniencing my family. As is typical of many of the participants in the program, I am responsible for many of the family responsibilities such as shuttling children to various sports practices and other activities, providing the basic household support, and keeping the family together. I truly appreciate being able to explore new ideas and still be able to "be there" for my family. If you checked the times of many of my postings, you would notice that many were done in the early morning hours. It is difficult to attend typical university courses at 5:00 A.M.!

My own experiences with the CCL have been very positive. I am in close touch with my students and regularly comment on their work. I also use the mailrooms from the CCL to write e-mail to whole groups and selected students on a regular basis. Thus, the use of the CCL allows me to stay in touch with my students and to establish a personal rapport with each of them. That rapport is substantive and focused on course goals through the use of critical reviews and conferences. Also, interactions are less focused on course goals but still professionally oriented through the use of notice boards, DJs, and electronic mail. Valuable interactions of a personal nature also occur through the use of e-mail. For me the CCL is now an integral part of all of my teaching.

There are two downsides to the use of the CCL. First, it is incredibly time-consuming. Since I am interacting with every student several times a week, and the mode of interaction is textual, I find myself spending a great deal more time on my teaching. Although I do not begrudge the additional time, it is necessary for anyone opting to use the CCL to realize that its functions set up an expectation of what the students' and teachers' roles will be. If either students or teachers do not fulfill those roles, participants can find their failure to perform to be a major source of dissatisfaction. A second issue concerns the ongoing development of the CCL. The skills needed to program the CCL are advanced, and I have had difficulty in retaining the services of computer scientists who have worked with us. As soon as they are able to write the PERL and CGI scripts needed in this work, they are very marketable and can command salaries beyond what we can afford to match. Accordingly, the turnover in personnel has led to the introduction of bugs that have been a source of instability in some features of the CCL. In addition, the goals of many of the computer scientists are oriented toward programming new functions, and it has been difficult to recruit personnel who will debug the CCL without wanting to rebuild it. Because all of the development has been done in a context of implementing a program of study in which large numbers of students are participating, there have been occasions when the level of frustration is high for both students and instructors.

Many faculty at several institutions now use CCL. It is used mostly in education courses but also has been used as a vehicle for courses in engineering, chemistry, and physics. As the development continues I envision the CCL being suitable for use in virtually any college level course.

The use of the CCL has enabled us to reach out to students in ways that respect their professionalism and the necessity to provide full-time employees with options that emphasize their autonomy and convenience. Our evaluations of the CCL suggest that we are successful in reaching some of our goals but our approach is no panacea. It is imperative that we realize that in creating any instructional delivery system we are not catering to the needs of all students. Our experiences have shown us that we have a long way to go in building a system that is dependable and allows students to learn and re-present what they know to benefit the learning of others and also to satisfy the requirements for obtaining grades. As we continue to develop the functionality of the CCL and expand its applications to include professional development for teachers dispersed over wide geographical regions, we are attending to the voices of the users and doing our utmost to create interfaces for asynchronous learning that are convenient, and secure and allow for the creation of professional discourse communities. It is important to listen to the voices of the users while keeping our theoretical frame of learning by co-participating in professional practice communities in the forefront of our minds. Despite the challenges, our brief encounters with the learning communities we have created are exciting and offer a glimpse of a possible future of cross-national communities interacting to learn from one another using the superhighways of cyberspace. We have far to go, but the beginning of our journey is such that, for me, the teaching of science education will never be the same again. The CCL is an integral part of my approach to the teaching and learning for all learners, whether they come to learn on campus or study from remote locations.

Afterword

The genre of research in this chapter is best described as a teacher-as-researcher study. As a teacher I reflected on my uses of technology and described a history of evolving practices that incorporated my own perspectives. In describing a landscape of applications of technology I used excerpts from students' written texts to illustrate the points I wished to make. The telling of events in this chapter does not reflect the perspectives of all stakeholder groups, and no effort was made to undertake a study in which my descriptions were regarded as assertions, each of which was to be supported by evidence. Instead, the chapter is a telling of what I considered to be the most salient impressions of a five-year program in which interactive computer technology was used to enhance the learning of how to teach science.

During the time in which this chapter was written as a draft and revised to reflect peer reviews and my own changing perspectives I have undertaken two large interpretive studies of the use of the CCL (Goh & Tobin, in press; Muire, Tobin, & Davis, in press; Tobin, 1997, 1998b, 1999). These studies employed the methodology advocated by Guba and Lincoln (1989) and emphasized authenticity criteria and dialectical and contingent selection of participants so that diverse perspectives of all stakeholder groups were considered. The findings of these interpretive studies are consistent with the more impressionistic accounts (Van Maanen, 1988) that are incorporated into the teacher-as-researcher account presented in this chapter. Such accounts may have applicability to contexts encountered by those who read this chapter.

Kenneth G. Tobin
Graduate School of Education
University of Pennsylvania, Philadelphia, Pennsylvania, USA

EDITORS

METALOGUE

Enhancing Student Participation

KT: The use of technology to facilitate learning has really opened my eyes as far as teaching at the college level is concerned. To begin with, the use of e-mail made it possible for students to show a side of their knowing that previously was not visible to me. As I have learned to use Connecting Communities of Learners (CCL) to advantage I have been able to build a system of asynchronous interaction that allows students to re-present what they know in a variety of ways and to do many things in a course that are just not possible in other ways. The biggest advantage, as I see it, is that students can interact more with one another and the teacher on substantive matters than is possible in the synchronous environments associated with verbal interaction.

PG: I learned to use the CCL through graduate courses that I took as a special student at Florida State University, while I was a professor of chemistry. At first, I was resistant to use it, but I learned its value and have used it extensively in my own teaching since then, in classes in physical science; science, technology and society; and biochemistry. In each setting I have utilized different aspects of the website. I think the most powerful part of it, however, is the part where students peer assess each other's work (see metalogue of Chapter 10). They get feedback from their peers, but they also give feedback to others. While they read each other's work, they get ideas of what they can do to improve their own work. They become more critical of their own writing and thinking, and thereby they learn and grow. I think they also become more in touch with how we each learn differently.

KT: The rapid advances in technology are such that international boundaries are regularly crossed during the teaching and learning in a course. Via the World Wide Web we can have asynchronous and synchronous interactions with teachers and students from around the world. It is no longer necessary to be without relevant expertise, and it makes no sense for one teacher to be a gatekeeper on what is learned, when, and how. Technology affords students the opportunity to be autonomous and responsible. However, there are challenges in organizations such as universities being able to break free from the bonds of tradition.

PG: One great advantage of the CCL is that students can use it when they have time to do their learning. Many of our students work part-time, or are parents themselves, and do not study at traditional times. These students can access the Web as their schedule allows. On many occasions students post in the middle of the night or early in the morning. One feature that some students utilize is the Dialogue Journal. This is a location within the CCL where I place students in collaborative groups, so they have a meeting room where they can post information that they want the other members of the group to read. Therefore, if there is a collaborative project due, the group can meet in this virtual space, relieving the problem of everyone trying to get together physically at the same time and location.

Encountering New Barriers

PG: Taking my first graduate course with Ken Tobin was like baptism by fire, reading Guba and Lincoln's (1989) *Fourth Generation Evaluation*. We were in the first stage of using electronic mail communication (before the CCL started). What CCL brought was organization to my own comments and those of others. When we utilized only e-mail correspondence, you didn't know how to sort it as it arrived. It was only after the course was finished that it was possible to sort. The advantage of CCL is that it sorts the student and teacher entries as the course progresses. You can find what you wrote or what someone else wrote that you want to read again more deeply. It is an excellent reflective tool, because it is organized so you can find what you want to reexamine and rethink.

KT: When a tool like the CCL is used in teaching it is essential that the teacher and students *negotiate new roles*. I found it necessary to step back and do less interacting on the CCL because there is just not enough time to respond in detail to everything and to participate in every conversation. Even though it is good fun and good education, the amount of time needed might catalyze someone to say "Get a life!" So as I cut back on the frequency of responding to everything that was posted I began to notice a significant number of students feeling that their posts were less valued because I had failed to comment on them. It becomes necessary for students to actually believe that they can learn from "listening" to their peers and then to construct their peers as teachers and themselves as learners with respect to their peers as well as to me.

PT: In my own Web-based teaching of distance learners (i.e., science teachers) over the past two years (using Internet and e-mail), I too have experienced the pedagogical problem of when to join in an ongoing conversation among students. I started from the opposite standpoint: minimal intervention by the teacher in (highly) pre-structured electronic discussion activities. After some time, I found that students were certainly listening actively to their peers; empathy was rife. However, my valued goal of students engaging in critical discourse was not going to be realized unless I intervened. I wanted students in this course to learn to reflect self-critically on the assumptions and values underpinning their own pedagogies, and I had intended that the readings and discussion activities would direct students toward realizing this goal. But this learning community had created a self-perpetuating support system that directed critique outward at others (curriculum policy makers, authors of articles). I found that I needed to enter the conversation with well-focused questions in order for students to understand what I meant by critical discourse and how to engage (empathically) in it with fellow students.

I also found the technology very time-consuming, especially monitoring complex electronic threads of discourse. And I also found that, although students wrote each other well-crafted and thoughtful pieces, the impersonality of asynchronous electronic text constrained students' conversations and precipitated serious misinterpretations. These teething difficulties aside, the technology enabled us to create a flourishing community of learners and challenged students and teacher alike to adopt new and richer educative relationships in which students are able to learn from each other's relatively unique professional expertise.

PG: I do agree that working on the CCL takes considerable time, both for the teacher and the students. It can also be frustrating, as a student waits for a person to post, so she can do a critical review once the writing is posted. Still I think the format is worth the effort, because it opens new avenues for learning and communication that were not even visible before.

KT: The two disadvantages that I can see of using a tool like the CCL in regular college classes is the additional time needed to interact to the extent that is possible and desirable—more time goes into teaching. A related issue is that more time also goes into learning, and students have complained to me that having the CCL in their lives is like taking a second course, one by turning up for the live classroom and the other via the CCL. The second problem relates to some students not liking to engage using the Internet. Although these students are in a minority, they seem to steadfastly refuse to participate regularly and this affects their performance. I wonder if an analogous situation occurs in class where some students do not like to participate orally and their failure to participate in class leads to lower learning and grades.

BIBLIOGRAPHY

Bourdieu, P. (1992). *Language and symbolic power.* Cambridge, MA: Harvard University Press.

Erickson, F. (1998). Qualitative research methods for science education. In B. J. Fraser & K. G. Tobin (Eds.), *The International Handbook of Science Education* (pp. 1155–1173). Dordrecht, The Netherlands: Kluwer Publishing Company.

Eudora, (1997). *Eudora Pro^{TM} 3.1.* San Diego: Qualcomm Incorporated.

Fairclough, N. (1992). *Discourse and social change.* London: Polity Press.

Gee, J. P. (1989). What is literacy? *Journal of Education 171,* 18–25.

———. (1990). *Social linguistics and literacies: Ideology in discourses.* New York: Falmer.

Goh, S. C., & Tobin, K. (in press). Student and teacher perspectives in a computer-mediated learning environment in teacher education. *Learning Environment Research: An International Journal.*

Guba, E., & Lincoln, Y. S. (1989). *Fourth generation evaluation.* Newbury Park, CA: Sage Publications.

Kincheloe, J. L. (1998). Critical research in science education. In B. J. Fraser & K. G. Tobin (Eds.), *The International Handbook of Science Education* (pp. 1191–1205). Dordrecht, The Netherlands: Kluwer Academic Publishers.

Millar, R. (1989). Constructive criticisms. *International Journal of Science Education 11*, 587–596.

Muire, C., Tobin, K., & Davis, N. (in press). Constructivism on-line: Improving mathematics and science teacher education through distance learning. *Florida Technology in Education Quarterly*.

Schön, D. (1985). *The design studio*. London: RIBA Publications Limited.

Tobin, K. (1997). *Use of technology to connect communities of learners*. Florida StateUniversity (unpublished manuscript).

———. (1998a). *Connecting communities of learners*. Philadelphia: The University of Pennsylvania.

———. (1998b). Qualitative perceptions of learning environments on the world wide web. *Learning Environment Research: An International Journal*, 1, 139–162.

———. (1999). The Internet as a tool for the reform of science teacher education: Transformative agent or catalyst for cultural reproduction? [Internet como instrumento de formación de los maestros de ciencias: ¿Agente transformador o catalizador de la reproducción cultural? *Enseñanza de las Ciencias* 17(2), 155–164.

———. (Ed.). (1993). *The practice of constructivism in science education*. Hillsdale, NJ: Lawrence Erlbaum & Associates.

Van Maanen, J. (1988). *Tales of the field: On writing ethnography*. Chicago: The University of Chicago Press.

CHAPTER 14

 **How Do I Express, Communicate, and Have Legitimated as Valid Knowledge the Spiritual Qualities in My Educational Journey?**

Ben Cunningham

In order to provide a professional context for myself as an action researcher, I explain the nature of the professional conflict I experienced while at a college of education, where I worked in a community of brothers in a religious order, and how I gradually came to a resolution of the conflict. This resolution was influenced by my emergent form of spirituality, which is dialectical, comprising both a contemplative and an active dimension. In its inner dimension of contemplation, my spirituality is one in which I continually seek to "face myself in the lonely grounds of my being without fear" (Merton; in Stone & Hart, 1985, p. 39). This enables me, I believe, to move toward reconciling my inner and my outer self, how I behave and how I want to behave. And I cultivate the active, "practical" part of my spirituality—my interpersonal relationships—by attempting to answer questions of this kind: "How does my life affect those I work with as I attempt to improve what I am doing?"

In saying that my spirituality is a dialectical one I am also accepting that I exist as a living contradiction. By that I mean that I subscribe to values, deny them in my practice and, in reversing the denial, I change and improve the situation. In describing and explaining my developmental dialectic within the contemplative and active dimensions of my spiritual journey, I believe I produce a "living" spirituality (Whitehead, 1993).

And let me also briefly explain the purpose of an earlier paper, on which this chapter is based, which enabled me to address the professional conflict I experienced at the college of education (Cunningham, 1996). Because the professional conflict felt so complex to me, I didn't entirely understand it. I didn't fully understand the responses to me of the people concerned and mine to them. If I wrote up my experience in a paper to be made public, I felt I would at least be explaining myself to myself as part of my research. That is, I would "know" that what I had said and done throughout the conflict would help me to learn to have some confidence in my knowing and how I could begin to change and improve in the future. Because I felt the experience of the conflict had seriously undermined me, I made a decision to leave the college of education. And one of the readers of the paper, whom I shall call Etty, found my description and explanation of these issues inspiring and useful as I describe here.

I address the utility of my writing, expanding it to a consideration of Lather's "ironic validity" (1994) as it relates to my evolving educational knowledge and to my narrative-reflexive writing style, in particular. In doing so, I address explicitly the issue of standards of "representation" and "legitimacy" (Denzin & Lincoln, 1994). The style of my writing, as a form of representation of the spiritual meanings I am trying to create, is influenced by a creative dialectic between my intellectual and my emotional rationality.

Some of Your Thoughts Mirrored My Own!

Etty wrote to seek my permission to use one of my papers (Cunningham, 1996). And she wanted to use it with a small number of women managers who had come together to discuss their own personal and professional development. Here is some of what she said:

> What struck me was how closely some of your thoughts had mirrored my own at various times in the past....When I read your paper today I saw that they [the women managers] may find some comfort in it and indeed some strength....I really did feel as if my own various frustrations were somehow acceptable and therefore were helped to be put into place. So thank you for writing it.

Explaining Myself to Myself!

The paper Etty read was an account of the professional conflict I experienced and my explanation of it to "Joe," one of my religious congregation officials, who recalled me to explain myself. In addition, the paper discussed what I learned from a visit to my dying friend, "Larry," and from visits I paid, at the time, to my sister and to two communities of brothers in religious orders, and how my feelings of alienation from the latter began to dissipate.

To enable you to understand the nature of the professional conflict I experienced at the college of education, I invite you to join me as I reflect on what becoming leader of an action research project for two years meant to me.

I was appointed the leader of the action research project. It was an appointment made by the previous leader informally, with the agreement of the other staff members, within the department in which I worked. Because this job wasn't listed in the college's governance board, it had no job description. The previous project leader, who was also leader of the department in which I worked, had a formal job description of that role. This was easily transferable to a description of her role as action research project leader. If only hindsight could be foresight, I would have sought a formal job description for leadership of the action research project! Doing so at the time might have enabled me to resolve at least some of the subsequent difficulties I experienced.

If I didn't have a job description, I had a leadership style, however. And my style was the opposite to that of the leader I replaced. I am not authoritative in style. That is, I am not instantaneously certain about my knowing; I believe it grows and develops over time. Regarding my dealings with people, I prefer to manage the process of the relationship rather than people themselves, and so I tend toward being non-directive. People rather than tasks have a higher priority with me, even though I also pay attention to the latter. I can tolerate a high level of ambiguity and a certain amount of disorder and instability. Rather than directing or controlling people, I prefer to believe that, given time and goodwill, I can enable them to become self-motivated and self-directed. As leader, I wished our meetings to be more reflective than problem solving, while not neglecting the latter. I was more interested in knowing how we each facilitated the teachers we were supporting in action research in the schools than in how many visits we paid to them, for example. The qualitative rather than the quantitative aspects of the action research project were more important to me.

What I hadn't reckoned with, though, was the personal stress and confusion caused to at least some of the staff by the change in leadership styles. They were faced with me, a leader whose style was the direct opposite to that of the previous leader, a very charismatic personality, who was still a fellow staff member as well as head of my department. I soon felt myself becoming destabilized by my perceptions of the staff's dismay at the change, by my own secret lack of confidence in myself, and also by the fact that another staff member, "Iris," began to act independently of decisions made at our meetings. Unknown to me, in her visits to schools, Iris used to take on tasks for which there was no mandate from our meetings, nor money to support them. She used to also discuss with other staff members, outside of meetings, decisions she intended taking that had not received prior approval at our meetings.

Eventually, having found out what was happening, I started worrying about the financial implications of Iris's independent decisions and how I felt her independence was undermining me and my leadership. I wondered also about how I might find a suitable way to reproach her in as gentle a way as possible. In the end I never did find a way. And then a further complication began to make itself felt. Iris increasingly went over my head to the principal of the college to seek support for her actions, support that she received. I now realized that my role as leader was rapidly eroding and, indeed, becoming untenable.

My stubbornness, however, wouldn't allow me to resign—at least then. It appeared to me that Iris and the principal were ranged against me and that some other staff members, at least, were somewhat unhappy with my style of leadership. In my efforts to understand why the principal had supported Iris, I wondered if her greater experience as an action researcher and her higher academic qualifications were reasons. I secretly grieved, though, that Iris had never attempted to discuss with me how she would have liked to act independently, even if within financial constraints.

Despite our difficulties, the principal, Iris and, I decided to seek a franchise for a master's award from a university for the action research work being done by the teachers. Having written the bulk of the appropriate documents, I was told by the principal that the university had said that these were, in his words, "woefully inadequate." Even though I had the university's formal franchise documents as aide-mémoir, I had let it be known to the then-principal that personal meetings between the university authorities and me were also necessary. My request had been turned down without explanation. Unusually, the principal then began to hold numerous successive meetings with me. I felt I was being "tested," but for what purpose I didn't know. Unexpected memoranda from him, consisting of complaints, began to appear in my mailbox.

At a meeting with him I was asked how I viewed action research. My answer, underpinned by my values, caused him to say that I would experience continual disappointment for the rest of my life! I told him I was willing to face that, however, because I wanted my life and those of the people I was working with to be lives of quality. To which he made no reply. The principal told me that I wouldn't now be attending a research conference for which I had a paper accepted; he would be replacing me, and Iris would be attending also. In addition, he wondered why Iris and I couldn't relate agreeably with one another, as we were both interested in action research and in working with teachers. I reminded him that things were constantly being done over my head without consultation and that I was constantly being undermined. To which he made no reply.

At a second meeting with Iris, the principal, and me, I argued with Iris about my right to my style of leadership. At her insistence that I change, I refused because I couldn't "turn myself inside out and pretend to be somebody I am not. If I did, I would lose all sense of my own integrity and identity" (my journal, 15th October). The principal said nothing. Later, however, I noted that "I also need to be reflexive—but on my own terms!"

At another meeting the principal told me that some staff in the action research project had informally complained about me, but when challenged, admitted that only one person had actually done so. He was not prepared, however, to explain the nature of the complaint. And switching topics, he suggested that, because I had recently passed the Ph.D. transfer procedure at the university, I should think seriously about taking study leave to finish the Ph.D. full-time.

On the one hand, I had been greatly angered by what appeared to be the "drip, drip" war of attrition against me, and if I left the college I felt I would somehow be admitting culpability. On the other hand, leaving would be a way of "solving" the plethora of complex difficulties with which I had been faced and to which, at that time, I couldn't see a resolution that would do justice to me. I decided to leave.

Reflecting at leisure on my experience of professional conflict and the ethos that I felt pervaded the college, I wrote to the principal and to two officials of my religious congregation about that experience. One of my complaints was that the two consecutive college newsletters published after I had left the college failed to mention me or my work at the Institute. My work was, in fact, attributed to others, including the principal, who knew nothing about action research. I felt infuriated and now insisted in my letter that remembering, paying attention to, noticing, reaching out to people, was essential to their human lives—and to my life. I lambasted what I considered to be the hierarchical and bureaucratic nature of the college where, in my opinion, nothing was decided at the appropriate lower level. Everything, no matter how minute or insignificant, was apparently decided at the higher level of department head or at the level of principal. And in the case of some decisions, they were sometimes decided on a whim. I even had the experience of a decision concerning myself being reversed (attendance at a research conference), and when I questioned the reversion, was told that the requisite authority had power to do so. I excoriated what I considered to be the centralization of power and said that whenever power is centralized in a few hands, there is the danger of corruption. And my career experience had shown me, among other things, that the corruption inherent in the "truth of power" (Foucault; in Gordon, 1980, pp.109–133) often leads to arrogance.

The college structures weren't helping me to achieve the freedom necessary for my personal growth. Hierarchy and bureaucracy were hampering my freedom. Unless I had freedom, I had no choice. If I couldn't choose, I couldn't grow. There was and is the question of who matters, of who is respected and who isn't. It was there that the essence of Christianity lay for me—the respect I was entitled to but hadn't experienced at the college. I had experienced a lack of freedom, of independence, and of respect when I was young and was now anxious to seek and to work toward living out these values in whatever work situation I found myself. My paraphrase of a paragraph from McNiff (1988) indicates how strongly I feel that this is what my learning, my educational development is about:

> I am convinced of the need...to appreciate the power of my "I," when that "I" engages in the process of my own development; of the power of my "I" to create my own understanding. This power allows me to apply my educational practices to the process of transforming my life and the lives

of others. If I am not happy with a situation, I change it. My educational knowledge is the process whereby I know why and how I transform my life...In order to do this, I must be free, socially, intellectually and spiritually. (p. 186)

And because Joe, one of the congregation officials to whom I wrote, was concerned about what I said, he recalled me for "consultations." At the meeting we had, I explained quietly but unequivocally who I perceived myself to be, what my values were, and how I perceived that I had been ill-treated in my previous job situation. Having heard what I had to say, he wrote shortly after our meeting:

Thank you very sincerely for our meeting last night. It was truly a graced occasion. When someone shares from the depths of personal experience as you did, Ben, it is always sacred....I want to wish you every blessing on the sacred journey of finding yourself at a new and deeper level. May your desire to serve others flower and blossom.

Was the Data in My Paper Valid?

I wrote to Etty about the validity of the "data" in my paper thus:

Actually you're the first to give me positive feedback on that paper. When I produced it, all I remember is people saying: "Well, who's to know that these things really happened? It wasn't taped, etc., etc." And, of course, it wasn't taped. It just wasn't appropriate to do so. What I actually did was to write field notes immediately after each event. Anyway, to my knowledge, these become data when you bring them to an audience to share and they feel they're authentic and give you comments on them. Now, you've given me feedback and I now have data that are useful! Up to this I had buried the paper and thought no more about it until you said it had use value. That, by the way, is the greatest compliment you could have given me! How many action researchers can say that some one-off paper they've done has use value? Very few, I think. So, Etty, you've done me a signal honor. That's for sure!

Is Anger Useful and Justified?

Etty explained in her letter that she was glad I had found a way of explaining myself quietly, yet determinedly:

Firstly let me say that for me it was a very happy paper. Happy in the sense that you got the better of things and succeeded, even though there seemed to be some very painful and upsetting experiences that you were relating. That's perhaps a contradiction in itself, but what I mean is that the overall feeling that I got was, "Good for Ben, at last he's saying what he wants to say!" No need for shouting or any song and dance, just quietly explaining himself, but at the same time I had this feeling that you were very determined to get your message across. It was as if your quiet determination (as opposed to your anger) was shouting at me, "Now listen!"

And concerning anger, Etty wondered "if you would have ever addressed these issues without anger?" And she continued:

Isn't this the tension that Jack [Whitehead, my Ph.D. supervisor] refers to which leads you on to put things right when your values are negated in practice? Sometimes I think it's good to stay angry, angry about the things that really matter to us, that way we can never forget what it was really like. At the same time we manage to accommodate the experience, remove the bitterness and in some way gain a different kind of peace, perhaps that's your spiritual peace?

And so Etty noticed and was encouraged by the fact that I was making an effort to deal with the causes of my anger:

All too often it seems that we just moan about things but do nothing about them, we're afraid to take the risk. We find fault, saying that someone has hurt us, but never move on to change the situation. That leads to quite depressing reading, although I accept that the sadness is a stage that we might stay within for quite a long time before being able or having the strength or support to do anything about it. I just found it encouraging to read your paper which suggested to me that you were moving on and away from your anger.

When I wrote to Etty I also added, however:

I stayed with the anger and the sadness you're talking about for a very long time before trying to do something about it. And then I finally plucked up courage and told some people what I thought of their treatment of me. It took some time before anybody took it seriously. Finally, somebody did.

And she commented on what she felt was spiritual in my paper: "It gave out a lot of hope and joy (Spiritual qualities?)." And Etty also felt that my paper provided a balance between the love and caring that you were experiencing in your communication with certain people that you mentioned [my sister and the two brothers of the religious order], and the anger that you felt at the injustice and lack of attention from others who should have been showing more care of you.

She also sensed, however, that I was frustrated, as she put it,

at your inability to communicate with those who were failing to show you the respect that you deserve both as a professional and as a fellow human being. I don't know if this is how you really felt but it's how it made me feel, it struck a chord with me and related to experiences I have had in my own life at certain times.

"I Was Very Moved by Your Description of Larry"

Etty went on to speak of how I described my friend Larry, and, in passing, paid me a handsome compliment, "I think you are becoming real and showing us what that is like." And she continued, "I was very moved by your description of Larry. He didn't have to pretend. He was just himself and at peace with himself." It seems to me also now that there was nothing controlling him. Perhaps that was another reason for my trust in him.

When I replied to Etty, I felt a need to add what I had said a year ago about my dead friend, Larry, who, like me, was a brother in a religious order (Cunningham, 1995):

Larry:	Yes, I have cancer. I have been given three months to live but I may die sooner. Funny, it wasn't until a past pupil commented that I had lost a lot of weight that I found out.
Ben:	And how are you now?
Larry:	Well, I have accepted it. The way I see it is that I am lucky to have had fifty-one years; many others don't. So I feel I haven't done too badly at all.
Ben:	And when you look back, what do you feel about your life?
Larry:	I enjoyed it. And I did what I was asked to do in school and I'm happy to think I did it well. What more could you ask?
Ben:	And what is it like for you at the moment?

Larry: Well, I eat a little at 8:00 A.M. and get up at 10:00 A.M. It's good to be able to look after myself still. I take a rest at 1:30 P.M. in the afternoon. You see, I have to be at my best when many people call to see from about 4:00 P.M. onwards!

I dreaded meeting my friend Larry. I didn't know how I could engage him in conversation as he was near death. How would I find the proper words, say the right things? I needn't have worried. In fact I was astounded by the matter-of-factness with which he had accepted the inevitability of his own imminent death. He was ready. He was satisfied with how he had led his life. He had accomplished what he had set out to do. His main worry now was that he would be in a fit state to receive his friends in these latter days. His mind was on others rather than on himself.

I told Etty how important Larry's apparently nonchalant help had been to me in the past:

> Yes, Larry has been an inspiration to me. I often think about his comparative "godlessness" with great affection....I felt he had literally saved me from death, many years before, when he was the one person whose help I sought when I was in the last throes of alcoholic despair and I couldn't go any lower, except out! And without fuss he did what he had to do and I was looked after. All done quietly. All part of a normal day's work. "No sweat," as we'd say nowadays.

I told Etty, too, why I considered Larry to be great. His total equanimity in the face of death seemed to me to be irrefutable proof of that. And I added:

> In a strange kind of way, my thesis and most of its themes are now drawn from what I feel I've learned from him. And yet when he was alive I never felt there was anything special about him. Neither did anyone else. I think what really electrified me and everybody else who knew him about his greatness—yes, I'm using that term now—was the way he accepted his death sentence. And that's what his form of cancer was. He was under imminent sentence of death. Total equanimity was his response and that was the way he remained right up to the moment he died.

Larry, I told Etty, was apparently able to act out of his own convictions without reference to others. It seems to me now, in retrospect, that he had found the delicate but important balance between independence and interdependence:

> Brothers in all our communities didn't know what to think of him. Here was this guy who seemed to neglect all the ritual things we were expected to do every day. He just acted totally independently. He attended ritual if he felt like it, he didn't when he didn't feel like it. That was that. And he was happy in his independence. He also knew he was being criticized *sotto voce*. That didn't matter either. He went to school every day, did his work, came home and did the crosswords. Had an occasional drink and a smoke. Was happy and that was it. On one occasion when he was asked to become a principal of a secondary school, he said yes. Six years later when he was asked to step down—he apparently wasn't exciting enough—again he said yes. That was that. Back to full-time teaching he went. And so life went on until his death.

I hold him in highest esteem—and I'm still trying to understand him!

"Why Were You Reticent about Speaking Out?"

In a later letter, Etty wondered if my reticence at speaking out to this point had to do with the fact that I am a brother of a religious order and that "you may feel that you are expected to give out kindness all the time with no return, and that you are not expected to complain." In replying to Etty, I felt a need to see if there were any other reasons for my reluctance to speak up on my own behalf, and I felt that a negative attitude toward myself was at least one, and here is what I said:

> Regarding my previous reluctance to speak up perhaps yes, being a brother of a religious order may have inhibited me. Perhaps I felt I should take whatever was doled out to me without complaining. That was the task of my faith. However, a more deep-seated problem was there I think also, masked by this role I felt a need to play. And it was that deep down I didn't feel good enough to object to any treatment meted out to me. Although I don't now want to apportion blame for who I had become—it's a useless and immature pastime—I had brought negative self-attitudes with me from my youth. When you are constantly told when you are young that you are useless, that you have nothing to recommend you, it takes a long, long time to get over it. I often feel I've spent a lifetime up to now having to compulsorily re-create myself.

Getting over my negative attitude toward myself, I told Etty, took a long time. And I was at pains to point out that I am no different from many other people, who undergo similar difficulties. Then I continued:

> It might sound melodramatic to say this but I've often felt that either I had to re-create myself from the ground up as it were, or disappear! I made several involuntary efforts through alcoholism to "disappear." I think my paper and the actions I took were a part of this long haul of the recreation of myself. It's unfortunate (and dishonest in my view!) that religious orders and congregations allow the perception to prevail that they are at peace with the world, "gaining strength from their silence," as it were. I wish that individuals among them at least would tell others that they have to struggle like everyone else to become whole, to become themselves; that they have similar problems in the workplace, etc. That is why I was fascinated by Larry's life. It was just "ordinary," like anybody else's and apparently extraordinary only at the end! It is some of these stories, yours and mine included, that we should be telling. It would uplift others, tell them that they're not alone with their concerns and personal and professional problems.

"Your Values Break Through, But What Do You Mean by Spiritual Values?"

In her letter, Etty detected that my passion about values had broken through, ones to do with "offering gestures of approval, of affirmation, of remembering, of intimacy." And she continued:

> And, Ben, you do that. I received the usual Christmas cards from friends and family, but there was one that I wasn't expecting—it was from you. I didn't realize the relevance at the time but now I do, thank you for remembering me.

Etty began to wonder how you explain action research and especially how you explain the action part:

> However, as I saw it, the action part in your paper was your speaking up and beginning to explain yourself, followed by trying to make sense of the experience in accordance with your values.

She also wondered about what I mean by spiritual values and began to speculate for herself about what they might be:

> Could the spiritual be the closeness you have with other people as you
> improve your communication with them and enable them to communicate
> with you?…the word conjures up for me something of wonderment, joy,
> peace, understanding, something bigger than us but part of us, something
> that is good and honest.

Etty thanked me for my paper and finished by telling me that I had reached into the spiritual in her, "Yes, Ben, through your sincerity, determination, and honesty you have reached out to the spiritual in me."

"Who Am I of Value in This Text?"

At this juncture in my chapter I am going to attempt to draw together what I think the meanings of my spiritual values are, and I wish to hold myself accountable for them to myself and to whomever may read this chapter. But just before I do so, I believe an extract from Fromm (1956) following on his question, "What does one person give to another?" is some of what I have been attempting to do in my practice and in this writing:

> He gives of himself, of the most precious he has, he gives of his life.…He
> gives him of that which is alive in him; he gives him of his joy, of his
> interest, of his understanding, of his knowledge, of his humor, of his
> sadness—of all expressions and manifestations of that which is alive in
> him. In thus giving of his life, he enriches the other person, he enhances
> the other's sense of aliveness. He does not give in order to receive; giving
> is in itself exquisite joy. But in giving he cannot help bringing something
> to life in the other person, and this which is brought to life reflects back to
> him; in truly giving, he cannot help receiving that which is given back to
> him.…In the act of giving something is born, and both persons involved
> are grateful for the life that is born for both of them. (pp. 24–25)

In attempting to draw the threads of this chapter together, I have to say that I have never found searching for "truth" easy. And yet I have felt a great need to search, to inquire, and to agonize in this search and in this enquiry. And to interrogate everything and everybody I meet, overtly and covertly about it, even though perhaps not naming it as such. Something or somebody, somewhere, must hold the key to my inquiry. But what or who is the key to my inquiry, to my truth? Does it reside in goodness, authenticity, integrity, the sacred, the mysterious? In appreciation, hunger for fulfillment, neediness, love, remembering? I know my search for truth, with a small *t*, is to do with all of these things. Not as abstract entities, but as embodied values. And I have been looking toward Etty and others for the embodiment of them. However, if I am to be authentic, a person of integrity—which I am claiming I want to be—then I have to admit that I too embody these values and that they can be seen in the form of life I am leading. I am practicing these values in my educative relationships with others—at least to some extent. And when I find I am not, I start over again. And that's okay. And because I am learning to own my own goodness, I can't any longer evade the question Jack Whitehead, my doctoral supervisor, asked me in his office: "Who am I of value in this text?"

And in taking Jack's question seriously I am now finally applying some of the words I used about my dead friend, Larry, to myself. "I think about his comparative 'godlessness' with great affection," I had said. And I am saying now:

> Yes, that's me. I don't want to be taken for granted by anybody, even by God. Threats, or even blandishments, won't do. God has to be gentle, understanding with me. I want to be accepted, to be appreciated, not for anything I've done but because of the simple fact that I am, that I exist....I want people, instead, to sit up and take notice, to be amused if they want to be. But also to know that I've said something that I'd like taken seriously, not especially because of its merits maybe, but especially because I've said it. For me it's a form of honoring and being honored, of remembering and being remembered.

And I sincerely want to bring goodness into the world in my dealings with others. And how I do it doesn't have to be grandiose, grandiloquent, noisy. It can be the quiet, small gesture—like mine. But, make no mistake, I've had to learn how to do it, but in doing it there is an unexpected but delightful pay-off. Others appreciate what I am trying to do, and in appreciating my actions, they are appreciating me.

And I also want and need to be independent. Larry's way of living brought it forcibly to my attention as did the way I felt I was treated in my previous employment. And it is why I now want to empower, to enable—but only at their behest—others whom I meet in the course of my action research work.

And as Larry didn't unduly disturb himself about anything, so I believe I am now able to say honestly that I "face myself in the lonely grounds of my being without fear." (Stone & Hart, 1985, p. 39). I am valuing myself and I am working at being as authentic as I can be. My solitude and my silence in my contemplative moments are also, I believe, moving me toward achieving a reconciliation between my inner and my outer self. If you see me being peaceful in interaction—and I believe you do—it is happening because I am now loving myself inside and out.

And I am learning to acquire Larry's equanimity, which for me is one of the rewards of learning to be authentic. I believe I am moving toward reversing all the years of internal despair at my own unworthiness. Sometimes now I want to shout with joy that my life has changed. But I won't shout, I'm happy to say it quietly: My life has changed and for the better. And in saying this I believe my story can help others.

But I am not naive either. Because I want the world to operate in a certain way doesn't mean it is going to do so. In my experience of working in organizations I have seen people being considered as units of production. I have experienced it myself. And sometimes I may succeed in being able to fight for others and also on my own behalf from inside organizations. Other times I may have to admit that retreat may be the better part of valor.

Writing to Etty, I told her I was now looking at the present but also at the future, referring to a recent job offer I got from the same college of education as my previous place of employment. And here is what I said to her:

> I accepted the overture very coolly but in a not unfriendly fashion either. I suggested a series of negotiations and…I think what I mainly want to do is to get to the point where I have negotiated what would satisfy me in terms of academic freedom if I were to take the job, knowing at the end that I am now worth more than I ever thought I was.

And so, in my negotiations about a future job, how can I ensure that I protect my independence, my hard-won level of self-esteem, my desire to be appreciated? And how do I negotiate so as to get my needs met and end up still respecting the other person in the dialogue while not agreeing with that person or with how he or she has negotiated? And I supply a partial answer when I pen the following and, because it is so important to me, I italicize it as follows:

> *And whatever happens when all the present negotiations are over, I want to end up being at peace. Holding to my equanimity. Feeling good about myself. Feeling appreciation for myself. My self-esteem, authenticity, and integrity intact. And maybe having to find a different arena than the college in which to work for others!*

Even though the negotiations turned out to be "discussions," what have I learned? In what I took to be mutual negotiations by a series of letters with the principal of the college of education about a job as academic coordinator of a university master's research program, I had made four requests: first, job definition and description; second, job protection; third, "negotiations that would satisfy me in terms of academic freedom"; and fourth, that "we continue to create an open, honest, and transparent dialogic relationship."

Writing to the principal, I said that I now realized that no negotiations had actually taken place and I concluded:

> It is with no little sadness and without animosity of any kind that I am
> now saying that I will not be a candidate for any job nor accept a job offer
> that may arise now or in the future at the college of education.

In his reply the principal said that we had not got "to the point where we would have been negotiating on a job offer." He categorized what had been happening as "discussions," and thanked me "for your willingness to engage in [them]."
And now I have to find work in a different college of education within which to work for others!

Could I have resolved the professional conflict I experienced at the college in any other way? I'm sure I could. However, I did it in the way I have just described because I am who I am. And now in retrospect, I am trying to understand my actions—or lack of them—and explain that understanding to you. In this, Elliott (1991) helps me when he tells me that:

> Spiritual development proceeds by resolving problems of living wisely.
> Wisdom can be defined as the achievement of a sense of unity of purpose
> in the multiplicity of decisions and acts which constitute a human life. (p.
> 147)

And some of the values that matter to me, too, are embedded in how I described my leadership style and how reluctant I was to let go of that when challenged by Iris. I like, for example, being uncertain about my knowledge. I keep my future open to new learning, new educational development. And there is my care for process in interpersonal relationships rather than the creating of products. I prefer reflection on and in action rather than action for the sake of it.

My decisions and acts then were largely influenced by my values, by my leadership style as well as by how I described my spirituality initially when I said it has a contemplative and active dimension. It seems to me that some at least of my instinctual gut reactions to my professional conflict were of the contemplative kind where being still, being peaceful, as I faced myself "in the lonely grounds of my being without fear" (Merton; in Stone & Hart, 1985, p. 39), was more important to me than a particular action I might have taken. I often thought and felt that reconciling my inner and outer self was paramount for me and that I could best do it by starting with what was inside me.

So what have I resolved? I have resolved that I now value knowing myself and being able to explain that knowing both to myself and others as I am doing here. I also accept that, despite inevitable—and necessary—contradictions (Whitehead, 1993), I often bring an aura of peace, joy, and fun to my interpersonal relationships, as with Etty, together with the values I articulated above, occasionally darkened by inevitable misunderstandings arising because of our mutually vulnerable human condition. And all of this goes into making up what my spirituality basically is—my autobiography! (Padovano, 1995). I reckon that, by Elliott's standards as I explain them, I am "resolving [the] problems of [my] living wisely."

Is My Research Useful?

I want to have my spiritual qualities legitimated as valid knowledge. And I do this partly through "self-validation." My criterion (standard of judgment) of self-validation is based on Polanyi's (1958) statement that "I am a person claiming originality and exercising his personal judgment responsibly with universal intent" (p. 327). I am offering meanings of my values revealed in my text and making decisions about living them out more fully in my ongoing negotiations regarding a job I have been offered. As part of my effort at legitimization, I am engaging in intentional critical reflection which, according to Whitehead and Lomax (1987), is

> the way in which a naive understanding of practice is transformed; where the practitioner reflects upon instead of merely experiencing practice; and where the process is made public and shared so that others gain an understanding of the practice. (p. 186)

My action research has "use value." Etty, for example, acknowledged that my paper indeed had use value in that she identified with it. Furthermore, she believed it would be useful to her women managers. I now believe that she and others have realized the importance for themselves of my struggle to understand myself and know that they can succeed in doing likewise. And they know that they can change themselves if they consider themselves to be important enough to be appreciated; that everybody deserves to be appreciated simply because they are human. That nobody should have to earn appreciation. I finally learned that lesson!

Others, too, realized the use value of this chapter. Peter Taylor, one of the editors of this book, felt that it would "inspire others with undreamt-of possibilities for renewal." Here is the full text of what he said:

> I have just read your...paper and have come to appreciate something of your struggle to be authentic and alive as you inject your spiritual values into an action research which breathes a refreshing breath of life into your pedagogy. And I want you to join us in our book. To have your story there to inspire others with undreamt-of possibilities for renewal. And I hope that, like Etty, my invitation pours energy into your sense of worthiness as a person and an educator. My invitation to you is couched in the spirit of celebration. In this book I want the celebration of life to burst forth and dazzle the reader with joyous anticipation of forming rich educative relationships with others. Will you join me/us in this endeavor? I hope you can.

And here is what "Rex Fischer," a participant at a Spirituality Conference at Roehampton Institute in London, wrote and gave me after hearing my presenting based on this chapter at a Spirituality Conference at Roehampton Institute in London:

> I wept copious tears last night as I was reading your paper. I was moved and powerfully gripped by the way in which you spoke about real live human beings grappling with issues that occur every day. You were, for example, trying to deal perhaps with an abuse of authority within hierarchical structures in an institution. Yet you were also able to empathize with someone who was dying. And Larry showed you and me how to die. A stranger, Etty, came upon these issues unexpectedly and saw how you and others dealt with them. And she was inspired—as I was. For the first time I can see the importance of research accounts which include the affective areas. They can tell you about the reality of the human condition and about what is educational. And I was delighted too

that it wasn't sanitized, it was as messy as life is. Thanks, Ben, for your humanity and honesty and for opening them up to us.

Having the Spiritual Qualities in My Educational Journey Legitimated as Valid Knowledge

But now it is time for me to formally address issues that I have informally been considering above under the rubric of "Is my research useful?" And the issues are those to do with legitimacy, representation and validity, particularly "ironic validity" (Lather, 1994), as that relates to my narrative-reflexive style of writing.

Legitimacy

For my research claims to be legitimate or for my text to make claims for its own authority it would, in the past, have used foundational arguments based on concepts such as reliability, generalizability, validity. However, Denzin (1994) argues for a more "local, personal, and political turn" (p. 504) and Seidman (1991) agrees when he proposes that "we be satisfied with local, pragmatic rationales for our conceptual approaches" (p. 136). In accepting a local and personal rationale for my case study, I base it on Winter's idea (1997) that "theory in action research is a form of improvisatory self-realization, where theoretical resources are not predefined in advance, but are drawn in by the process of the enquiry." Where I disagree with Winter, however, is in the necessity for drawing in theoretical resources in the process of an enquiry. Rather, I believe that in being improvisatory, I am creating my own descriptions and explanations for my own self-realization, for my own educational development.

Representation

Representation means how I transform what's in my consciousness into a public form so that I can analyze and share it with you (Eisner, 1993). I know I will never be able to capture directly what's in my consciousness. I can only attempt to create its contents in some form for you. That there is a problem in directly linking my experience with text is known as the "crisis of representation" (Denzin & Lincoln, 1994). Nevertheless, I attempt in this paper to construct my form of representation in a way that will act as a lens through which you are able to perceive my meanings (Laidlaw, 1996). So I construct my form of representation in a subjective way, not in the "disparaged" sense of that term but as my "aware subjectivity." That is, I am both in and observing myself in my research. It is a rigorous process in which I am seeking to be critically aware of my own perspective and am hoping to keep it open to change as I move through my research and track how it affects how I make sense of how I am acting—and reflecting.

Validity

Validity, simply understood, means, Does my research really do the things it claims to do, and can you, reader, believe the results (McNiff, 1988)? The explanation for my research, however, becomes more complex when I add to it the adjective "ironic." Basically, "ironic validity" (Lather, 1994) means that the validity of my account can best be judged in relation to its failure to actually represent what I know. Because my intrapersonal and interpersonal spirituality, which I have been describing and explaining in this chapter, is living, dynamic, and constantly changing, the irony is that I can never communicate fully this knowledge that I embody in my living and lived reality. I am always falling short in my representation of it. And the same is true of how I use language in this chapter to represent what I know. My use of it fails to communicate fully what it is I know. But what forms of language am I using? I am using both propositions and descriptions of the felt state, principally my own. And let me explain what I mean by propositions and by the felt state.

Propositions are statements, and statements have been central to our traditional view of knowledge (Eisner, 1997). The traditional view of knowledge then is linked with matters of verification, and verification is linked to matters of truth. Truth is linked to claims, and claims cannot be made without assertions. Assertions, in turn, require propositions, and propositions return us to a traditional view of text—which is that it is about propositions (p. 7). In fact, this section of the chapter is close to the propositional position, which is principally to do with my intellect. And let me tell you where I place my intellect in the conduct of my life.

For me, the intellectual and the conceptual are always secondary to the affective, are secondary to my felt experience, to my emotions. My intellectual ideas develop over a period of time, and they appear to me to go through a series of events before they become mine, before I can appropriate them. And in that time I am trying to be a person of integrity, to be honest and sincere in my appraisal of my intellectual thoughts. If my emotions, my feelings, my instincts accept them as fitting in with my experience then I can, and do, accept them. If, however, they still remain strangers to my emotions, my feelings, and if my experiences do not know them, well then, I cannot accept them, I cannot entertain them further. And if on occasions I appear to accept them, I do so only because of my respect for the source from which they emanate. In the fullness of time, I will wish to reject and overthrow them when my sense of integrity appeals to me for arbitration at the peril of losing it.

And so it is with no little appreciation that I embrace the notion of being a "living contradiction" (Whitehead, 1993). I appear to accept some external conceptual ideas, some external intellectual ideas as values, as small truths if you like. I can entertain them even when I know they are negating who I believe myself to be. Eventually, in order to maintain my integrity, I reject and replace them with my own form of existential truth, be it in a self-appropriated intellectual form or in an affective form or one form accompanied or interpenetrated by the other.

Then there is the "felt state," which is to do with the emotions, my emotions. I believe with Macmurray (1991) that my "feeling the world" is more basic than my thinking it. Feeling is the touchstone of reality, my reality. In using my emotions in dialogue and in my descriptions in this chapter I am showing my concern for myself and for others as persons. My emotional intelligence (Goleman, 1996) also embraces zeal and persistence, the ability to motivate myself, and that leads me toward my moral instincts in my relationship with others. While I am sure that the intellectual and conceptual can enlighten me about the pros and cons of the adoption of these stances, can offer me propositional knowledge, I am not persuaded that it can be the moral agent that motivates me in my caring, compassion, and empathy, which are essential ingredients of my spiritual values. And so I believe my emotional intelligence is a surer guide for me to my exercise of spiritual values than intellectual and conceptual intelligence.

And let me bring you on a tour of my psyche and heart so that you can see how I treat emotional and affective ideas. I rarely hesitate to absorb, to appropriate emotional, affective ideas, because I feel I have lived with them, interiorly and exteriorly, all my life. They are a lifetime's houseguests, guests of my interior, which I call home. They are familiar, you see. I don't have to doff my hat to them, be polite in their presence. And it's not that they own me or that I am beholden to them even when I allow them to disport themselves, as they sometimes will. My instincts trust them. They have always been my touchstones to reality, the real guides to my life. And while I will not be entering a full-scale defense of my emotions, suffice it to say that one of my gifts that I prize highly is empathy. And in my work with others I use empathy, which is at least partly the source of my altruism, "the ability to read emotions in others; lacking a sense of another's need or despair, there is no caring" (Goleman, 1996, p. xii). Without caring, compassion, and empathy I believe my explanations of my spiritual values in this chapter would be seriously deficient. And because I am a researcher who respects rigor in its intellectual and other garbs as a form of validation, I check my thoughts, feelings, and instincts with others as I have done, for example, with Etty.

At the same time, I am not attempting to oppose one form of rationality with another, the intellectual with the emotional. Rather, I am attempting to use both, linking them with the synthesizing capacity of my "I" through using a propositional and felt form of language. And I believe I have succeeded to some extent. However, I also acknowledge with ease that I am still failing to actually represent what I know. In fact I will never really be able to do, so as my knowing keeps developing even as I am describing and explaining it, and so every effort I make to represent and explain it fails. But it is not a failure beyond which I cannot go. No, in fact, it causes me to redouble my efforts to go on trying to see if I can, in my form of representation, approximate more closely what I tacitly know.

Ben Cunningham
School of Education
University of Bath, Bath, UK

EDITORS

METALOGUE

Self-Inquiry: Room for Others?

KT: Ben's descriptions of himself as a leader reminded me of the fallibility of self-descriptions. I would like to be like Ben, and I wondered if this chapter was placed here so that I would reflect on his description in relation to myself. I hoped there was room on that pedestal for more than one of us. My colleague Fred Erickson has cautioned us not to rely too much on interviews in interpretive research, and Paul Churchland has also addressed the viability of oral accounts for what has happened. He cautions that the human quest to be rational sometimes leads to the construction of plausible narratives that are seen as reasonable and become objects for reflection but are actually not correct explanations for the actions they seek to describe.

My thought as I read over Ben's description of self was how others might describe his leadership. Would they use the same categories or others? There is a political agenda here. How would the previous leader describe him/herself? Perhaps s/he would use some of the same categories as Ben. I find it useful to think not only of the differences but also the similarities. It might be that through different theoretical lenses the commonalities were more pronounced. Would Ben then have acted in the same way with respect to the circumstances of interactions within the community? Ben's actions with respect to others may have been framed by his oppositional perspectives on what was happening and why it was happening. Did his action research allow him to put his theories to the test? Is there value in putting them to the test?

PT: In action research or other forms of professional self-inquiry, critical self-reflection can be a two-edged sword, then? On the one hand it requires moral courage to deconstruct the demons in one's own closet; to identify the hard questions whose pursuit can take us through a (tortuous?) autobiographical journey from which we emerge refreshed with a deeper understanding of our own subjectivity (integrity, values, spirituality, emotionality) and its defining role in our sense of what it means to become a better professional. On the other hand, if we become too self-absorbed we can lose sight of the cultural complexity of our relationships with significant (especially troublesome) others, and run the danger of roaming our inner landscapes, tilting quixotically at windmills.

Ken, you seem to be suggesting that a balance can be achieved by bringing multiple perspectives to bear on our critical self-reflective inquiry. And that in the case of our relationships with others, we should endeavor to adopt theoretical standpoints that emphasize both our differences and similarities. This seems like a good strategy for preserving the dialectical nature of the relationship between self and other which, otherwise, might collapse into an unhelpful singularity. On the other hand, we sometimes need to act decisively in order to protect ourselves from repression, especially of a political nature. Context and circumstance seem to me to be the important criteria in this equation.

PG: I think the value of this chapter is to see Ben struggling to write of his emotions coupled to the intellectual aspects of his being. He has chosen to do action research, to validate himself as a person of worth. He selected an example of a situation in which he was questioned as a leader, and he wrote about it. Etty wrote him after reading his study on values. There was much with which she could identify, and she felt joy and hope. Ben also recounts his meeting with Larry, shortly before he died of cancer. Ben connected to the spiritual and focused on what we can give one another. He asked, "Who am I of value?"

Becoming More Fully Human As a Teacher

KT: Ben provides so many good ideas for teachers at any level to follow in respect to the way they interact with others. Learning can be enhanced if "I" can, when invited, provide respect, trust, and opportunities to do science through the exercise of autonomy and, when things do not work out as intended, to be on hand as a co-learner, co-teacher and co-researcher. Each of these roles is an essential part of "being" an effective teacher. Yet there is more. Ben refers to it as spirituality. I see it as connecting the learner to what is to be learned and, in Ben's way of saying things, appreciate all of your students not just because of what they know and can do, but because they are human.

PG: Ben wants to empower others as they learn to be authentic through action research. Ben quotes Elliott (1991), saying, "Spiritual development proceeds by resolving problems of living wisely" (p. 147). Ben feels that we learn by intentional critical reflection. Ben bravely tells his story, to give others courage to critically reflect on their own lives. I think with time, we can go deeper into our reflections, to learn more.

PT: If, as science educators, we take a narrow view of constructivist theory, then we tend to focus only on the traditional teaching goal of enabling students to construct scientific knowledge. Thus, our pedagogical sights are fixed firmly on deconstructing students' conceptual impediments (e.g., misconceptions) to their achievement of a scientific way of knowing. And any concern for students' co-constructive learning activity tends to be subsumed by this single-minded goal. Ben alerts us to a richer view of constructivist theory, one that compels science teachers to empower their students as more efficacious learners: as socially responsible, collaborative, and self-critical inquirers. Thus, our pedagogical responsibility extends from the traditional domain of knowledge construction to include the domain of construction of self as learner, a vital role also for scientists. And, of course, as science teachers, we are implicated in this concern for self-development.

I suggest that the ultimate goal of *being* an effective science teacher needs to be predicated on the initial goal of *becoming* more fully human, a goal that can be realized (in large part) by fostering the mutual development of both one's self and one's students. Thus, a fundamental concern for the science teacher (as co-learner) should be with the question of how to improve the quality of his educative relationship with students, a quality that is defined as much by what it means to be fully human as by a (traditional) concern for reproducing a scientific way of knowing.

Passionate Teaching, Passionate Learning

KT: Ben talks about his knowledge going through a series of stages before he can appropriate it as his own and use it to pursue his goals. He also talks about the connection between his knowledge and affective issues. To me this is an important wake-up call for science teachers. Students should be able to study science that has strong interest to them and the connections between values, beliefs, and interests should be highlighted. I call this linking science and passion as a way to enhance the quality of college science teaching.

PT: For too long science teachers have injected into their teaching the so-called scientific virtue of value-neutrality: a powerful cultural myth dating back to the Enlightenment, when faith and reason were officially separated. This myth enshrines the belief that scientific knowledge is revealed by an impersonal process of inquiry (e.g., theory-free pattern recognition in the data) that has been purged of human values and emotions: The data "speaks for itself." In the thrall of this myth, many a teacher of science has adopted a transmissionist classroom role and portrayed science as an indubitable "body" of knowledge capable of speaking for itself, a teaching role whose uninspiring dullness is due in no small way to the teacher's lack of personal research experience in generating scientific knowledge claims. Notwithstanding the dullness of much repetitive laboratory analysis, scientific research can be an exciting adventure, especially when significant social or environmental problems are addressed in local contexts.

As Kielborn and Gilmer (1999) illustrated, elementary school teachers undertaking scientific fieldwork for the first time can become so engrossed in their value-laden experiences that they passionately transform their own teaching practices. Perhaps this is an argument in favor of college science teachers being engaged in scientific research. In the absence of this compelling experience, it is difficult to imagine how they can appreciate the role of the values, beliefs, and interests of the scientists whose knowledge (theories, laws, principles) they teach about.

The Irony of Assessing Understanding

KT: Ben also makes the point that it is never possible to represent what is known. How then can assessment of students be undertaken with anything other than ironic validity? What are the implications for teaching science if it is not possible for students to write and speak all they know about science?

PT: The deeply problematic issue of knowing "objectively" another's knowledge should alert professors of science to the need to refer to science education for guidance when it comes to assessment of student learning. Ben reminds us that because the task of making clearly explicit our own understandings is fraught with unassailable difficulties, we should take care to treat cautiously our claims to understand others' understandings. When we give expression to our own understanding, whether it be in writing, orally, diagrammatically, or in dance, etc., the best we can achieve is a "re-presentation" via symbolic form(s) that then must be interpreted by others. Science teachers should be careful not to (over)interpret a student's ability to reproduce faithfully the teacher's symbolic expressions as unimpeachable evidence of the high quality of their scientific understanding.

If the best we can do is to make an interpretive judgment of a student's understanding, then we are obliged to seek guidance from educators who, themselves, practice the ancient art of hermeneutics. Challenging pedagogical questions then arise. What types of interpretive evidence are required? When best should evidence be obtained (before, during, after learning)? By whom (teacher, student, student peers)? What criteria are appropriate for judging the adequacy of claims to know a student's understanding? Can these criteria be used equally to assess the quality of a student's conceptual understanding, participation in a discourse of critical inquiry, and progress toward the goal of becoming a critically reflective learner? What level of uncertainty about our knowledge claims is acceptable, and does the answer depend on the purpose it serves and the contexts in which it is elicited?

BIBLIOGRAPHY

Cunningham, B. (1995). *Valuing the spiritual.* Paper presented at the CARN (Collaborative Action Research Network) Conference, Trent University, Nottingham, UK.

————. (1996, February). *Accounting for myself: Building towards a joyous anticipation.* Paper presented to Action Research Group, School of Education, University of Bath, UK.

Denzin, N. K. (1994). The art and politics of interpretation. In N. K. Denzin & Y. S. Lincoln (Eds.), *Handbook of qualitative research* (pp. 500–515). Thousand Oaks, CA: Sage Publications.

Denzin, N. K., & Lincoln, Y. S. (Eds.). (1994). *Handbook of qualitative research.* Thousand Oaks, CA: Sage Publications.

Eisner, E. (1993). Forms of understanding and the future of educational research. *Educational Researcher 22*, 5–11.

————. (1997). The promise and perils of alternative forms of data representation. *Educational Researcher 26*, 4–10.

Elliott, J. (1991). *Action research for educational change*. Milton Keynes, London: Open University Press.

Fromm, E. (1956). *The art of loving*. London,: HarperCollins Publishers.

Goleman, D. (1996). *Emotional intelligence: Why it can matter more than IQ*. London: Bloomsbury.

Gordon, C. (Ed.). (1980). *Power/knowledge: Selected interviews and other writings 1972–1977: Michel Foucault*. London: Harvester.

Kielborn, T. L., & Gilmer, P. G. (Eds.). (1999*). Meaningful science: Teachers doing inquiry + teaching science* [Monograph]. Tallahassee, FL: SouthEastern Regional Vision for Education.

Laidlaw, M. (1996). *How can I create my own living educational theory as I offer you an account of my educational development?* Unpublished Ph.D. thesis, School of Education, University of Bath, UK.

Lather, P. (1994). Fertile obsession: Validity after poststructuralism. In A. Gitlin (Ed.), *Power and method: Political activism and educational research* (pp. 40–42). New York: Routledge.

Macmurray, J. (1991). *The self as agent*. London: Faber and Faber.

McNiff, J. (1988). *Action research: Principles and practice*. London: Macmillan Education.

Padovano, A. T. (1995). *A retreat with Thomas Merton: Becoming who we are*. Cincinnati, OH: St. Anthony Messenger Press.

Polanyi, M. (1958). *Personal knowledge: Towards a post-critical philosophy*. Chicago: University of Chicago Press.

Seidman, S. (1991). The end of sociological theory: The postmodern hope. *Sociological Theory 9*, 131–146.

Stone, N. B., & Hart, Bro. P. (Eds.). (1985). *Thomas Merton: Love and living*. London: Harcourt Brace.

Whitehead, J. (1993). *The growth of educational knowledge: Creating your own living educational theories*. Collected Papers. Dorset, UK: Hyde Publications.

Whitehead, J., & Lomax, P. (1987). Action research and the politics of knowledge. *British Educational Research Journal 13*, 1175–189.

Winter, R. (1997, October). *Managers, spectators, and citizens: Where does theory come from in action research?* Keynote address to the 1997 CARN Conference, Watford, London.

CHAPTER 15

 # Spirituality in the Classroom

Margarita Cuervo

What is so special about teaching? What do teachers think about themselves? About what they do? About their students? And why talk about spirituality if, after all, this is a book about college and university teaching, where students are adults, know what they are doing, and are completely responsible for their lives? We higher education teachers just have to teach the content, right? Well...let's talk about what is really happening in our college classrooms. Let's talk about what we do. Let's talk about how things are and how things could be if we shifted our energies more toward spiritual matters.

A good teacher is an efficient, organized, accessible professional and active scholar who is, most of all, concerned about the students. The ideal professor relates to the total student, academically, emotionally, socially, and psychologically; establishes a mutual zone of comfort in and out of the classroom; and exudes enthusiasm, energy, empathy, and excellence. My goal is to be like that.

God gives us talents; He places us in situations in which we can use those talents. He expects us to use those talents in those situations. I know that God has created me to do some definite service. He has committed some work to me that He has not committed to another. I have a mission; I am a small link in a very long chain. Therefore...I shall do good. I shall be a person of peace and truth in my own place.

So what does the spirit have to do with college teaching? Everything! Do your students know you? I don't mean your name, title, publications, and professional accomplishments; I mean, do they know you as a real-life person with joy and happiness, trials and tribulations, successes and defeats? Do they think you exist outside of the classroom? Do they know your life stories, or at least, some of them? Do they know how you got to be there? Tell them, don't be afraid! By being vulnerable you become strong. There is a special spiritual strength in connecting and bonding. Let the real you come through. You will feel more accomplished in your work and will experience the inner peace of knowing that your job is making a difference in the lives of those around you.

Education helps to shape character and conduct. Our colleges and universities are expected to teach virtues such as hard work, responsibility, honesty, loyalty, respect for authority, respect for property, and so on. A lot has been written about the rights of professors, but not enough attention has been given to our corresponding responsibilities. The original "profession" was the religious (clerical) profession. So, before the year A.D. 1500 one made a profession by taking the religious vows of poverty, chastity, and obedience. The term "professor" grew directly out of this monastic tradition. A professor has something to profess, has to be dedicated to the preservation and development of knowledge. So we ask ourselves, What are the standards of conduct appropriate in classrooms, in grading students, in meetings, in conducting research? Professors are not used to many kinds of constraints, but morality imposes limitations. It is well known that we, the professors, have much power over our students. We do not like to be reminded of our potential for doing serious harm to our students; we can frustrate them, mislead them, or destroy their will to learn. Or, we can help greatly. Many teachers act conscientiously, prepare their classes with care, grade carefully, offer students extra help, and participate constructively in a variety of faculty activities. But, unfortunately, there are others who are not so positive, who are more concerned with their own advantage and well-being, without regard for the welfare of others. Students in our colleges and universities are entitled to presume that their teachers are dedicated conscientious professionals. As ethical professors we should take care never to betray that trust. Let's decide to be special, to be people committed to spiritual values.

I am responsible for what happens in my classroom. I know that I set the mood. I have a captive audience; therefore, I have a moral obligation to use my time and talents efficiently. Ultimately, if my students do well, they win, I win. If they don't, they lose, I lose. My students know I am on their side. Their success is my success; their happiness is my happiness.

Then and Now

I have been a mathematics college teacher for over twenty-five years. I think it all started way back when I was a little girl and we used to play school, and I was always the "teacher." My little sisters and my friends were the "students." I remember having imaginary friends so as to have more students in the class. Of all the subjects, mathematics was always my favorite. While most of my classmates dreaded it, somehow it was easy for me. Was it because my mother was a mathematics teacher? Was it the genes or the environment or both? I love mathematics. It is just so logical, so basic to all scientific thinking.

I have a good life now, but it wasn't always like that. It was very tough for many years. I am the oldest of eight children (six sisters and two brothers) and in 1961, together with our parents, we immigrated to the United States from communist Cuba. I was fifteen years old at the time, and my youngest sister was two. We arrived in Miami on a cold January day and after some bureaucratic formalities, in the middle of the school year, I was enrolled in the eleventh grade at the neighborhood high school. It was difficult, it was depressing, and it was overwhelming.

There were many difficulties, including financial problems. We had no money. We had to do without many basic things. At that time, a hot lunch at the public high school was thirty-five cents. I remember as if it were happening now, how I used to eat just a hot dog for fifteen cents at the school cafeteria, and save the rest of my lunch money. No French fries, no soda, just a hot dog and water. With my savings of twenty cents per day, at the end of the school week I would have $1.00 to spend on the weekend.

The next year, at the age of sixteen and a senior at high school, I got my very first job, and I have been working ever since. I became a cashier at a neighborhood theater at seventy-five cents an hour, where my mathematics skills came in very handy (there were no computers, and all the mathematics was done manually). I worked every day after school and on Saturdays and Sundays, so by then, I was making around $15 a week, and I had some money of my own. I spent and saved wisely, which made it possible for me to buy the graduation ring, the yearbook, and a dress for the senior prom.

Things were getting more and more difficult for my family. And then, thanks to the relocating efforts of some community agencies, we moved to Puerto Rico, where both my parents started all over again, a second life as working professionals as they originally had in been Cuba.

I had in my possession a high school diploma and $15 in cash. A new chapter had started.

There were still so many difficulties. But I knew that education was my passport to freedom, to a better life. I enrolled at the local community college with $50, money borrowed from a granduncle. He is dead now, but I am eternally grateful that he opened the door of higher education for me. I did not waste the opportunity. I studied hard, transferred the next year to the University of Puerto Rico on a merit scholarship, and, a few years later, graduated *magna cum laude*.

As they say, "nothing succeeds like success," and as with everything in life, one good thing leads to another. So I went on to graduate school at New York University on another merit scholarship. Life was still very difficult, and particularly my life as a student in New York City was very tough at all levels. But I knew the value of education; I persevered and succeeded. I graduated, earning a master's of science degree in mathematics. Right after graduation, I became a full-time mathematics college teacher in New York, at the age of twenty-five. Three years later, I moved to Miami, where I have been a full-time college mathematics professor at Miami-Dade Community College since 1974.

I can tell you that things are very different now. I am not talking about the technology, the changes in the curriculum, the shift to distance learning, and everything else. I am talking about the student body and how dramatically it has changed throughout the years, not necessarily for the better. Just to start someplace, when I graduated from high school in 1962, the three "big problems" were talking in class, chewing gum, and getting out of line. When my son graduated from high school in 1992, thirty years later, the three big problems were sex (pregnancy and AIDS), violence (including carrying guns to school), and drugs (including alcohol). Big difference! What I am saying is that we are teaching to a different population in a different place. Our students live in a more difficult and complex world. Most of the time their circumstances are overwhelming. Their lives are full of serious problems.

As I was recounting my school and college years, full of difficulties as they were, I have to admit that, for the most part, my world then was a safe place to live. Family and friends were always there, the community was protective and supportive. The environment was not so dangerous. I had fewer monsters to fight. But the struggles helped build my character and made me, I hope, a better person.

How is that related to what happens in my classroom now, so many years later? And what is the spirituality I am talking about? *Funk and Wagnalls Standard Desk Dictionary* defines spirit as "the vital essence or animating force in living organisms, especially man, often considered divine in origin; the part of a human being characterized by intelligence, personality, self-consciousness, and will; the substance or universal aspect of reality regarded as opposed to matter." Spirituality is defined as "pertaining to the spirit, as distinguished from matter; affecting the immaterial nature or soul of man; marked or characterized by the highest moral or intellectual qualities." Spirituality refers to the nonmaterial essence of people.

Therefore in what I say and do, purposely or not, I am affecting the souls of my students. I want to understand their needs and feel compassion for their sufferings. I want to do all the good I can. From my own teachers I learned the behaviors I wanted to copy and the ones I would never imitate. I believe that the principle of service is what separates special people from the rest. We are always making choices, and our selfish human nature will push us not necessarily in the right direction. It is a matter of awareness and personal decision, but it takes effort to choose right.

The first challenge of my daily life as a teacher is finding a parking place, but I am usually very lucky in that respect. Somehow, I always find a parking spot at the last minute, just when I need it. You can laugh if you want, but I will tell you that I have an angel in charge of parking. So, my first feeling of the day is of thankfulness. I say, "God, thanks, I owe you." So I enter the classroom in a grateful mood and with a spiritual debt. An important problem was solved for me, so it is my turn to do for others.

If you are an extrovert like me, you are always talking and sharing your personal experiences with others. As I said before, I enter the classroom feeling good because I found a parking place and made it on time. And it goes from there. But what if some bad things happened to me or to people I know the day before? Well, I just talk about them. And don't argue that there is no time, there is plenty of time for everything. You just have to be efficient and integrate the whole teaching-learning process. Throughout the semester, a process of discovery is taking place. Not only about the content and academic knowledge, but also about ourselves. It is in this self-revelation, self-exposure, that the teacher becomes a real human being in the eyes of the students. When I am happy or sad, celebrating life, or carrying a big emotional burden, I talk about it and become vulnerable. Does that make me weak? I don't think so. I have experienced so many wonderful moments, full of testimonials and confessions from my students. I have learned so much from their lives, from their experiences. Imagine what I would be missing if there were this "distance" between us that would not allow this emotional interaction. I do not have an imaginary sign that says DO NOT ENTER. My invisible sign says, WELCOME, DOOR IS OPEN, JUST COME IN.

Faith, Hope, Love

For the last six years, as part of my regular load, I have been teaching two bilingual (English and Spanish) mathematics courses every semester, with excellent results. This delivery method is very much appreciated by many students who are not native English speakers. I remember grading algebra tests for one of my bilingual classes, and being very upset and frustrated with the terrible results. After all, I had explained the material thoroughly, I had reviewed the content, I had practiced a sample test with them, but all to no avail. So, when I started returning the tests, their faces were unhappy, and just a general feeling of doom was permeating the room. And then I started thinking, "What's happening? What's wrong with them? What's wrong with me? What's wrong with the world?" And three words came to my mind. I wrote on the blackboard "FAITH, HOPE, LOVE." I remember saying something like, "This is not really part of the syllabus, but we are going to talk about spiritual ideas, without which life, maybe, is just too difficult and has not too much meaning." After a while, their eyes were wide open, and they had hopeful expressions on their faces.

I said that, first of all, we need faith in God; that I respected whatever their religious beliefs were, but that I sincerely believed that God is our creator, and He loves us. All blessings come from him, and our lives need to be rooted in having faith in someone bigger than ourselves. Then I went on to talk about having faith in ourselves, faith in our relatives and friends, faith in our teachers, faith in the institution, faith in the "system." I told them that I find living without faith and trust so difficult. And a very interesting discussion opened up. The expression on their faces started to change.

Well, then I went on to talk about hope. It is sad, but I think many of us live in a state of despair, of hopelessness. There is also the problem of this generation, which is "instant gratification." Most of them are not used to waiting for anything. We know that our society gives us everything we want *now* whether we are ready or not, whether we can afford it or not, whether we deserve it or not. So to have hope is against the grain of our materialistic lives. We don't like to wait, and hope requires patience, trust, and faith. Maybe the results are not there yet, but we can believe that things can turn around, that things will work out, that there is light at the end of the tunnel.

Then I talked about love. We started discussing the different meanings that we give to the word *love*, but how unfortunate it is that many times we just don't love ourselves at all, or not enough. It is common to hear students say, "I hate math," "I hate school." How damaging those expressions are to their emotions! So, I continued to talk about love, and how it could dramatically change their lives as students.

I have to tell you the story of a beautiful girl who was in that class but was failing. She had a very pretty face, perfect skin, perfect features, delicate carriage, but such a sad expression! I remember approaching her a couple of times and asking what was wrong. Of course, the answer was always, "Nothing, I am fine." Then one day, after class, she came up to see me and started crying. She was saying, "Yes, Dr. Cuervo, you are right, I am very sad, I have many problems and I am in a state of despair, that is why I am doing so badly in this class." She went on to tell me that she had a very small baby, with no father. The baby had fallen out of the crib two times because her emotionally sick mother, who was baby-sitting so that she could go to school, could not do things well. She kept talking about her hopeless life, her dull job as a store cashier, and how she just did not know what to do. I almost started crying myself. This was an opportunity to reach out and maybe make a difference. In all sincerity I told her that an F in my class was not really important because that was no crime or reason to feel shame. We talked about priorities, about being a young single mother to a newborn baby, that maybe what she needed to do was to take care of her baby first, that school could wait. What was important was that a connection, a bond, an "I care about you," happened.

We don't want to be guilty of arrogance, rigidity, insensitivity, dullness, and hypocrisy. Let's make an effort to practice, on a daily basis, right in our classrooms, the virtues of humility, flexibility, sensitivity, spontaneity, and sincerity. I keep asking myself, Which qualities do students use to describe me? I hope they are from the virtues list.

Lessons for Life

After reading the book *The Measure of Our Success*, by Marian Wright Edelman, I got the idea of using her 25 Lessons for Life (written advice she was giving her three sons who were going to college) to inspire and motivate my classes. I edited her Lessons for Life into "20 Lessons" and use them as an introduction to every class meeting. These are the lesson titles, as I rewrote them:

Lesson 1	There is no free lunch.
Lesson 2	Set goals.
Lesson 3	Assign yourself.
Lesson 4	Never work just for money or for power.
Lesson 5	Don't be afraid of hard work.
Lesson 6	Don't be afraid of taking risks or of being criticized.
Lesson 7	Don't ever stop learning and improving your mind.
Lesson 8	Take parenting and family life seriously.
Lesson 9	Your wife is not your mother or your maid, but your partner and friend.
Lesson 10	Choose your friends carefully.
Lesson 11	Slow down and live.
Lesson 12	You are in charge of your own attitude.
Lesson 13	Be a can-do, will-try person.
Lesson 14	Be honest.
Lesson 15	Be reliable.
Lesson 16	Be faithful.
Lesson 17	Never give up.
Lesson 18	Finish what you start.
Lesson 19	Believe that you can make a difference.
Lesson 20	Always remember that you are not alone.

What I do in my classes is write on a corner of the blackboard the Lesson for the Day and discuss it for a few minutes. We talk about it and relate it to our mathematics class. For example, the first day we discuss the first lesson, "There is no free lunch." I ask students to raise their hands if they want to get an A in the class. (Can you believe that not all do?) So we start a philosophical discussion about how there is a price to pay for everything. We just cannot expect to get something for nothing, we should not feel entitled to what we have not worked and struggled for. The amazing thing is that many of our students are unfamiliar with that concept. Many believe that somewhere there is a free lunch. I tell them, it might be free for you, but somebody else is paying. We are adults, we should pay our own. We continue talking about efforts and results, doing what it takes to win at what we are doing. From the very first day I am teaching in a way that can help change a life.

When we discuss the lesson, "Assign yourself," some important concepts emerge. We usually don't like other people telling us what to do. But teachers are always telling other people what to do: "read this, write that, do this assignment for next week," etc. Why is it important that I assign myself whatever work is needed? Because that is one of the secrets of success. A doer, a self-starter plans his own strategy for winning. I tell my students they don't have to wait for me or anybody else to tell them what to do and when. The idea is to assign yourself whatever extra work you know you need to achieve your results.

When we talk about the lesson, "Believe you can make a difference," lots of interesting discussions take place. On a bumper sticker I read DON'T MAKE A BUCK, MAKE A DIFFERENCE. And when you think about it, unfortunately that's what all of us are mostly concerned with, making money. Nothing wrong with that, but we need to be reminded of our potential and our obligation to make a positive contribution to our community. The financial rewards will follow. Students certainly need the encouragement. They all have special family and work situations where they can make a difference. This is connected to another lesson, "Never work just for money or for power," because that is not going to make anybody happy. The key bad word there is *just*. We know that many people in positions of power abuse it and are corrupted. There is moral corruption in low and high places. And it is not okay to do the wrong thing, even if everybody is doing it. So we talk about what it means to have a job with authority and power, and how responsibly we would have to act at all times. During these discussions, students are talking about their present and future lives. They are getting a college degree to move up in the world; all these ideas are very relevant to their lives.

At the beginning, students are very skeptical; after all, that's not what mathematics teachers do, maybe philosophy or humanities teachers...but a mathematics class? After a couple of weeks, the first thing they usually ask is, "What is the lesson for the day?" I assign students to write the list of lessons and, at the end of the semester, select the three that have helped them the most. Then, they discuss them in class with other students, and talk about valuable personal experiences. Many wonderful things always happen as a result of this activity. Students tell me how they have grown emotionally and spiritually, how differently they feel about themselves, how differently they deal with some controversial issues, and how they see the world in which they live with new eyes.

Community Spirit

Another thing that I do sometimes is to ask students to undertake some kind of community service, as little as five hours per semester. They may do it individually or in small groups. Of course, I do my hours too. This type of activity always generates a lot of interesting discussions about life and people. And it always amazes me the wonderful things that these ordinary students think of to help other people. The condition is that they do not get paid for whatever they are doing, and that they do not get anything from me either. They earn "brownie points," though.

I remember one semester when there was a severe blood shortage in our city and the Red Cross was having an aggressive campaign asking for volunteer blood donors. I had just returned a test, and it occurred to me to offer five points on the next test to anyone who would donate blood. There was some silence and, after a while, a young man raised his hand and said, "Make it ten points!" and I said, "You got a deal!" Well, a very interesting thing started to happen. The next day, two or three students came up to me and showed me a paper proving that they had donated blood. After a few more days, more students did the same, but then the bloodmobile was gone from campus. Then, more wanted to know of other places where they could donate blood. They were confessing to me that they went to do it just for the extra points but, in the process, they had lost some of the physical fear and had begun to appreciate the sense of well-being they experienced knowing that they were doing something good for their community.

"Being" a Teacher

The essence of our lives is to be happy, productive individuals. We can make the world around us a better place for everybody. "I like what I do, and I do what I like" is a statement that expresses a life of harmony and peace.

In one of my travels, I found an ancient parchment, of unknown author, with very wise words and inspirational thoughts. I have it hanging on a wall in my office, and this is what it says:

> If you love your job more and more as time goes by...
> If your punishments are fruit of your love and not vengeance...
> If you try to renovate yourself in every class...
> If you can follow a method without becoming its slave...
> If you can learn besides teach...
> If you can study again what you already know...
> If you know how to teach, but even better, how to educate...
> If your students want to be like you...
> ...then you are a TEACHER

In closing this chapter, I think about Mother Teresa's words to the sisters of her order, Missionaries of Charity: "Don't let anybody that comes to you leave empty-handed."

The same thing applies to us, in and out of the classroom. Have a friendly expression on your face, and students and colleagues will approach you with all kinds of requests. Be there, be genuine, be generous, and say "Yes." Be a giving person, don't let anybody go empty-handed.

Margarita Cuervo
Department of Mathematics
Miami-Dade Community College–Kendall Campus, Miami, Florida, USA

EDITORS

METALOGUE

Connecting with Students' Lifeworlds

KT: I do think it is a good idea for the students to know their teacher as a person. It helps them to put what they have to learn into perspective. Too often science and the knowing of science are seen as activities separate from teaching and I think this leads to many of the problems to which Kate alluded in her chapter (Chapter 5).

PT: That is why I think it is important for science professors to include in their teaching a focus on their own research activity. I recall only one interesting science professor from three years of undergraduate physics lectures. He taught us about his research on the sun's solar wind and how it impacted on the earth's magnetosphere. That guy talked with great enthusiasm that revealed his love of physics research. His teaching brought physics to life for me. Even if the professor is a teacher rather than a researcher, it is still important for him to communicate a sense of personal enthusiasm about his discipline and why *he* believes that it is relevant to his own *lifeworld*, the world outside of college and the lifeworlds of his students.

KT: Margarita is much like Ros (Chapter 8) who teaches science in a community college. She alludes to the different problems she now faces with students of today compared to when she began to teach college mathematics. Teachers need to know about the lifeworlds of students, what they know and can do, and how to use their *primary discourses* to make sense of mathematics. As students enter the classroom they do not leave their problems at the doorstep. These problems constrain learning and must be considered by a teacher as the curriculum is enacted.

PT: By getting to know their students, college professors are better able to make connections between their discipline and the contexts of students' lifeworlds. By understanding better how their discipline can contribute to the world outside of college—relevance to other professions, usefulness for everyday life, pertinence to a community's environmental problems—the professor can situate learning in contexts of relevance and value to students' current and future lives.

PG: One of Margarita's great strengths is that she connects to the students, to their real lives, to their struggles and joys. When speaking with Margarita, it is obvious that her students are very important to her. Her students know that Margarita cares whether they learn, about what is happening in their lives, and how they feel about themselves and about mathematics. She shares with her students what helped her find her way as a Spanish-speaking immigrant, arriving in the United States from Cuba as a teenager. Margarita had to learn not only English and our culture, but also the language of mathematics. Her students are also native Spanish speakers, so they can identify with her struggles. They see Margarita as one who has excelled, one who has become a professor of mathematics. Her students want to succeed too, so Margarita's caring and guidance are meaningful.

KT: Margarita is able to teach in Spanish and English, thereby making it possible for her students to use their native language to access all of what they knew as children, prior to learning to speak English. By teaching in both languages she makes it possible for them to connect what they know in their different languages to their emerging understandings of mathematics. The native language of a learner is perhaps the greatest form of capital that a learner can have and it only makes sense for teachers to make sure that students can use all of their language resources to learn mathematics.

PT: Very much like Dr. Mary Buenos (Chapter 1), Margarita creates a rich linguistic environment in her classroom that optimizes students' opportunities to become co-participants in a learning community. As Noelle explains in Chapter 3, students' own *language register* is a powerful resource for them to use in negotiating mathematical meaning with the teacher and among themselves. Margarita's ethos of care for her students means that learning is very much a discursive activity that occurs inside her classroom.

Spirituality, Not Religiosity

Critic: It is true that we need to transform our community via cultural production, but I think that the ideas in this chapter are located too far away from our community, and too much in communities that are almost the antithesis of science and mathematics education.

KT: Earlier this week one of my student teachers was reprimanded by the school principal for pulling out a Bible and speaking to his mathematics students about God and advising them to act more like God's children. There was concern that his teaching and learning should separate his religious beliefs from the mathematics he was to teach. The student teacher did not agree with the ruling but decided to comply because he was teaching in a public school. I have seen Margarita interact with her students, and there is no doubt that to know Margarita is to recognize that she does not separate her spirituality from her teaching of mathematics, eating of dinner, or being in the world. In his chapter (Chapter 14), Ben made the case for professional actions that have integrity and authenticity. In Margarita's case that would involve expressing her spirituality in ways that are not offensive to her students.

PT: Indeed there are many who would be offended by a teacher's blatant religiosity, whether it be expressed in earnest evangelical zeal or in muted tones. But should we equate religiosity with spirituality? Might we be in danger of throwing out the baby with the bathwater? This is a complex issue in much the same way as the issue of research ethics discussed earlier, and it is likely to evoke a greater emotional response, especially among those who advocate secularism in public education. Nevertheless, it would be a great pity to reject outright any consideration of the value of spirituality as a referent for teaching and learning. In saying this, I am aware of the centuries-old divide between faith and reason, a distinction that underscores a popular definition of science as being concerned only with that which is empirically verifiable and not with the metaphysical. Putting aside the inadequacy of this as a definition of theoretical science (especially the "New Physics"), we have seen in many chapters of this book that the cold logic of science (or mathematics) is not an adequate logic of teaching. We have been presented with compelling evidence that if science teachers are to decenter from transmitting the logical facts of their discipline and take into account the humanness of their students then they are in need of other ways of thinking, knowing, and valuing. Although I'm not really sure what Margarita means when she refers to her "spirituality," I find it fascinating that it seems to provide her (and many others throughout the world) with the passion, energy, and selfless commitment to care deeply about the nonmaterial well-being of others.

I wonder what other referent one might invoke to create a similar driving force in one's teaching? I find that, in different ways, Ben Cunningham (Chapter 14) and Penny Gilmer (Chapter 17) seem to be talking about the importance of spirituality in their professional lives. Could it be that we are reaching a watershed in Western society where the pursuit of material well-being and individual freedom is in need of being counterbalanced by a greater sense of meaningfulness in our professional and personal lives? And perhaps, for many, that comes from a stronger connectedness with others in family or local community or workplace. Could it be that *spirituality* is a temporary label for this need?

PG: I do know what Peter Taylor means by increasing the meaningfulness in our professional and personal lives. When one can bring the forces together in one's life, they are more than additive, they can enhance each other in a multiplicative sense. I sense this meaningfulness in Margarita's life and teaching as well. Margarita's students sense this meaningfulness too, in how she tries to help them overcome their (perceived and real) barriers to excelling in their lives, in learning mathematics, and in moving toward realizing their dreams.

A Legitimate Story for Science Education?

Critic: The audience for this book will *not* find the chapter interesting or even the kind of story that they admit as an allowable story. If the audience had been a church club, a Bible group, a group helping academic survivors (e.g., Academics Anonymous), the story would certainly have had a considerable resonance. But even from my understanding of the stories told among another anonymous group, Alcoholics Anonymous, the stories they tell have recognizable components both in their semantic and syntactic makeup. That is, both the form and the content of an AA story has to conform to the community norm. As part of the process of coming off an addiction, members learn to tell stories in the particular way of the community. This chapter violates the norms of the science education community and the community likely to read this book on college science teaching both in its form and content.

PT: These are strong and perhaps not surprising claims. I disagree with them on two counts. First, I do believe that many who read this book will be attracted to the range of provocative ideas that challenge existing orthodoxies. But these ideas are not simply provocative; they also are well grounded in theories of epistemology and actual practices of teaching and learning. There is much of relevance to those who are interested in interpretive research methodologies, especially at *the leading edge*. The range is broad, including the neotraditional empirical approach of Guba and Lincoln (1989) and recent narrative methods of writing as inquiry and critical reflexivity (Richardson, 1994; Steier, 1994). Margarita's chapter is located in the field of narrative inquiry and sits best with the genre of *autobiography*, especially *evocative autoethnography* (Ellis, 1997). The self-descriptive nature of the tale will be of interest to researchers studying teachers' *personal practical knowledge* (Clandinin & Connelly, 1994). Little of this research has thus far penetrated the conservative field of science education.

And, second, for reasons discussed above, I believe that a consideration of alternative referents associated with the *nonrational* (but not irrational) are of increasing interest to educators. The provocative referents of dreams and spirituality are both well justified in empirical and theoretical terms in the final two sections of the book, which aim to establish new trajectories for the near future of undergraduate science teaching in order to serve better the interests of prospective teachers of science as well as future scientists.

PG: These nonrational referents are part of our human experience, so to ignore them would be a loss. It is difficult within our rational culture to bring together elements such as spirituality and mathematics (this chapter), or dreams and computer education (Chapter 16), or spirituality and action research in education (Chapter 14), or dreams and chemistry (Chapter 17). However, these nonrational aspects of our lives can be powerful referents for us.

I shared our (developing) table of contents for this book, while Peter Taylor, Ken Tobin,and I edited it, to science educators and others around the world. Some felt threatened by the titles of chapters on spirituality and/or dreams, but many were fascinated with these ideas of spirituality and dreams as referents in teaching. I feel too, like Peter Taylor, that we may be approaching the watershed of change from a dominating rationality in science teaching to a melding of the nonrational with the rational. This can be dynamic, as at the triple point, in which all states interconvert in rapid equilibrium (Chapter 17). We can touch our students' lives, goals, and struggles, by expanding the vision of ourselves from mere rational beings to dynamic individuals who not only love to think about science and mathematics but who also savor our humanness and our rich inner lives.

BIBLIOGRAPHY

Clandinin, D. J., & Connelly, F. M. (1994). Personal experience methods. In N. K. Denzin & Y. S. Lincoln (Eds.), *Handbook of qualitative research* (pp. 413–427). Thousand Oaks, CA: Sage Publications.

Ellis, C. (1997). Evocative autoethnography: Writing emotionally about our lives. In W. G. Tierney & Y. S. Lincoln (Eds.), *Representation and the text: Re-framing the narrative voice* (pp. 115–139). Albany: State University of New York Press.

Guba, E. G., & Lincoln, Y. S. (1989). *Fourth generation evaluation.* Newbury Park, CA: Sage Publications.

Richardson, L. (1994). Writing: A method of inquiry. In N. K. Denzin & Y. S. Lincoln (Eds.), *Handbook of qualitative research* (pp. 516–529). Thousand Oaks, CA: Sage Publications.

Steier, F. (1994). From universing to conversing: An ecological constructionist approach to learning and multiple description. In L. P. Steffe & J. Gale (Eds.), *Constructivism in education* (pp. 67–84). Hillsdale, NJ: Lawrence Erlbaum Associates.

CHAPTER 16

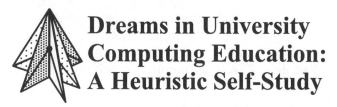

Dreams in University Computing Education: A Heuristic Self-Study

Mark Campbell Williams

Was I to concede and begin to formulate my questions with clarity, or was I to stumble on? In a sense there was no choice, since my uneasiness about seeing the world in tight, empirical, rigid and academic terms was reaching acute proportions.
—O'Connor, *Understanding the Mid-Life Crisis*

As with my research described in this chapter, Peter O'Connor uses empirical research material from dreams. Like O'Connor, I find it very difficult to escape the feeling of "being silly." As he mentions in a later book, "The mainstream prefers predictable, conforming men who not only play the rules but never question them, usually as an outcome of what has euphemistically been called education" (O'Connor, 1993, p. 65). Yet I find myself increasingly respecting this type of research material (Williams, 1995). And, as with O'Connor, I find that, as I share my research and listen to comments from friends and colleagues, I obtain confirmation and support.

Overview of My Research

Over a period of five years, I conducted an interpretive study of the relationship between *technicism* and *open discourse* in my university undergraduate course entitled Business Computing (Williams, 1996a). I understood technicism as an overemphasis on technical, instrumental, or strategic techniques or actions, to the detriment of wider human communication, human values, or human purpose (Habermas, 1972). Persons thinking in technicist ways tend to define human problems in terms of rational and technical solutions, thus leading to an undue emphasis on the science, instrumental rationality, and technology of the modern age (Habermas, 1972). I coined the term *open discourse* to represent a process of communication concerning personal and social meaning, purpose, orientation, values, goals, concepts, ideas, feelings, and emotions relevant to, but distinct from, the narrow subject matter, techniques, and administration of the curriculum (Williams, 1995; Taylor & Williams, 1993).

During the early part of the study, I introduced a number of teaching-learning strategies in an endeavor to promote open discourse in my Business Computing classes. These strategies included requiring students to keep personal learning journals and to participate in personal introductions in early classes, "check-in" sessions at the beginning of classes, and a "buddy system" of working in pairs on the computers. My initial research focus was on the interplay of open discourse and technicism in the outer world of human relationships, technology, and the subject matter of the course.

In the later part of the study, I engaged in a *heuristic self-study* (Moustakas, 1990) of my own teaching-learning-research practice. This activity led me to a critical awareness of ethical shortcomings stemming from my imposition on students of learning activities such as writing about their personal values and feelings and my collection of research data without due written permission. I came to recognize an *inner technicism* operating within both my research and teaching. I realized that I had used unwittingly what Habermas (1972) terms *strategic action* to confirm my implicit hypothesis "that open discourse could counterbalance technicism" in my Business Computing classes.

During the heuristic self-study, I strove to manage my own subjectivity through self-reflective examination of my inner subjective drives. I had weekly talks with the University chaplain, the Reverend George Trippe, who has had over thirty years of experience in Jungian dream interpretation and psychological *inner work*. This contact, combined with my existential interest in Jung's (1964) notion of *individuation*, led me to explore unconscious ways of knowing in the research process. I surmised that an *inner open discourse*, a form of what Habermas (1972) terms *communicative rationality*, might help me to address my inner technicism. Thus, I reflected on, analyzed, and interpreted a number of my dreams that had relevance to the research.

It should be noted that I did not discuss my dreams in lectures or computer laboratory tutorials, even if they included symbols or images of technology. It was only through my conscious mind reflecting, analyzing, interpreting, and extending dreams that they became part of the open discourse of the research. In this sense, my use of dream material was in line with what Moustakas (1990) terms the *indwelling* aspect of heuristic inquiry.

Heuristic Inquiry, Dreams, and Self-Dialogue

Moustakas (1990) states that in heuristic inquiry, "I may become entranced by visions, images and dreams that connect me to my quest. I may come into contact with new regions of myself" (p. 11). Recognizing that intellect, emotion, and spirit all have their place and a unity in heuristic investigation, he asserts that, in the heuristic process, self-understanding and self-growth proceed together as the phenomenon under investigation becomes intimately understood.

Moustakas emphasizes the dialectic involved as the researcher, in *self-dialogue*, moves back and forth from individually subjective understandings to general understandings in bringing forth insights. In the endeavor to understand intimately the experience or phenomenon under inquiry, self-dialogue is a process of speaking to oneself in an open but recordable manner. Moustakas provides an example of self-dialogue by quoting from his own self-dialogue, between Varani as "John" and Varani as "John Paul," in his heuristic study of the psychological dimensions of mystery.

Although Moustakas describes the importance of tacit knowing and intuition, he does not mention the Jungian idea of the unconscious. However, by emphasizing terms such as *subliminal* to describe Polyani's notion of tacit knowing, he does not exclude the notion of the unconscious. Indeed, he asserts that in the indwelling process the heuristic researcher seeks whatever possible avenues, from narrative to dance, that will bring to light subjectively valid insights into the nature of the phenomenon or experience under investigation. In my case, I sought to address my inner technicism through my inner discourse of Jungian-based dream work.

The Unconscious and Science Education

The unconscious aspect of research is only thinly referred to in the science education literature, even though it was two and half millennia ago that Socrates founded our rational tradition with the compelling injunction "know thyself." While conducting my heuristic research, I came to understand that knowing myself means not only finding out what I think in a narrow conscious sense. It includes also knowing my emotions, feelings, dreams, hunches, intuitions, loves and hatreds, passions and delights, taboos, and the myths and metaphors that I live by.

Von Wodke (1993) urges computing users to keep a dream journal and use Robert Johnson's (1986) Jungian dream interpretation methods as a spur for creativity. As part of human balance and development, von Wodke also advises the use of meditation, visualization reveries, mental and physical relaxation procedures, and the writing of daily journals. In this work, I found support for exploring unconscious wellsprings of knowledge in computing education.

Krathwohl (1994) advises educational researchers to "use your unconscious." I consider it unfortunate that Krathwohl seems to be advocating a strategically rational use of the unconscious in the sense that one can use the unconscious instrumentally without respecting its processes. The beginning point for respecting the processes of the unconscious is to acknowledge, as does Erickson (1986) in his work on qualitative research, that one of the insights of Freudian psychology is that "people knew much more than they were able to say" (p. 123). This seems like common sense, but the fascinating question for me as a researcher is just how to access that unconscious knowledge.

Dieckman (1993) reports using psychoanalysis in educational research in a recent ethnographic study. A form of psychoanalysis was used "as an initial procedural guard to check for researcher bias and raise the awareness of the researcher before entering the field of study" (1993, pp. 49–50). Robert Early (1992) actually collects and interprets students' dreams as an optional part of the curriculum in his college mathematics classes.

Although Clandinin and Connelly (1994) describe a dream scenario, they do not interpret the dream but use it as part of a story, because "we believe it is important for those who study personal experience to be open to a rich and sometimes seemingly endless range of possible events and stories" (p. 417). Although my use of dreams is different from their usage, their central point about being open to personal experience is crucial to what follows.

The Unconscious and Dreaming

Although we can choose not to remember, we all dream—several times a night (Sanford, 1989; Segaller & Berger, 1989). Jung (1964) understood that many people discard dreams as irrelevant or even dangerous to consider. Others agree with the last sentence in the section on "Dreaming" in the *Oxford Companion to the Mind*, "Despite the symbolism and fascinating condensation of ideas to be found in dreams, there is no evidence that a more useful understanding of personality can be gained from them than can be divined from the realities of waking behaviour" (Oswald, 1987, p. 203). A senior psychiatric professor from a prestigious university states later in the work that, although lacking grammar or structure or any real communicative function, dreaming "merits attention as a modest manifestation of human creativity" (Zangwill, 1987, p. 275). Zangwill rather "damns by faint praise," but it is interesting to note that, even with this perspective, dream interpretation would seem to merit some attention in science education research.

For Jung, the character of most dreams reveals vital inner forces or psychological energies of the personality in the interface between consciousness and the unconscious (Sanford, 1989). Jung accepted, but moved beyond, Freud's somewhat negative perspective to understand a creative and communicative function in unconscious communication. This is especially so in the way the unconscious uses dreams to communicate with the conscious mind. Different from Freud's problem-oriented dream work approach, Jung saw a creative function within the unconscious. While Freud saw the unconscious as a region of the mind where consciousness dumped unwanted impressions and memories, Jung viewed the unconscious as an autonomous second psychic center with its own mysterious authority, capable of its own processes and communication with consciousness (Monick, 1987). His focus was less on the psychological maturation problem (which is also encountered in dreams), being more on the development of mature individuals, with their aims, aspirations, and racial/cultural historical contexts giving potential meaning and dignity. Ultimately, Jung understood dreams as part of the interplay between psychological energies in the conscious and the unconscious, as part the human journey of finding a meaningful place in a purposeful universe.

In Jung's approach of development toward wholeness, *individuation* results from an engagement, a listening to the unconscious, and an appropriate modification of our conscious point of view (Moffett,

1986). Finding meaning and purpose in the creative inner workings of the psyche, within human community, individuation is "the urge or instinct towards wholeness. This is a universal quality of human beings and continues to push or pull us towards psychological growth" (O'Connor, 1993, p. 47).

There is no lazy way to achieving balanced psychological development. Diligence and attentiveness are required to keep on the path (von Franz, 1968). Partly influenced by von Wodke's (1993) recommendation, I used Robert Johnson's (1986) Jungian-based *inner work* process to value and to listen to the unconscious. This process involved my inviting the unconscious to communicate through dreams, making a written record of any significant dreams, interpreting the dream symbols through making associations and understanding the dynamics of the message, conducting rituals based on the interpretation, and continuing the dream process by using *active imagination* in writing out a continuing dialogue with the dream characters.

In this chapter, I describe several dreams with my interpretations and my dream inner work. I treat the dreams as messages or elucidations of various aspects of my self, which were relevant to the research. Briefly stated, at certain points in the research process, the unconscious was encouraging me to continue in certain directions or was warning me that I was being driven by an inner technicism, a strategic rationality, rather than acting out of balance and wholeness. By listening to these dreams as part of the voice of the unconscious in the inner open discourse of the research, I believe that I was able to respond in my teaching, learning, and researching, by acting and thinking in ways more congruent with an ethical sense of virtue and wholeness. One way of warranting my belief is by describing autobiographically the way in which I responded to the dreams.

When describing my dreams, the first person "I" represents my *ego*. According to Chetwynd (1974), the "I" in a dream usually represents the ego, that aspect of the psyche that coordinates the energies of the personality. He described the ego as the psychological energy for assessing reality and coordinating and communicating with deeper energies of the psyche to allow the individual to live and work appropriately. When I am describing the content of the dream, I use the first person "I" to represent my ego, because that is who I am in the dream. However, when a person lives and works and relates to others, it is not only the ego but other aspects of the psyche that are present in the person's personality.

Chetwynd (1974) contends that a common theme in dreams is the concern that the ego is too frenetic, overstating certain approaches and not listening or allowing other deeper aspects of the psyche to flourish and contribute to a balanced psychological wholeness. It is often the case that many people have a weak or unbalanced ego that tries to dominate the psyche or is, itself, dominated by only certain inner energies at the expense of the other energies. However, when a person lives from the Self, the ego is only the background coordinating energy of the psyche that allows the deeper, more powerful energies of the psyche to emerge in a balanced manner in the person's life and work.

For example, in my last dream, my ego is depicted as a listening and circumspect academic at a colloquium at which other, more adept, academics are contributing. When I describe the content of the dream, I mean my ego when I say that "I am listening at a colloquium." However, in commenting on the dream, I say later that "I reflected on this dream"; it is my whole personality, hopefully my inner energies (depicted as separate characters in dreams) working together in harmony as the true Self, that is now the I. While I am not asserting that I indeed live out of the true Self at all times, that is my aim, and I write from this hope.

Chetwynd (1974) thus asserts that a healthy psyche would allow every aspect of the personality to thrive in its own area without getting out of hand and dominating the other inner energies. In a similar vein, O'Connor (1993) asserts that, particularly for male academics, there is a conflict between the logical rational intellect aspect and the feeling imaginative aspect of the psyche (often represented as the anima) that facilitates a sense of connection.

In my research, I came to understand my need for psychological growth toward maturity through a demonstration of balance. One way of understanding this balance may be through an appropriate blend of technical, practical, and emancipatory interests, or an appropriate blend of instrumental and communicative action without strategic action. Strategic action and technicism could be understood as an unbalanced immaturity. I will describe how this theme of balanced psychological development is apparent in the following series of dreams. Before I relate these dreams, however, I need to discuss some methodological and ethical considerations.

Methodological Considerations

Moustakas (1990) asserts that in heuristic research, the essence of validity pivots around a question of meaning. In presenting my dreams and my inner work based around those dreams, there were no others experiencing these phenomena, so mine is the sole meaning. Hence, I can only refer to my own experience as I reflect on Moustakas's statement, "Does the ultimate depiction of the experience derived from one's own rigorous, exhaustive self-searching, and from the explications of others present comprehensively, vividly, and accurately the meanings and essences of the experience?" (p. 32). In this work, I depict two major experiences: (1) that of my inner technicism, a type of strategic rationality; and (2) a form of inner open discourse in the Jungian inner discourse that helped me in my struggle to address my inner technicism.

The main challenges of including dream texts, dream interpretation, and active imagination dialogues based on the dream material is that the research data can be experienced and recorded by only one person. For example, one of my dreams (Williams, 1996b, pp. 49–50), which I entitled "Academia," revealed a dramatic scene in which some of my inner energies were represented as a group of academics talking around a table, with the final scene one of abusive verbal opposition. Why would the unconscious have given such a warning? I wrote down the dream, reflected on it, and interpreted the meaning. Using active imagination as part of the process of inner work, I wrote out a continuation of the dialogue of the dream as if writing the script of a play. I discussed my interpretation and inner work with Trippe, the interfaith chaplain at the university who is an accredited Jungian analyst, who led me to new insights and a deeper understanding.

As the above example implies, a form of rigor can be demonstrated if a properly documented and systematically recorded dream journal is maintained. In the data analysis, I found it helpful to involve a person experienced in Jungian dream interpretation and active imagination.

I found that Richardson's (1994) idea of "crystallization" was more appropriate for my heuristic research than the notion of triangulation. Using her metaphor of a crystal, my research writing is a postmodernist mixed-genre text that reflects and refracts ever-changing pictures and images of my central themes. I did not triangulate rigorously to prove objective truths about technicism and open discourse.

Further research may be able to supply other frameworks for providing warrants for heuristic inner-discourse knowledge claims. I am thinking especially of what Guba and Lincoln (1989) term *catalytic* and *ontological authenticity.*

Ethical Considerations

In recounting in this chapter my dreams and inner work, I do not deal primarily at the level of the personal unconscious. I do not deal with deeply personal issues of my own emotional, psychological, and spiritual development. Although I see this self-evaluation as important for the research, there are limits and boundaries to be respected. I skim over the intense inner work I have done, even with the reported dreams, to understand my feelings and emotions revealed in the dreams.

In a similar vein, it should be noted that I have *mythologized* certain elements of some dreams by providing a degree of anonymity in order to safeguard myself and others. In Trippe's view (personal conversations, September 21, 1994, and July 1, 1996), mythologizing is writing the dream and inner work in such a manner as to make it more accessible to a wider group of people, without getting distracted by an individual psychology. The meaning of the dream can become clearer when expressed on a more general or cultural level. For example, if I were to have a dream about my departmental head, mythologizing may result in the meaning of the dream becoming more accessible or general if I were to write about the vice chancellor of Cambridge University.

Mythologizing also provides a means of providing anonymity. Other methods, such as introducing fictive elements (Oakley, 1982) or fictive characters, or reducing a dream to the bare minimum of content, can detract from the power of the message of the dream. In mythologizing a dream, no reader will be able to say which elements of my dreams, in detail, have actually been experienced. This ensures that any academic debate including the content of my dreams can only be tentative and circumspect. I have chosen this method as the best way to safeguard and protect my Self. Boundaries have to be set. In writing this report, I am willing to expose only some of my Self to educational debate. Following Clandinin's (1993) advice, it would be unwise and naive, perhaps immoral, to be utterly self-disclosing and to remove ambiguity in this situation.

My practice of mythologizing my dream material may be helpful to educational debate in that it may encourage communicative action. In educational debate, persons can use instrumental action from the technical interest to control, dominate, or manipulate others. Individuals can thus gain a type of strategic advantage. This tendency can be alleviated if elements of the research can be said to have been edited for a general audience. The debate may then concentrate more on sharing insights, on bringing out meanings from one's own perspective. The debate would thus be more oriented to reaching intersubjective understanding and mutual agreement, rather than scoring an academic point or focusing on a specific interpretation of a previous experience.

Projection

I think it is important to note the psychological tendency to project one's inner energies onto significant others (Sanford, 1989). It has been useful for my interpersonal relations in the research to be aware of projections, and thus guard against the ever-present danger of treating others based on my projections onto them. In other words, I strive to listen and communicate openly and honestly with people without assuming in them qualities that I project onto them. My assumptions of their motives and communication could easily become entangled with my own projections.

In my experience, unconscious projection can be a major challenge in research teams, as in any human group. The more someone is unaware of his or her inner life, the more the probability of this unconscious projection (Sanford, 1989). For example, while working closely in a research team I could easily project my inner archetypal soul-woman *anima* onto a woman and be liable to a romantic affair. Or I could project my archetypal inner "wise old master" onto a senior academic and expect of him far greater wisdom than may be the case. The worst projection (that seems to me to happen far too often in acrimonious academic debate) is that one's inner archetypal *shadow* is projected onto another colleague so that one tends to overemphasize and confront the colleague due to a perceiving of one's own negative and underdeveloped aspects in that person (Moffett, 1986).

However, even with attention and foreknowledge, such projection is inherent in the immense potential of full human communication and contact (even in a research project). It is interesting to speculate that, if managed and held in balanced creative tension, such projections can be channeled toward wholeness for a research group. If positive interpersonal skills and open discourse were to be encouraged, then the negative possibilities could be mitigated, or resolved, or even transformed to positive results.

The Voice of the Unconscious in Inner Open Discourse

O'Connor (1993) relates a dream with reference to academics:

> The dreamer, having tied the boat up, discovers on the wharf a sack containing a number of decapitated male heads. Heads that have been severed, disconnected from their bodies and of course their hearts. This is such a clear image of so many over-intellectualized men, all head, no body, no substance, no pulsating feeling heart. Academics tend toward this beheaded image; all in their heads, often without substance or feeling. What better image could one have of an over-intellectual man, nothing more than a head! (p. 61)

If a researcher were to have a similar dream, could it not be that the researcher's own psyche is giving a warning that the research is ignoring other ways of knowing—the emotional, for example, or the somatic, or the intuitive, or the mystical? The crucial point is whether or not the researcher has an in-depth understanding of human beings that acknowledges the importance of the affective domain, of the body, of the senses and of the unconscious. If not, then the researcher may indeed be in danger of unbalanced and unhealthy one-sided psychological development. This may be in line with the dream image of decapitated heads. With that scenario, it is little wonder that the voice of the unconscious is often silenced.

Dreams can be experienced as a kind of inner open discourse in which the unconscious communicates with the conscious mind as a partner in the process of individuation. As Sanford (1989) puts it, "We can describe this relationship between the conscious and unconscious parts of the psyche as a dialogue, or discussion (*auseinandersetzung*, in German), a word that implies taking apart, clarifying, as well as confrontation" (p. 126).

In one of my dreams, I noted that some of my inner energies were depicted as a group of academics talking around a table, with the final scene one of abusive verbal opposition. I took the message of this inner open discourse as a warning that I needed balance and correction in the manner in which I was teaching and conducting my research. The inner open discourse of dreams can act as warnings and compensations for neglecting certain aspects of ourselves in our everyday activity, but also can give motivation.

Due to my inner discourse, I came to realize that during the ethically flawed initial stage of my research, my ego was not able to coordinate a proper balance between the rational intellectual side and the relational, creative aspects of my psyche. As demonstrated in the first dream, I needed to become more systematic and steady. The second and third dreams indicated that I was being dominated by a talented and quick thinking, but frenetically driven, inner energy that actually appeared in the image of a slave-driver. I was not working out of the balanced Self, that is, out of a balanced wholeness in which the ego steps back to communicate with and coordinate all the deeper energies of my psyche to flourish in their own domains and combine together to form a united personality. These dreams led me to do a great deal of self-management, inner work, and reflective writing to facilitate the development of a balanced Self. My final dream indicated that I was some way on the path to a balanced wholeness.

A Series of Dreams

The first dream, which I entitled "Music," occurred when I had just finished a two-year period of gathering research texts while lecturing full-time, on top of my personal life with a young family. I was exhausted and certainly did not feel equal to the task of writing up my doctoral research. I was insecure, being uncertain about my rather unconventional writing style, my novel approach to qualitative research, and my unusual approach of including the voice of the unconscious in the research. I was in need of motivation and purpose and direction.

First Dream: "Music"

I was to compose a piece of music for the University Choral Society. Although I did not feel adequate to the task, I attempted to begin by humming a refrain from Beethoven's Fifth Symphony. Carl Haus, the wise old master of music, portentously wandered by and commented that I should watch my pitch. I was somewhat unnerved, but then saw a vision of a princess on a throne amidst blue hills and green dales, and heard a sweeping melodic musical composition.

In reflecting from a Jungian dream-work approach on the Music dream, I understood that the music I was to compose was a symbol of my doctoral thesis (I had just started writing up in earnest). My ego was represented as the "I" in the dream. By composing for the University Choral Society I took it that my ego was to act as a coordinator of my deeper academic energies, who would "do the singing." In the image of Carl Haus, the wise old music master, I detected elements of the archetypal inner wise old man, from the collective unconscious. By his commenting that I should "watch musical pitch," I understood the unconscious as advising me to remember fundamental research and writing skills, being systematic and careful.

My feeling inadequate to the task of composing represents my feelings of inadequacy in writing the thesis. At this emotional low point, my psyche dug deep down into the racial or instinctual and cultural stock of symbols in the collective unconscious to give me a picture of a princess on a throne amid beautiful scenery and evocative music. She symbolized the archetypal *anima* or inner "soul woman" energy that acted to inspire me. This inspiration was emphasized with the addition of the sweeping melody. While my humming may be seen to represent my initial attempts at writing the thesis, the final melody may be seen as a glimpse of the finished article.

Following Jung's insights and in the process of remembering, recording, and interpreting this dream, I understood that my conscious mind was in dialogue with unconscious aspects of my psyche. I detected symbolic figures from the collective unconscious participating in the inner open discourse. I understood that the dream inner work process gave me a graphic awareness of a crucial stage in the research. The warning was for me to be rationally systematic and careful while, at the same time, being open to the imaginal and relationally oriented part of my psyche. In terms of my research approach, I began to take very seriously Kathryn Ahern's concerns (personal conversation, March 21, 1994) about the ethical and procedural shortcomings of my recently completed qualitative research writing.

That acknowledged, I noted also the message that I should not feel embarrassed or insecure by my heuristic research approach and postmodernist writing style but should continue in this unconventional manner (i.e., "keep humming"). I would be flooded and sustained by a deep inner strength and motivation that would be akin to a powerfully evocative symphony or a sweeping landscape of journey and delight. The unconscious reinforced this message not only by the lovely landscape image and surging music, but also by picturing the archetypal symbol of the golden-haired princess as an inner soul-woman anima energy. This archetypal anima aspect of my psyche can be understood, from a Jungian perspective, as a potentially guiding and motivating inner energy. On the path of individuation, the anima energy can inspire and guide the psyche to balanced wholeness of the personality in life and work.

After reflecting on this dream for a considerable time, I determined eventually to continue my research with boldness. I decided to continue to investigate postmodernist writing styles and ways to include my use of the inner open discourse centered around my dreams as part of my doctoral research thesis. I decided to continue with the idea of writing the research as an autobiographical narrative and not to be afraid to use a rhetorical style. In my Business Computing classes, inspired by my readings in curriculum reform (von Wodke, 1993), I decided to continue introducing in mass lectures and computing laboratory groups new strategies such as meditation, rites of beginning and rites of ending, and guided visualisation. In the subsequent research data collection and analysis, I began to use the more extreme aspects of Colaizzi's (1978) method of *phenomenological reflection* by being more than moderately subjective.

However, I had not heeded the warning about being systematic in my research approach, nor had I taken the ethical questions of care and rigor with enough seriousness. I next had a dream that I entitled "Quality Assurance," which warned me that perhaps I had been overly bold in my research.

Second Dream: "Quality Assurance"

I was at a teaching quality assurance meeting at a university with Jane (supportive and interested associate dean), Michael (a senior supportive academic), Andrew (marketing business wheeler-dealer), Anthony (young enthusiastic co-researcher in creative approaches in computing), and Arnold (old-fashioned expository teacher type). Arnold walked around expounding the virtues of traditional expository lecturing. I brusquely intruded by saying that approach is poverty-stricken. Arnold swore vehemently. I was taken aback but asked him to go on so that I could listen. He talked for a while. I drifted off and become confused but when he was finished I tried to repeat the main points of his approach.

He angrily stated that I was "all wrong" and then sat down indignantly. My supportive co-researcher, Jill, said, "What's happened to you, Mark?"

The setting of this dream was in the rationally intellectual atmosphere of the traditional university, but this setting was balanced positively by the presence of the wise woman anima energy (Jane, the associate dean). The meeting around the table for "quality assurance," and the ego asking Arnold to talk more, held promise of progress toward a balanced wholeness, with all the energies of my psyche flourishing in their own area and communicating together coordinated by the ego around the round table (an archetypal symbol of the Self). Represented are the four (archetypal number of the true Self) masculine energies together with the feminine anima energies and the ego. Arnold represented the heroic intellectual energy; Anthony, the emotional youth prince energy; Michael, the intuitive magician or priest energy; Andrew the sensate businesslike energy; Jane and Jill, the anima energy; and "I" was the ego.

The trouble that the unconscious was bringing up to consciousness was that "I" as ego was being confrontational and was polarizing my inner energies. Instead of listening to and appreciating the steady old-fashioned intellectual part of me, I rejected his potential contribution with a judgmental remark. Even when this energy spoke again I was too busy or confused to listen.

Part of the message from the dream may have been a warning from my unconscious about being careful and attentive to interpersonal university department politics in my outer life. (Indeed, the next week I was careful in departmental politics, to the betterment of the department and myself.) However, I took the prime message to be a call for me to do some inner work, to allow myself time and space to listen to all the energies that make up my psyche, especially the logical rational intellectual energy. My ego was being dominated by the wildly creative aspect of my character.

As a result of working with this dream, I toned down my more adventurous teaching reforms. By thinking seriously about questions of ethics and rigor, I finally discontinued my analysis of the early qualitative research data in which I had been concerned with proving that open discourse could balance technicism in university Business Computing classes. Instead, I began in earnest a more systematic and careful heuristic research approach based on Moustakas (1990), rather than on the more extreme aspects of the phenomenological reflection approach commended by Collaizi (1978).

Some time later, again exhausted from working rather too furiously on my doctoral thesis and lecturing full-time, I received the following dream that I entitled "The Slave-Driver."

Third Dream: "The Slave Driver"

In the dream, an Egyptian-looking fellow, with a short beard, was slave-in-charge of affairs on a property (represents that slave-driver part of me, forcing me to overwork frenetically on writing my doctoral thesis). He typed far into the night, writing a massive project of research into the names of a type of Biblical idea of the human race (representing my thesis). His wife (representing an inner soul-woman anima) reminded him that he missed his promised appointment to be with the young prince who owned the property (represents the Self).

I understood the dream as giving me a graphic picture from the unconscious of that part of my psyche that was forcing me to overwork on my doctoral thesis, a veritable slave-driver. Due to my working too hard and too long, this slave-driver part of me missed out on meeting with the Self. Due to reflection on this dream, I took a rest from working on my thesis, to relate more fully to the significant people in my life and give more attention to my students and colleagues in my university faculty. I took time to review my life's direction, reflecting whether the activities I was doing were balanced and focused in achieving the important commitments, responsibilities, needs, and desires I had in my life and work. I revised my self-management and interpersonal skills development in areas such as time management, relaxation and stress management, exercise, diet, communication, listening, assertiveness, negotiation skills, mental and physical ergonomics, and workspace management.

I spent a considerable amount of time working through books such as Paul Timm's *Successful Self Management* (1987), Stephen Covey's *The Seven Habits of Highly Successful People* (1989), and Bob Montgomery's *You and Stress* (1984). In general, the authors encourage comprehensive self-management and self-motivational skills such as values clarification, goal setting, self-contracting and self-rewarding, as well as brainstorming and consensus synergistic group activities, visualisation and imaging and inner awareness.

Related more to computing usage, I followed Mark von Wodke's (1993) advice to use regular ergonomically designed exercises and relaxation techniques, personal creative diaries, inspiration from dream and meditative reverie images, and wide experimentation with multimedia, to revitalize vision and energy in business computing usage. When I returned to my research, I endeavored consciously to balance the imaginal and relational aspects of heuristic research and postmodernist writing styles with the more systematic and rigorous approaches to research and writing. About six months later, perhaps as a result of my endeavors, I received a dream that I entitled "Academia." The previous few days I had felt confident and balanced. After getting my doctoral supervisor's comments, I had done some productive work and was feeling excited about my doctoral thesis.

Fourth Dream: "Academia"

I was late to an academic seminar colloquium, but I was able to catch on to what had been happening. Auguste Comte (representing an inner masculine academic energy—mathematically talented but hyperactive and frenetic) was speaking with Aristotle (another inner masculine energy—a balanced academic researcher and manager, mature and talented). They were speaking about a mathematical problem called the "segment of the circle" (the circle as a symbol of potential balance and wholeness, the true Self). Plato (another inner masculine energy—intuitively talented and brilliant academic, but sometimes too extreme, wild, or esoteric) talked about some wild ideas. We all went for a walk, during which Aristotle confided to me that Plato was rather wayward, even though he had unexpected value for the team.

I understood the dream as giving me a graphic picture, from the unconscious, depicting several inner energies that were working within me as I worked on my research. There are signs that, from a psychological perspective, I was working well, and making progress toward a balanced wholeness (note the reference to the "segment of circle"). I took it as a good sign that my ego, the "I" in the dream, was not fighting my other inner energies, but communicated with them as a team. Note that the ego was talking with and listening to the balanced academic leader energy in me, Aristotle. Aristotle was wary of Plato, a symbol for my wildly intuitive academic inner energy, but acknowledged his potential value.

I understood the general message of the dream as an indication that I was proceeding in a fruitful manner in my thesis, and also toward general psychological health. As a form of dream-based inner discourse, using the method explained by Robert Johnson (1986), I imagined myself back in the dream. I imagined that I waited for Plato and Comte, and we went into the seminar room together. In my inner discourse, I asked the four of us to sit around the central circular table. I put a proposition to all that we needed to work together to finish the doctoral thesis and that Aristotle should be in charge. I thanked Plato and Comte for their crucial contributions. We all agreed to make Aristotle the leader.

In reflecting on my dream inner work imagination, I noticed that the round table and the fact that there were four of us in number were two further symbols to reinforce the idea of a balanced wholeness. My ego was acting in a healthy way, not acting out of strategic rationality, nor being aggressive or overly dominated by any one of my inner energies. My ego acknowledged the value of my inner balanced academic leader energy and, through communicative action, suggested that the other energies acknowledge this as well. I saw here an example of a balanced rationality producing both instrumental and communicative action without the error of strategic action. I took this dream inner work as an indicator that I had made some progress toward addressing my inner technicism through the process of my inner discourse.

In this process, I had begun to address my problem that the ego had not been communicating with other inner energies, and my being driven by the inner technicism of a frenetic slave-driver that had no contact with the Self. Rather, this last dream of Academia was an indicator that the Self was beginning to emerge in my life and work, with the result that communicative and instrumental action were appropriately balanced, without strategic action.

Conclusion

In this chapter, I report on a dialectic interplay between some outer-oriented aspects and subjectively inner-oriented aspects of technicism and open discourse in the broad context of teaching, learning, and researching in university computing education.

After reflecting on some methodological and ethical considerations, based mainly on Moustakas's (1990) notion of validity in heuristic research, I describe the content of and interpret the relevance of four dreams that (I thought) were relevant to my research. In so doing, I touch on questions of motivation and academic rigor, and the self-development and interpersonal skills I employed to address my inner technicism, which was driving me into error. I make no empirically warranted assertions about the ways, if any, that my psychological reflection changed my teaching, learning, or researching. However, I did give some personal anecdotal comments about the way in which psychological reflection, evident in my dream inner work, influenced my conduct in teaching and research and in dealings with academic colleagues.

My overarching theme is to chart some progress toward a kind of psychological wholeness or balance. In this process of the interplay of conscious and unconscious knowledge, I became aware of a needed balance and correction in my early research—the more creative, intuitive, and relational side of my nature was in conflict with the more rigorously intellectual analytical side. The result was my adoption of a psychologically oriented heuristic research approach in which I endeavored to balance subjective intuition with systematic care.

I do not claim rigorous empirical warrant for my suggestions, comments, or descriptions to do with my inner work self-study. I take it that in my last dream, about an academic colloquium, the unconscious gave me a picture that I was progressing toward a kind of wholeness or balance. How could one possibly warrant such an assertion? What do I mean by "a kind of wholeness or balance"? How, other than my existential *Eureka* experience, could I warrant that my interpretations of my dreams are correct? I leave these, and the host of other such questions, for further research.

As a result of my heuristic reflection based around multimodal thinking, I did change my teaching practice in my university Business Computing courses. I now endeavor to provide rich learning environments in our classes by continuing the initiatives I had incorporated in 1992 and 1993, but in a more relaxed and gentle manner. I also incorporate the stretching and mental relaxation exercises suggested by von Wodke (1993) into the computer laboratories. Von Wodke asserts that ergonomic, relaxation, and imagination activities, even to the point of including Jungian dream interpretation and journal writing, encourage creativity and mental and physical health while using computers.

For example, as the computers are turned on, I encourage the students to sit straight, close their eyes and breathe slowly and deeply while silently repeating "calm body, alert mind" while breathing in, and letting go of tension while breathing out. As the applications are loading from the menu I ask them to join with me in stretching across to the left and then to the right. I stress the value of ergonomic approaches and the resting of eyes and body. In my postgraduate seminars, and even at times in the undergraduate mass lectures, I guide the students in brief mental relaxations. At times, I include visualizations such as imagining a splendid flower blossoming on the computer screen or imagining a rural walk with sun and grass and flowers and breeze. Robert Flood (verbal answer to a question at the keynote address of the Australian Systems Conference, 26 September 1995) refers to these sorts of practices as physical and mental ergonomics.

I emphasize that my work is largely tentative. I have described the inner journey that I undertook as a consequence of my heuristic analysis of my early flawed qualitative research. This inner journey could be understood as similar to what Habermas (1972) describes as the process of educative *self-formation* (or *Bildung*, in German). In this process of self-building, Habermas contends that subjective self-reflection on unconsciously produced constraints, which can be encouraged by psychoanalysis, brings to consciousness blockages to self-development. The next step is *rational reconstruction*, which revolves around more objective reflections on the conditions of how one knows and acts to assist critical social action in the world.

Jung (1964, 1977) contends that the dialectic between the outer and inner worlds, the conscious and the unconscious, although of vital importance in almost every area of human pursuit, is often neglected, usually at a cost. I explored this dialectic in the context of reflecting heuristically on a teaching reform in my university Business Computing course. I think that the conscious–unconscious dialectic is present in any research and is pervasive in academia, but largely ignored, usually at a cost. For some researchers the cost may be high. What price is the Self—or the soul? For science in general, some would contend that lack of reflectivity, including lack of awareness of the unconscious, lies near the heart of some failures, perhaps *the* failure, of the scientific project of modernity.

Mark Campbell Williams
Department of Management Information Systems
Edith Cowan University, Perth, Western Australia, Australia

EDITORS

METALOGUE

Connecting with Students via Open Discourse

KT: I see Mark's chapter as being significant in many respects. The issue of *open discourse* is appealing as an instructional tool. I have used this approach frequently this year in my new position at the University of Pennsylvania. I came to the city of Philadelphia to address the problems of teaching and learning in urban settings in the United States. I have a social commitment to making a difference through my professional activities in an Ivy League university that looks out onto an area of the city that is in many respects run-down and poor. It seems as if all that I know about teaching and learning is out of alignment with what I see in these schools. So when my prospective teachers come to me for advice on how to teach in urban settings, I find I am frequently unable to connect with their concerns.

After more than twenty-five years in universities I find myself having to build a community in which we develop a process of communication concerning personal and social meaning, purpose, orientation, values, goals, concepts, ideas, feelings, and emotions relevant to, but distinct from, the narrow subject matter, techniques, and administration of the curriculum. The community has to extend beyond the walls of my classroom as I make a sincere effort to connect my knowledge to the circumstances in which they are to practice. Teaching has never been such a struggle for me, and my students might make much the same claim. But the open discourse makes the learning worthwhile. We have shared goals. As co-learners we have shared pain too. In some regards this might be analogous to what Hal referred to as "managed frustration."

PT: I, too, aim to establish open discourse in my teaching, especially my Web-based teaching of a Curriculum course for science teachers taking master's degree studies in the distance education mode. The epistemology of my course is shaped largely by the metaphor of *curriculum as personal experience*, a metaphor that places a major focus on the science teachers as reflective and collaborative learners. Our website provides an electronic Discussion Room for teachers to engage one another in semistructured discourses about curriculum issues. A pedagogical challenge for me (as it was for Mark) is to promote an open discourse in which science teachers "speak" richly (albeit asynchronously in written text) about the contexts of their own professional practices, and act dialogically by inviting others to provide alternative viewpoints on issues that they experience as authentically problematic. In this way, I hope that the course provides a highly relevant focus on science teachers' professional development needs and makes use of the range of practical expertise that they bring with them to the course.

Open discourse allows also the course tutors to speak from the heart about their own professional values, rather than feigning a value-neutrality. As you can imagine, it is not always easy to establish open discourse among science teachers who have been conditioned to speak/write as a learner in an objective voice that is dispassionate, analytical, and passive. They need explicit instruction about the nature and legitimacy of using metaphor and narrative to speak about their educational history and current learning experience. I am very keen to see professors of college science encourage open discourse in their own classes as part of an endeavor to reconstruct the image of science (in part) as a discursive activity in which learners (scientists, teachers, students) wonder out loud about their inquiries of the natural world and their processes of making sense of it.

(Re)Connecting with Self

PG: Mark's chapter was a critical one for me. This was because I had read a portion of an early draft of Mark's doctoral thesis when Peter Taylor spoke in Ken Tobin's 1995 Evaluation class. This was the first course I had taken since receiving my doctorate in biochemistry from the University of California, Berkeley, more than twenty years previously. Hearing how Mark used his dreams to help him get in touch with his teaching triggered my powerful triple point dream, which I discuss in Chapter 17. I was one of Mark's outside referees for his thesis, and had the challenge and joy of getting inside his head as he analyzed his dreams in an effort to understand and influence his role as a teacher.

KT: I have no skepticism at all. Mark makes a strong argument for reconstructing dreams and interpreting them in a process of getting to know your own "self." I see the reconstruction of dreams and subsequent analyses as being similar to *storytelling* and subsequent *reflection* on the stories. My reason for saying this is that the dreams come to us as *images* to which we assign language, and these stories then become objects for reflection and associated analysis. Since this is not a part of my professional practice, I find it difficult to recommend to others that this be done. I can recall dreams and I can do the analyses that are suggested. Perhaps with practice I will learn some things about myself that I do not presently know. However, for the time being I see the main value of the activity as *building concrete referents* on which one can reflect about self and others. This is a psychological or "between the ears" analysis and may underplay the social and cultural self—which can be understood in different ways.

PT: Like you, Ken, I have had very little personal experience with the practice of dream analysis, although I now tend to reflect on the significance of my more powerful dreams (the ones I can recall). So why am I enthusiastic about the professional use by Mark and Penny of dream analysis? First, because it reminds us that creative thinking is not just a rational activity involving propositional deductive logic. As Penny illustrates (Chapter 17), dreams connect us with our *unconscious*, the place where inspiration, intuition, and imagination are located, and can inspire our creative use of powerful images and metaphors. In science classes, an account of the dream-inspired discoveries of famous scientists (e.g., Kekule's benzene ring) might inject a much-needed sense of the importance to science of the creative process of discovery and the central role of the scientist's subjectivity.

And, second, dream analysis reminds us of the possibility of tapping into the (Jungian) *collective unconscious*, the place deep within us that is populated by cultural archetypes that direct us to enact in our waking state taken-as-natural cultural myths. As Mark illustrates, dream analysis can be a powerful tool for conducting an *inner discourse* that can raise our critical self-awareness of deep-seated beliefs and values. For critical reflective learners seeking an enhanced sense of personal agency, dream analysis might enable us to understand the extent to which we have been colonized by cultural myths (e.g., rationalism, objectivism, empiricism, masculinism) the predominance of which restrains severely our understanding of the dynamic, contingent, and constructed nature of science.

Connecting with What Is to Be Learned

KT: The tendency to disconnect what is to be learned from the learner is a fatal flaw in my mind. Mark's chapter reminds me of the need to connect our emotions and inner self to what it is to be learned. Learning is not only an objective value-free process that leads to objectified truths about the world. To the extent possible in science we might show what is to be learned in terms of its values for us as learners and teachers and to the sociopolitical milieu in which we are learning. Knowing oneself as a part of a sociocultural milieu is just a part of connecting that knowledge to what we know and can do in science. Rather than coming to know science as objectified truths that are value-free, it is possible to imagine at least a form of science that is linked to the knowers and the social and cultural contexts in which they live their lives.

PG: I too think there is much we can learn from Mark's chapter, concerning who we are, what our experiences have been, and how they influence our learning. The sociocultural milieu in which we reside influences our very being and the constructions we make. Getting in better touch with who we are, whether through dreams, as Mark did, or by reflection, empowers us as teachers and learners.

PT: On the issue of the connection between the learner and what is learned, was it not Richard Feynman, the Nobel Prize–winning physicist, who said that scientists do not reflect directly on nature; rather they reflect on their understandings of nature? And Ernst von Glasersfeld makes a similar point in radical constructivism when he argues that we make sense of the world by recursively abstracting patterns from our experiences of the world. And feminist theory adds a congruent perspective in promoting a *connected* way of knowing; one that involves an empathic relationship with the object of inquiry, whether it be self or non-self (animate or inanimate). What comes through strongly in Mark's and Penny's chapters is that a strong connectedness with one's self can generate a deep understanding of one's relationship with the world, in both the social and physical senses. And surely science teaching is concerned with the connection between both the social world (of the learner) and the physical world?!

Can you imagine a *connected science classroom environment* in which students are encouraged to engage in a way of knowing that involves becoming emotionally well prepared for communing with nature and weighing their learning experiences in (inter)subjective (aesthetic, ethical, emotional) terms? How did it *feel* to be involved in that inquiry? Which of y/our lifeworld values was enhanced/challenged by the learning experience? What did you learn about your self and about others? Questions such as these are certainly not meant to stand alone but to complement the usual objective questions that are addressed currently in science classes. Once we acknowledge and give full credit to the embodied and communal nature of knowing scientifically we will be able to enrich immeasurably the learning experiences of students of science and, thus, the practice of science itself.

BIBLIOGRAPHY

Chetwynd, T. (1974). *Dictionary for dreamers*. London: Paladin Books.

Clandinin, D. (1993, November). Personal experience methods in research on teaching. Paper presented at the 1993 International Conference on Interpretive Research in Science Education, National Normal University of Taiwan, Taiwan.

Clandinin, D., & Connelly, F. (1994). Personal experience methods. In N. Denzin & Y. Lincoln (Eds.), *Handbook of qualitative research* (413–427). Newbury Park, CA: Sage Publications.

Colaizzi, P. F. (1978). Psychological research as the phenomenologist view it. In R. S. Valle & M. King (Eds.), *Existential phenomenologist alternatives for psychology* (112–137). New York: Oxford University Press.

Covey, S. (1989). *The seven habits of highly successful people*. New York: Simon and Schuster.

Dieckman, E. (1993). A procedural check for researcher bias in an ethnographic report. *Research in Education 50*, 14–21.

Early, R. E. (1992). The alchemy of mathematical experience: A psychoanalysis of student writings. *For the Learning of Mathematics 12*, 15–20.

Erickson, F. (1986). Qualitative methods in research on teaching. In M. C.Wittrock (Ed.), *Handbook of research on teaching* (3rd ed.) (119–161). New York: Macmillan.

Glasersfeld, E. von. (1995). *Radical constructivism: A way of knowing and learning*. London: The Falmer Press.

Gleick, J. (1992). *Richard Feynman and modern physics*. London: Abacus.

Gregory, R. L. (Ed.). (1987). *The Oxford companion to the mind*. Oxford: Oxford University Press.

Guba, E. G., & Lincoln, Y. S. (1989). *Fourth generation evaluation*. Newbury Park, CA: Sage Publications.

Habermas, J. (1972). *Knowledge and human interests*. London: Heinemann.

Johnson, R. A. (1986). *Inner work: Using dreams and active imagination for personal growth*. New York: HarperCollins.

Jung, C. G. (1964). Approaching the unconscious. In C. J. Jung (Ed.), *Man and his symbols* (1–95). London: Aldus Books.

———. (1977). *Memories, dreams, reflections*. New York: Random House.

Krathwohl, D. (1994). A slice of advice. *Educational Researcher 23*, 29–32.

Moffett, P. (1986). *The applications of analytical psychology to educational thought*. Unpublished master's thesis, Murdoch University, Perth, Australia.

Monick, E. (1987). *Phallos: Sacred image of the masculine*. Toronto, Canada: Inner City Books.

Montgomery, R. (1984). *You and stress*. Melbourne, Australia: Lothian.

Moustakas, C. (1990). *Heuristic research: Design, methodology and applications*. Newbury Park, CA: Sage Publications.

Oakley, A. (1982). *Taking it like a woman*. London: Flamingo.

O'Connor, P. (1993). *The inner man: Men, myth and dreams*. Sydney: Pan MacMillan.

———. (1981). *Understanding the mid-life crisis*. Melbourne, Australia: Sun Books.

Oswald, I. (1987). Dreaming. In R. L. Gregory (Ed.), *The Oxford companion to the mind* (201–203). Oxford: Oxford University Press.

Polyani, M. & Prosch, H. (1975). Meaning. Chicago: University of Chicago Press.

Richardson, L. (1994). Writing: A method of inquiry. In N. Denzin & Y. Lincoln (Eds.), *Handbook of qualitative research* (138–157). Newbury Park, CA: Sage Publications.

Sanford, J. A. (1989). *Dreams: God's forgotten language*. New York: Harper & Row.

Segaller, S., & Berger, M. (1989). *Jung: The wisdom of the dream*. London: Weidenfeld & Nicholson.

Taylor P., & Williams, M. Campbell. (1993). Discourse towards a balanced rationality. In J. Malone & P. C. Taylor (Eds.), *Constructivist interpretations of teaching and learning mathematics* (135–148) Curtin University of Technology, Perth.

Timm, P. R. (1987). *Successful self-management*. Los Alton, CA: Crisp.

Williams, M. Campbell. (1995, April). *Including the unconscious in qualitative research*. Paper presented at the National Association for Research in Science Teaching Conference, San Francisco.

———. (1996a). *A self-study of teaching reform in a university business computing course*. Unpublished doctoral thesis, Curtin University of Technology, Perth, Australia.

———. (1996b). *Personal dream journal*. Unpublished manuscript, Perth, Australia.

Von Franz, M. I. (1968). The process of individuation. In C. J. Jung (Ed.), *Man and his symbols* (157–254). London: Aldus Books.

Von Wodke, M. (1993). *Mind over media: Creative thinking skills for electronic media*. New York: McGraw-Hill.

Zangwill, O. L. (1987). Freud on dreams. In R. L. Gregory (Ed.), *The Oxford companion to the mind* (274–275). Oxford: Oxford University Press.

SECTION IV

 ...Being Realized

CHAPTER 17

Opalescence at the Triple Point: Teaching, Research, and Service

Penny J. Gilmer

Typically, there are three domains of responsibility for higher education faculty: research, teaching, and community service. The proportion of time that one spends in each domain varies with one's institution and one's own interests and goals. In my experience in a college of arts and sciences, faculty tend to work on only one domain of responsibility at a time: teaching, research, or community service.

Sometimes, however, university and college faculty work at interfaces between domains. When we are teaching an undergraduate or graduate student how to do research we can be simultaneously teaching and conducting research. Also when we teach, we may envision ourselves as doing service for the community, either at a local level or for the greater community of our profession. When we conduct research, we contribute to knowledge in our profession, so we can also view ourselves as serving at the research-service interface with a wider community of scholars.

For many of my twenty-four years of university teaching, I have worked within these three domains of responsibility, and sometimes at the interface between two of them, generally those of research and teaching. My research has included ethics in science, biochemistry, and science education. My teaching has ranged from Honors General Chemistry (for first-year university students), General Biochemistry (at graduate and undergraduate levels), Physical Science for Elementary School Teachers (for practicing teachers at the graduate level), and Science, Technology & Society (graduate and undergraduate levels). I have "served" in administrative positions and as a leader on national, regional, and local committees. I have served my community also when interacting with schoolchildren while presenting a chemistry show called *Off to See Ms. Wizard*, in which I share my love for chemistry.

I was driven to conduct this study because I had seen many middle and some elementary school teachers who were part of the Science for Early Adolescence Teachers (Science FEAT) program (Spiegel, Collins, & Gilmer, 1995) change their teaching dramatically as a result of action research in their own classrooms. I had not taken this brave step to examine critically my own teaching as they had. I was still constrained by the dominant culture within science, to teach using a "stand and deliver" lecture format in my chemistry classrooms. I was not getting the students as actively involved in learning as I might have. I tried some of the new ideas from cognitive psychology and sociocultural linguistics on how people learn. Part of what held me back from changing was the fear that were I to change my teaching, my students might not be prepared to go on to the next chemistry class and do well. If it didn't work, would my chemistry colleagues deride me for trying some of the new ideas from education? If it did work, would my chemistry colleagues feel threatened by my successes?

Using Symbols in Dreams as Metaphors

It was not until I had my "triple point" dream six years ago that I started to think differently about my academic life. On the evening before the dream, I had class with Professor Kenneth Tobin as teacher and Peter Taylor as our guest speaker. Peter, from Curtin University of Technology, was on a sabbatical leave at Florida State University for the fall 1995 semester. I was taking a graduate level course in Evaluation for credit, the first course I had taken for credit since finishing my doctorate in biochemistry at the University of California, Berkeley, 29 years ago.

Peter spoke with us in class of how he was just finishing writing his "tales of the field" of teaching mathematics and science at community colleges in Florida (Taylor, Chapter 1 in this book). He described how he used alternative genres as his way to communicate science and mathematics teaching in community colleges in the state of Florida (Taylor, 1996). He also shared with us how his graduate student, Mark Campbell Williams, had utilized dreams in his dissertation to help him get in touch with how to become a better university teacher (Campbell Williams, Chapter 16 in this book; Williams, 1996). Peter shared with us some portions of an early draft of Mark's dissertation.

This evening class triggered many thoughts in my mind. I had kept a dream journal for over ten years during my first marriage to a psychiatrist. I described in class that evening how, on one occasion, I had solved a scientific research problem through a dream. At that time more than twenty-five years ago, I was trying to make red blood cell "ghosts" and reseal them, so that I could test their reactivity when loaded with cyclic nucleotides. The experiments had technical problems at that point in time, and I was trying to figure out what parameters to change in the experimental protocol. My dream was that I had a blow-out while riding a bicycle. After recording my dream, I realized that the bicycle tire was a symbol or metaphor for the red blood cell and that the blow-out signified how I was manipulating cells too roughly during the experiment. I realized that I was releasing red hemoglobin molecules too quickly and was giving the red blood cells hypotonic shock. There was a "blow-out" in the cellular membrane, and the membrane would not seal up when I restored isotonic salt conditions to the preparation. When I realized this through the dream, I redesigned the experimental protocol, and the experiments started to work beautifully (Kury & McConnell, 1975). That experience epitomized how powerful a dream can be. For me it is a matter of tapping into my unconscious thoughts, a process that enabled me to solve problems and tap into a deeper energy source.

In the evening class with Peter Taylor as our speaker, I shared the bicycle dream and its interpretation with my classmates. Georgia Jeppesen, a student from class, remembered how another chemist, August Kekule, had solved the cyclic structure of benzene, C_6H_6, through a dream of a snake biting its own tail. Prior to Kekule's time, chemists considered molecules as either linear or branched but not as cyclic. As a chemist myself, Georgia's recounting of Kekule's dream resonated with me (as electrons in benzene do). I was the only Ph.D. scientist among a class full of science educators, and I felt rather daring even to mention the bicycle dream in class. What would they think of an analytical thinker who dares to share dreams with science education classmates?

That evening after Ken's class, I decided to put my dream clipboard out by the side of my bed, as I used to do years ago before my two children were born. On the morning after Peter's guest presentation in our class I awoke from a dream, hearing just two words, spoken three times: *"triple point, triple point, triple point."* Because I used e-mail with students in Ken's class to share understandings, I got up early before my family awoke and sat at my computer to write my classmates about the meaning of the "triple point" dream.

Many students in Ken's class enrolled only in distance education mode, which was available through the Internet, so they had not attended class the night before, nor did they know the context of my dream. Therefore, as I warmed my mind by writing the context for Peter's presentation and the ensuing discussion, ideas flowed (e-mail to <sce-6761-01@garnet.acns.fsu.edu>, 22 November 1995, 6:38 a.m.).

To a chemist like myself, triple point is a specific temperature and pressure at which a pure substance, such as water, coexists as three phases: solid, liquid, and gas. This is represented in a phase diagram (Figure 1), which demonstrates regions in which the molecule can exist in each phase. Triple point of water is at 273.16 K (or 0.01 °C) and 6.03 X 10^{-3} atm (Atkins, 1998, p. 143). At the triple point three phases are in rapid, dynamic equilibrium with solid becoming gas and liquid, gas becoming liquid and solid, and liquid becoming gas and solid simultaneously. In fact, there are no degrees of freedom at the triple point, because if you change the temperature or pressure just the smallest amount, you are no longer at the triple point. At the triple point each phase is rapidly becoming the other two phases. In fact, the phases are indistinguishable, because the process of interconversion is so rapid. Water at its triple point is beautifully opalescent.

What could be so beautiful in my life? What did triple point as a metaphor mean to me? For the previous sixteen months I had been struggling with a diagnosis and treatment for breast cancer, which included surgery followed by seven tough months of infusion with a three-poison chemical cocktail called "CMF" chemotherapy. When I did not bounce back after the end of chemotherapy and after a friend died of metastasized breast cancer one month later, I lost my zest for life for seven months. I lost confidence in myself, as a person and a chemist. I kept losing weight, and I, like my family and friends, feared that I was dying.

During the following summer I had chosen to submit my long overdue folder for promotion to professor of chemistry because I realized that I did not want to die as an associate professor. If I were to die soon, I wanted to be professor. Preparing that promotion folder was very difficult, because I lacked my usual self-confidence and energy. You must "promote" yourself as you prepare your own promotion folder. My colleagues in chemistry who knew me well recommended my promotion,

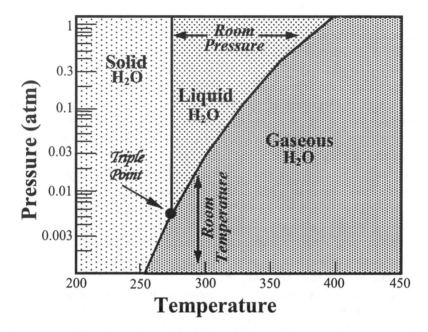

Figure 1. In the phase diagram for water, the pressure increases by a factor of ten at regular intervals along the vertical y-axis. This method of graphing allows large ranges of pressure to be plotted. Room temperature is 298 Kelvin (25 °C), and atmospheric pressure is 1 atmosphere. Water is solid in the dark gray region, liquid in the medium gray region, and gaseous in the light gray region. The triple point is indicated as the temperature and pressure at which water coexists in rapid equilibrium as solid, liquid, and gas. (Figure adapted from original found in *Chemistry: Science of Change*, Third Edition, by David W. Oxtoby, Wade A. Freeman, and Toby F. Block, copyright © 1998 by Saunders College Publishing, reproduced by permission of the publisher)

but science area and college promotion committees did not. I had to write appeals to each succeeding committee. In the process of writing appeals, I was finding my old self again. I was starting to explain how what may have seemed to others like disparate fields of scholarship (i.e., ethics in science, biochemistry, and science education) are connected to each other.

The morning I had the triple point dream was the day I submitted a final appeal for promotion to professor in a letter to the dean of faculties. Triple point was my metaphor for having brought my scholarly worlds together to a single point, where my domains of interest were interconverting with each other in dynamic equilibrium. After having the dream, I knew the letter I had just finished the day before was a convincing appeal and that I would be promoted (which I was).

In my dream I realized that the first triple point represented bringing together my three areas of research: ethics in science, biochemistry, and science education. As an example of that triple point, that semester I was starting to offer "contextual learning" experiences for biochemistry and science education majors to learn not only biochemistry but also science education and ethics in science (Gilmer, Grogan, & Siegel, 1996; Muire, Nazarian, & Gilmer, 1999). I encouraged students to work in teams to focus either on a specific autoimmune disease or a particular form of cancer, learn science that was known about that disease, work collaboratively in groups, interview patients (while addressing the issue of ethics in science by utilizing human consent forms and maintaining patient confidentiality) who have the disease and doctors who care for those patients, and share understandings on the Internet and through publications and presentations. This research overlapped with my biochemical cancer research in the laboratory where we study modulation of immune recognition of tumor cells. This contextual learning was an example of working at a triple point. Students were learning biochemistry, ethics in science, and science education in a way that interconverted, like three phases of a pure substance at a triple point. Learning can be more meaningful in such situations.

On a more personal level, the second triple point is a merging of family, friends, and self. Through my struggles with diagnosis and treatment for cancer, I saw myself differently, and came to understand my family and friends differently. My family, friends, and self have merged at a second triple point. We are all connected. We are all one humanity.

The third triple point brings together three responsibilities of university faculty: teaching, research, and service. Instead of working in separate domains, or even along interfaces between two domains, working at a triple point in which teaching, research, and service are one can be tremendously energizing. In contextual learning experiences, while I am teaching students, we are simultaneously conducting research and providing a service to the community (of patients, doctors, biochemists, World Wide Web users, and science educators) (Gilmer, Grogan, & Siegel, 1996; Muire, Nazarian, & Gilmer, 1999).

When all three of my triple points merge into one, all phases of my life become focused. You can think of each triple point as being at the centroid of an equilateral triangle. If you connect the centroid to three corners of an equilateral triangle, this forms three domains. Each of three equilateral triangles lies in a plane (Figures 2a, 2b, 2c). One triangle involves three areas of research (Figure 2c; in my case, ethics in science,

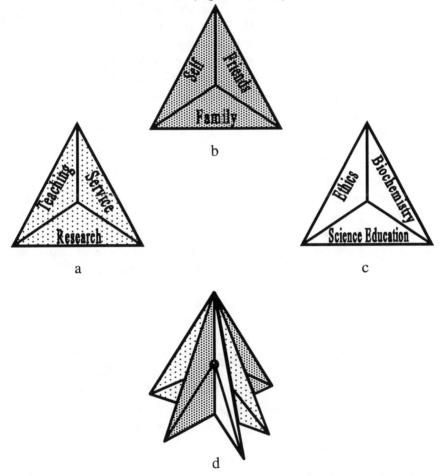

Figure 2. Representation of the triple point in each of three dimensions (a, b, c), and the superposition of three triple points at the centroid of each equilateral triangle to a triple "triple point" (d).

biochemistry, and science education), the second being family, friends and self (Figure 2b), and the third being research, teaching, and service

(Figure 2a). The triple "triple point" then forms at a single point where three centroids merge (Figure 2d).

A triple "triple point" is a point and, therefore, has no dimensions. From a single point, you can go anywhere. Nothing gets in your way. You are so focused. When you think of a triple point, consider the release of energy that occurs in an actual phase transition to change a liquid to solid or a gas to a liquid. I can tap into that energy by working at my triple point. I can feel the opalescence, the beauty of the triple point. I feel like a pluripotent stem cell. I can differentiate any way I want, or I can remain pluripotent at the triple "triple point," tapping into energy flow. When I am at my triple point, I can feel the opalescence of a beautiful Australian opal with its lightning-like streaks visible within a stone, capturing the energy of our universe. I can tap into my triple point as a seemingly infinite source of energy. This energy empowers me to be the person I want to be and to act in multiple domains, to reconceptualize myself in the act of actualizing.

This metaphor that I constructed through a dream has been powerful, even years later, because it linked me to my life through my experiences. This dream helped me through a difficult time of coming to grips with my own mortality, recovering from the physical trauma of surgery and chemotherapy, appealing a negative decision from my college for promotion to professor, trying to recapture my health, renegotiating my role in life. Triple point may work as a metaphor for someone else as well, but more likely, each person will have his or her own personal metaphor. If a person recognizes that metaphor, it can be empowering and a source of energy.

Metaphors are powerful referents for reconceptualizing roles and images of self (Lakoff & Johnson, 1980; Abbas, Goldsby, & Gilmer, Chapter 7 in this book). It is a matter of listening for your own metaphors and tapping into them, reflecting, and learning from them (Lakoff & Johnson, 1980; Williams, 1996). There are various ways to tap into metaphors, and capturing your metaphors through dreams is only one of them (Williams, 1996; Williams, Chapter 16 in this book), but it is one that works for me, which is why I share it here. The actualized metaphor can allow one to not only cope but to energetically tackle multiple roles of self, to tap into energy for self-actualization, to propel oneself forward to accomplish difficult goals.

The goal I chose was to change my teaching as I had seen the Science FEAT middle and elementary school teachers do so bravely.

They too worked within cultures that had constrained them, but they broke free from the constraints. Each teacher chose to examine some component of her/his own teaching. The Science FEAT teachers had a supportive environment within our program to do this. Although their schools still constrained them, the Science FEAT teachers were brave to take the steps that they did. I could see that the teachers became liberated from what had held them back previously. As they reflected on their teaching, they had great insights and I experienced their growth as teachers and learners. I wanted to see if I could change my teaching and, thereby, the learning of my students. I figured that since the Science FEAT teachers became brave within the culture of their graduate program, I too might become brave were I to do similarly, to become a graduate student again and immerse myself within the culture of science education, to support me on this journey.

Scientist Negotiating Her First Educational Research

In this time period while facing many of life's challenges, I enrolled in my first graduate course since I had received my doctorate in biochemistry from the University of California, Berkeley, in 1972. I did this as a special student at the institution at which I was and still am a professor of chemistry. It was Tobin's course in Evaluation, which used Guba and Lincoln's (1989) *Fourth Generation Evaluation* as the text. It was a tough book to read, especially by a practicing scientist like myself. Most students taking this course were distance learners, so we communicated by e-mail (Tobin, Chapter 13 in this book) on our insights and understandings of readings and dialogue, both in class and by electronic communication. I became part of a discourse community (Erickson, 1998) in which I could learn from others and others could learn from me.

During the fall semester in which I had the triple point dream, I was teaching a group of thirty students in my Honors General Chemistry course. When the course first started, I was still suffering from the ill effects of chemotherapy and recovering from physical effects and loss of confidence in myself because my body had forsaken me when the cancer had grown so rapidly. This malaise made me feel anxious about teaching, especially honors students. During the experience of preparing for and teaching the course, I reconnected to my former love for chemistry, and it helped me regain my self-confidence. What I love about chemistry is that a large part of the world makes sense through chemistry. The process of reteaching chemistry to myself and teaching it to students helped me recover from my illness.

The textbook gave me a fresh approach to teaching chemistry (Oxtoby, Nachtrieb, & Freeman, 1994) . What I particularly liked about this book was the high quality of problems at the end of the text that related chemistry to industrial and other applications. This introductory textbook was the first that I had used that included a chapter on condensed phases and phase transitions, including topics of phase equilibrium, phase transitions, phase diagrams, triple points, and critical points. A week after teaching that chapter I tapped into my triple point metaphor and had my dream.

As my body and mind recovered from chemotherapy, I tried some collaborative group work with these honors students. During class time, once every three or four class periods, instead of lecturing and my solving problems on the board, I had students work in groups on solutions to some tougher problems. A few student groups worked collaboratively on solutions, teaching each other how to solve problems. Members of other groups tended to work individually rather than collaboratively in their groups. At the end of the semester, part of their assessment was based on self- and peer-evaluation of how they did in their group work.

However, I felt that all students did not engage well in the discourse of canonical science. Most students would come to class and listen to lectures. I realized that I did not provide enough opportunities for students to have dialogue using the language of chemistry, with me or with other students. Some students learned through traditional lecture, but I felt I could have encouraged students to engage better.

One year later I taught Honors General Chemistry again and decided to conduct my first educational action research and try to tap into my metaphor, tapping into the energy at my triple point of research, teaching, and service (Figure 2a). I conducted this research in an effort to improve my teaching and provide a service to my students. I utilized the metaphor of triple point so I could tap into energy transformations at the triple point and use that energy productively to improve my teaching, conduct research, and simultaneously provide a better opportunity for my students to learn.

I chose to do an action research in my own university-level chemistry classroom. Kurt Lewin (1946) first coined the term action research. What is meant by action research is

> a spiral of circles of research that each begin with a description of what is occurring in the "field of action" followed by an action plan. The movement from the field of action to the action plan requires discussion, negotiation, exploration of opportunities, assessment of possibilities, and examination of constraints. The action plan is followed by an action step, which is continuously monitored. Learning, discussing, reflection, understanding, rethinking, and replanning occur during the action and monitoring. The final arc in the circle of research is an evaluation of the effect of the plan and action on the field of action. This evaluation in turn leads to a new action plan and the cycle of research begins anew. (Collins & Spiegel, 1995, p. 117)

Recently, action research has become a powerful genre in education (Kemmis, 1983; Kemmis & McTaggart, 1988). Educational researchers are working with K-12 teachers, bringing innovative teaching and learning into science classrooms at the K-12 level (Berliner, 1998; McDonald & Gilmer, 1997; Spiegel, Collins & Lappert, 1995), university level (Gardner & Ayres, 1998; Chapters in this book: Scantlebury, Chapter 1; Tobin, Chapter 13; White, Chapter 9; Williams, Chapter 16) and community college level (Humerick, Chapter 8 in this book).

I share this initial educational action research study now with you, the readers, as it may give courage to other scientists who may be considering how to improve teaching and learning in their classrooms. It may also be a window into the mind of a practicing scientist for science educators. Scientists not formally educated in education may feel that they need to rethink everything to change their classroom, and thereby put off making any changes at all. One of my points in this chapter is to show that you can start action research in your own classroom by making a discrete but (what may seem like) small change in your teaching plan, as long as you are involved in the action, monitor progress of research, assess learning of students, reflect on what has happened, and evaluate "the effect of the plan and action on the field of action" (Collins & Spiegel, 1995, p. 117). The process of doing this immediately suggests new areas of educational study for the next semester of teaching. Action research as a cycle begins anew each semester.

The purpose of this entire book is to highlight different approaches to educational research, to improve teaching and learning in university and community college classrooms. Much of the emphasis in our book has been on teacher preparation classes that involve teaching students who are preparing to become science teachers. Therefore, I will highlight one of my students, Sarah, who was planning to become a science teacher, but will also include other students who were planning to become scientists and engineers.

Next I will provide you with my theoretical framework, followed by results of my educational research, and finally, what I think it means, both to me and to students in my classroom.

Defining the Research Problem

The opportunity for my action research project came when I taught Honors General Chemistry again. I wanted to encourage my students to be more active learners. The purpose of this action research was sixfold:

- to increase university students' interest in chemistry
- to deepen students' understandings of chemistry and science, in general
- to encourage students to utilize canonical discourse of chemistry
- to improve communication between teacher and students
- to meet the needs and interests of students
- to learn how students come to learn and understand chemistry

In the "field of action" in my classroom undergraduate students generally did not fully engage in the discipline of chemistry. Most students in Honors Chemistry arrived as first-year university-level undergraduates, fresh out of high school. All students had taken chemistry in high school, but the way they learned it was more by rote than dynamic learning. Even though many of the students in Honors Chemistry lived in a special dormitory for honors students, before this semester they did not seem to engage each other in a discourse on chemistry. These honors students were among the best and brightest we have at the university, so they were able to learn on their own. I felt that they could learn more deeply if I gave them opportunities to engage in canonical discourse on chemistry. I wanted them to know more than memorized chemical facts. I wanted them to see the relationships of chemistry to their lives, to the world. I wanted them to understand different areas of chemistry and how they relate to each other. By having a discourse with co-participation, it could be dynamic and evolve toward active learning and construction.

The action plan for this educational research was to determine if encouraging first-year university students to attend chemistry departmental and divisional seminars, including special opportunities for open house and poster sessions, and to communicate with me by e-mail about what they learned, increased their interest and deepened their understanding of chemistry and science, increased communication with me, and met their needs. Through this process I wanted to learn better how my students "come to know."

Theoretical Frame

The theoretical frame for this chapter is sociocultural and examines learning communities in terms of discourse theory (Tobin, 1995). By discourse, Lemke (1995) means a "social activity of making meanings with language and other symbolic systems in some particular kind of situation or setting" (p. 8). Developing a common discourse community between teacher and student encourages student learning and is dynamic. Tobin (1997) states that co-participation in a discourse community "implies the presence of a shared language that can be used by all participants to communicate with one another such that meaningful learning occurs" (p. 371). Furthermore, Tobin reminds us that "[t]he best an outsider, such as a teacher, can do is to mediate in the process of constructing, reconstructing, and enacting knowledge" (p. 379). Tobin further states:

> The mediating role of the teacher is focused not only on what students know and how they can re-present what they know but also on the identification of activities that can continue the evolutionary path of the classroom community toward the attainment of agreed-on goals. Thus, the concern is beyond representation and also involves mediating the constructions of a discourse that becomes increasingly scientific in its character. (p. 372)

I wanted to engage my students in canonical discourse of chemistry, so that it would become "increasingly scientific in character" (Tobin, 1997, p. 372). Based on research in secondary students learning science, Driver, Squires, Rushworth, and Wood-Robinson (1994) encouraged teachers to do more than lecture, noting, "Experience by itself is not enough. It is the sense that students make of it that matters" (p. 7). A teacher needed "the ability to listen to the sense that learners are making of their learning experiences and to respond in ways which address this" (p. 8).

The Study and the Students

It is a challenge to get and maintain interest of a high proportion of students in an introductory college-level chemistry course. Some methods that are proving successful in teaching science include focusing on materials science in chemistry (Lyons & Millar, 1995); introducing organic chemistry into freshman chemistry (Coppola, Ege, & Lawton, 1997; Ege, Coppola, & Lawton, 1997); using collaborative learning groups (Bruffee, 1993; Johnson, Johnson, & Smith, 1991), and utilizing e-mail for communication (Dougherty, 1997). A book that discusses teaching and learning in university courses that I found particularly useful was published by the Committee on Undergraduate Science Education at the National Research Council (1997).

I conducted an interpretive action research study (Collins & Spiegel, 1995) using a qualitative research design described by Guba and Lincoln (1989). Gallagher's description of the uses of interpretive research in science education was most helpful in my study as well (Gallagher, 1991). My university-level introductory chemistry class met three hours a week for fifteen weeks. My action research plan evolved early in the semester. I involved students in a hermeneutic dialectic circle, by inviting students to attend extracurricular seminars on chemistry. They communicated their learning (or lack of it) to me by e-mail, and I responded. Using criteria from *Fourth Generation Evaluation* (Guba & Lincoln, 1989), I conducted member checks of preliminary results of my action research near the end of that semester with my students. In addition, during the study I shared my understandings with peers at a regional science education meeting (Gilmer, 1996). In addition, I expanded the stakeholder group to include my chemistry professor colleagues, by sending them electronic copies of the paper I presented.

The primary data was e-mail correspondence between first-year honors chemistry students and me as their university chemistry teacher.

In addition, students responded to a survey about the course midway through the semester and anonymously to a longer questionnaire at the end of the semester.

Early during the semester of this action research study, I invited twenty-four freshman from my class to attend a chemistry education seminar presented by a chemist from the University of Wisconsin–Madison named Professor Arthur Ellis. Dr. Ellis is a winner of the Pimentel Award in Chemical Education from the American Chemical Society, and he brings vitality into the university-level teaching enterprise. Basically, he has brought his research expertise in materials science into his teaching (Ellis, 1995). He and his colleagues from Wisconsin have a powerful website on teaching chemistry at all university levels (http://www.chem.wisc.edu/~concept). Ellis's colleagues, Lyons and Millar (1995), have published a case study in breaking chemistry "curriculum gridlock."

I proposed to the students that they could receive ten extra credit points (out of 700 total) if they attended Ellis's seminar and, within five days, send me an e-mail of at least one paragraph about chemistry that they learned.

Ellis delivered a particularly well-organized and interesting lecture, involving many instructive demonstrations of applications of materials science research in chemistry. Half of the students in class sent me an e-mail concerning the departmental seminar. Students enjoyed it so much that I decided to make this an action research study. I continued the practice of awarding extra credit for students who attended other divisional or departmental seminars each week, provided that students correspond with me by e-mail about the chemistry they had learned.

Each week, I sent an e-mail to alert all my students of upcoming seminars that they could attend. Some students chose to attend a physical chemistry seminar, while others went to a chemistry departmental seminar, biology colloquium (when it explicitly involved chemistry), teaching and learning colloquium (when it involved chemistry), a ChemiNole poster session (on the research of faculty and graduate and undergraduate students in the chemistry department), and an open house celebration at the National High Magnetic Field Laboratory.

Some students could not attend seminars due to other classes or work schedules. I allowed them to view previously taped seminars on chemistry or biochemistry or TV documentaries that involved chemistry.

This study focuses on e-mail correspondence that students sent me over a seven-week period and the results of two questionnaires distributed in class. Of 24 students in this class, 19 of them sent me a total of 63 e-mail messages with a response rate of between 1 and 7 e-mail messages per student, with an average of 3.3. The quality of science learned, as evidenced in each student's response, varied widely, depending on both the student and the seminar attended. For instance, some students had trouble understanding spectroscopy presentations in the physical chemistry seminar. However, one student who was persistent in attending the physical chemistry seminar series finally was able by the end of the semester to start to make sense of complex topics of spectroscopy.

I attended some of the same seminars, so I heard science content through the "lens" of a both a scientist and a science educator. I have corrected e-mail messages quoted in this chapter for syntax and spelling mistakes. Students, seminar speakers (other than Arthur Ellis), and others have been given pseudonyms.

I had to select certain e-mail messages to include and others to omit from this chapter. As I mentioned earlier, I have chosen to highlight Sarah's comments because she was planning to become a science teacher. In addition, I chose some examples of disconfirming evidence as well as confirming evidence, because I wanted you, the readers of this chapter, to see the range of responses I received.

I have organized student responses into four domains: students learning the content of science; students experiencing the culture of science: "a clash of chemists"; process of student learning; and process of teacher learning.

Students Learning the Content of Science

Learning Practical Applications of Chemistry

A visiting organic chemist gave a departmental seminar on zeolites, a class of clay minerals called aluminosilicates with open spaces in their interior. Students learned not only what zeolites are, but also some history of science and how zeolites are used in the world:

Dolores said:

> In the 1700s, [Alex] Cronstadt discovered a crystal called zeolite at a
> volcanic site. He did some tests on it and thought that it absorbed water
> because when it was heated up, it did not produce a vapor or turn into a
> powder. We know today that you can use zeolites to convert methanol into
> gasoline, make detergent, or use it in medical waste treatment, as a drying
> agent, for gas purification, bulk separation, gas chromatography, and
> liquid phase separation. There are different types of zeolites. One has just
> silicon in it, with oxygen bound to it, and one has silicon and aluminum
> with oxygen bound to it. Zeolites are very stable, which is why they are
> used in many industrial developments. Zeolite's shape is tetrahedral. It has
> a void space topology. Methanol goes into the zeolite, and depending on
> the size and shape, comes out with almost the composition of gasoline.
> This also works with peanut oil, which I thought was very interesting.

It is impressive how many technical words Dolores learned and
could use after hearing just one seminar: crystal, zeolite, vapor,
separation, gas chromatography, liquid phase separation, silicon, oxygen,
aluminum, tetrahedral, and void space topology. She used these words in
proper contexts as well. She was interested in how zeolites can be used to
break down compounds like methanol, forming gasoline. She also
addressed how zeolites are used in various analytical methods; she
probably would not have learned about those methods until her
sophomore or junior year of college, if then, had she not attended this
seminar. She also addressed the chemical structure of zeolites, with
references to the tetrahedral shape and their void space topology.

The part of Dolores's writing that did not make chemical sense
was when she said that Cronstadt "did some tests on it and thought that it
absorbed water because when it was heated up, it did *not* produce a vapor
or turn into a powder." In fact, Cronstadt, the discoverer of zeolites in
1756, found that heating a natural mineral named stilbite released water
that was trapped in void spaces within the structure. Therefore, I think
Dolores must have meant that stilbite *does* absorb water because when
heated it will lose water, becoming a powder. Depending on the size of
the void within zeolite, certain molecules can diffuse into the void,
depending on their molecular diameter. This allows zeolites to catalyze
cracking of molecules like methanol that can accommodate molecules of
a certain size in zeolite's void site.

Some students attended a chemistry departmental ChemiNole
poster day at which faculty and graduate students presented posters of

their current research in the Chemistry Department. Two students mentioned posters on biochemistry and physical chemistry that particularly interested them because of practical uses that chemical or biochemical research addressed.

Sarah commented:

> The poster about the harmless virus that they are trying to inject into people in order to get genes into the human body was very interesting. From what I understand, so far there is a little problem because there have to be several injections, but the body builds up an immunity to the virus, so the gene doesn't make it in to replace the defective gene. I thought that was the most interesting [poster] because I could see them in a few years discovering ways to phase out genetic defects like color blindness, which my brother inherited, and then even fatal gene disorders.

Chuck's favorite experiment was the project on lasers:

> The group talked about lasers and discussed some specific areas or types of lasers. The most interesting one was what is known as optical tweezers. A. Ashkin and J. M. Dziedic in 1981 found that they could be used to move around microscopic bacteria without any harm to the bacteria. The term "optical tweezer" really means single beam gradient laser trap. It works by applying a stray focusing beam, which applies a laser gradient to a particle giving it momentum to move the bacteria. Kind of neat, huh?

Sarah focused on how gene therapy might be used to solve genetic defects. At that time technology was not perfected for introducing a viral vector into humans, because humans can develop an immunity to the virus. With more research Sarah realized that a similar method might be used to solve genetic defects like her brother's color blindness.

Chuck enjoyed technological aspects of microscopic optical tweezers that work by using a laser which attracts transparent dielectric particles toward it focal region. He did not mention that to attract a reflecting particle, such as a bacterial cell or a protozoa, that a doughnut-shaped laser beam is needed. Ashkin and Dziedzic's (1985) pioneering method of using optical tweezers was in 1985, rather than 1981 as Chuck mentioned. Ashkin, Dziedzic, Bjorkholm and Chiu reported first work on using optical tweezers for biological samples in 1987.

Sarah and Chuck were particularly interested in practical applications of science at these poster presentations. Both of them were looking to the future, to innovations that may change the way people live. They were looking ahead to how they, if they choose to become professional scientists, may develop these methods further. Sarah's brother, who is color-blind, may motivate Sarah to go into scientific research. In general, these two students had the main ideas correct, but there were some minor misinterpretations of complex subjects for freshman undergraduates.

Engaging in Creative Thinking While Learning Science Content

When students indicated chemistry that they had learned, they sometimes extrapolated to creative solutions of how chemistry might be used in the future to solve problems in the world. Both Milton and Sarah wrote about what they learned from Art Ellis's seminar called "You Do Teach Atoms, Don't You?"

Sarah said:

> I liked the "memory metal" demonstration, I have never seen anything like that before. I shaped my two pieces of memory metal into hearts. I'm going to send one to my boyfriend, but I'm not going to tell him the trick so if he unbends it, which I know he will, I'll fix it back. I think that the models that Art Ellis showed would be very helpful in learning the chemistry, especially the models of the atoms that are like Lewis dot structures, because they make what is small and easy to be unaware of, big and easy to visualize.

Milton said:

> I did not know about induced dipolarity, in which an electric current forces molecules to align, making a liquid solid between the poles. I wonder what kind of force the solid can withstand, and what the properties of the solid are for the different materials used. The demonstration reminds me of muscle. Imagine a bionic arm with the liquid enclosed in long, thin watertight bags as the muscles. You could connect the electrodes so that impulses from the brain triggered the electrodes, flexing and unflexing the muscle.

These two students understood science well enough to be able to extend their ideas to practical applications. Sarah also pointed out the importance of showing models of atoms, because it is so hard to visualize them. Having seen demonstrations of how these materials can be utilized facilitated students' understandings of basic concepts of induced dipolarity and memory metals. Student-generated ideas of changing physical properties upon induction of polarity in a bionic arm and of returning a NiTi metal to its original shape by heating it with a hair blower are creative.

Students Experiencing the Culture of Science:
A "Clash of Chemists"

The students not only had a chance to learn science but also to see scientists interacting within their discourse community. Samuel spoke of argumentation in science that he witnessed at a chemistry divisional seminar.

> The seminar proved to be both educational and entertaining. Perhaps a better title would have been "A Clash of Chemists." A man better known to most of us as Nick constantly argued with the speaker on the validity of the speaker's research and his diagrams of certain compounds. Nick's, for the most part, foreign counterparts often chimed in with their own devious questions. This provided much entertainment in a lecture that under normal circumstances would have put me to sleep. They seemed to disagree with him about stimulating only one half of the molecule because of self-absorption. [The speaker] claimed it did not matter because of the direction facing the laser.

Samuel continued:

> Between all the arguing I did pick up on a few abstract things about lasers. There was a fifty-year gap between the first concept of a laser and the first one actually being built. This was due to experimentation with diatomic molecules, instead of more complex states, and also the difficulty of obtaining a population inversion. True frequency of the laser is obtained by avoiding resonance (or echoes)....That's about all I picked up from the lecture....Although, once again [it was] over my head, I had a much better time at this lecture than the last one.

I was struck by Samuel saying, "that's about all I picked up from the lecture." Samuel learned considerably both about the science and history of lasers and about the culture of science.

Encouraging Samuel to use the discourse of science allowed him to integrate content and culture of science into his extant knowledge and experiences. The process of reflection and writing encourages deeper analytical thinking and allows students to make connections. As Glynn and Muth (1994) have reported, "Through reading and writing, students can build upon their prior learning and make real-world connections" (p. 1069).

The seminar also made Samuel aware of the argumentative tone of discussions in science and that science is not always cut and dry. Scientists do argue about their understandings, and he witnessed this aspect of the culture of science and seemed to have enjoyed the debate. This process of having an argument about scientific data is an important part of the culture of science because this process helps scientists think about and understand data more deeply. I responded by e-mail to this student that science advances by scientists challenging each other's ideas.

Process of Student Learning

Learning About Learning

Sarah attended a seminar focused on having students use computers to think about what is happening to molecules during chemical and physical changes, such as in a process like boiling water, at the macroscopic, particulate, and symbolic levels (i.e., with equations), examining how a student integrates at least two of these conceptual levels. For example, there was a phase diagram with figures to demonstrate what is happening to molecules in different regions of a phase diagram. This was a particularly interesting topic to me and Sarah, because we had learned about phase diagrams in class. Sarah liked an approach to learning "of not only knowing the facts, but visualizing them and using them in real life."

Sarah wrote me her reactions to this seminar:

> I really liked what [the speaker] had to say about using computers to test and teach. Since a student's learning is the motivation behind teaching, it is very important that teachers find out if their students are learning. It is amazing for me to see the progress that has been made just since I was in elementary school. I remember when you had to stick the disk into the computer to make the program run...that was before there were hard drives. I think it would be very interesting to have the computers a part of our class, but I do have to admit that even though I was raised in "the

computer age," they still scare me!! While he was showing the demonstrations, I was trying to figure out the answers to his questions, and I was very pleased to say that I got a couple of them right. The ones that I got wrong, I wanted him to stop and have him explain why the answer was what it was. Speaking as a student, the approach of not only knowing the facts, but visualizing them and using them in real life is exactly what students dread, but it is also what teaches the most.

Sarah's response was very helpful for me. Her one sentence about how students' learning is the motivation behind teaching and how it is very important for teachers to find out if their students are learning really influenced me. Sarah also mentioned how she would like to use computers to learn. It made me much more interested in trying to find ways to determine what my students were learning.

Sarah mentioned how students often wanted a chance to ask questions during demonstrations. Her comment led me to realize that I could involve students in demonstrations in my classroom, rather than just doing the demonstrations myself, as I had in the past. By having students involved I found there was more chance for interaction and discourse, with students asking questions and explaining their understandings to others.

The use of technology really interested Sarah. She has seen great progress in computer technology since she was born, and she could also see how teachers could use technology to help students learn. In our classroom, the only technology we used during that semester was e-mail. This was in 1996, and then most freshmen did not have ready access to e-mail. Our classroom had only blackboards, chalk, and an overhead projector.

Accessing the Language of Science

An international speaker spoke about photophysical effects of homogeneous broadening on spectroscopy. Several students had quite different experiences of this presentation:

Samuel wrote:

> I learned that the separation of two components of spectral broadening is based on red edge excitation. Homogeneous broadening affects all molecules in a spectrum. It could expand up to hundreds of centimeters. There is also inhomogeneous broadening. Homogenous broadening is due

to electron pairs coupling and has a strong temperature dependence, whereas inhomogeneous broadening is due to statistical irregularities in the environment and is independent of temperature. I also learned the methods of broadening by photon echoes, fluorescence line narrowing, photochemical persistent spectral hole burning, or single molecule spectroscopy. Broadening is used in photosynthetic light harvesting systems, glasses doped with rare earth metals, protein and protein probe compounds, and amorphous semiconductors. I probably would have been able to learn more if the guy talking was not speaking in an international dialect of the English language.

Michelle had a different version of the same seminar:

This seminar made my jaw drop. I could not understand a word the speaker said. It was not because the material was too difficult, which I am sure it was. It was because I am not sure what language he was speaking. I could understand when he was talking slowly, but when he spoke faster I did not get it.

Thomas spoke about the culture of science he observed during the same seminar:

[The speaker] spoke on spectroscopy and photophysical effects of homogeneous broadening. I could not understand a single word he said, except for a couple of comments. A faculty member remarked once that something wasn't possible, and the speaker replied, "Well, I didn't publish it." Someone else asked him if he was going to give some information, and he replied, "I gave it, you just weren't paying attention." At the end of the seminar, someone said, "You have five minutes left," and he said, "No, no I don't," because he had nothing else to say. I have nothing else to say.

Despite an obvious difficulty in trying to understand this seminar, one of these three students, Samuel, was able to explain some chemistry learned from the spectroscopy seminar. However, all three students got a potent taste of the culture of scientific seminars, and they witnessed argumentation that can occur in this context. Scientific language was hard to understand because it was so technical and advanced. The seminar speaker spoke very quickly and with an accent. Students picked up that the speaker addressed questions with some hostility. Three students reacted differently to the same experience, based on their prior experiences.

Still, Samuel was able to access the discourse of physical chemistry and to understand the complex topics included in the seminar. He was able to make sense of the ideas in his mind and could write me about his learning. He was using the language of the discipline. By what these three students said, I could see what they could understand and what they could not yet decipher. This sort of information gave me clues as to what my students were ready to learn, their stages of intellectual development, and what I needed to do in my own classroom to facilitate their learning chemistry.

Participating in Scientific Discourse Community

One student, Michelle, attended a physical chemistry seminar every week. In the beginning of the semester she wrote that she understood little of what was said. I had suggested to her by e-mail that pieces of the spectroscopy puzzle would probably start to fit together if she were patient, if she kept participating in seminar series and sharing her current understandings in an e-mail discourse with me. In fact, the very next e-mail message from Michelle confirmed:

> The extra credit on Monday was one of the more interesting ones I have attended. The speaker had lots of colorful charts to show us. I understand the idea of exciting something in order to see a change, by twisting the molecule and by shaking it (in a sense). My chemistry laboratory teaching assistant had to explain some things after the seminar once again, but I did walk away from this one knowing some new material.

Michelle's teaching assistant was a physical chemistry graduate student, so he attended these same seminars. Quite often Michelle mentioned that her teaching assistant would explain some of the concepts from the seminar to her. Eventually, Michelle was able to put the pieces together, and seminars started to make sense, so she could write about her learning. My suggestion to her that her understanding would start to increase as she continued to make an effort may have given her confidence that she could start to understand tough concepts in spectroscopy.

Learning to Use Metaphors as Tools for Learning

I shared my dream of triple point with students in class and explained how I have used it as a metaphor to focus my professional and personal life. Samuel utilized metaphors as powerful means to learn more about chemistry. In his e-mail about the spectroscopy talk, he wrote me about a metaphors to make sense of what he heard, for example, using the term "middle men" as intermediates in chemical reactions.

> Once [the speaker] got down to the chemistry aspect of his talk, I actually understood him for a good period of time. He was discussing chemical intermediates. Chemical intermediates are like middle men in a reaction. They are created and consumed within the chemical reaction, and are unaccounted for in most cases. These intermediates, however, are very important, as they can give us hints about reaction possibilities and provide other important information about the reaction. There are three types of intermediates: organic free radicals, organic metallic radicals, and radical complexes. The way [the speaker] learned about these radicals was through high-resolution spectroscopy. I found that pretty interesting because spectroscopy is what I learned about last week [at a physical chemistry seminar]. I think it helped me understand a lot better. It is neat how all this stuff is starting to come together.

The idea of chemical intermediates is taught at a higher level than undergraduate freshman chemistry. It was interesting how much Samuel learned about both spectroscopy and reaction intermediates. His learning was built on his prior learning from other spectroscopy seminars during that semester. I responded to Samuel's "middle men" e-mail as follows:

> I'm glad that you learned about chemical intermediates. Once you know the structure of intermediates, you can design transition-state analogs (i.e., compounds that resemble the transition state) that can inhibit enzymatic reactions. I used one of them very handily in my doctoral dissertation. As you mention, intermediates can tell you a lot about a chemical reaction.

Process of Teacher Learning

Getting "Into Demonstrations"

Sarah had mentioned in an e-mail I cited earlier in this chapter that she really learned from seeing demonstrations on chemistry. Also other students indicated from Arthur Ellis's seminar that chemical demonstrations helped them to learn chemistry:

April declared:

> I am a much more hands-on type of learner, and it makes things a lot
> clearer for me when I can actually see what is happening. Each time Art
> Ellis did a different experiment, I would think that I really like it, then
> when he did the next one I would like it even better. I thought that the
> demonstration in which he showed oranges being cut [into sections] really
> explained the whole idea of a unit cell.

Robert said:

> At the beginning of the [Ellis] seminar, I thought it was going to be rather
> boring, because I'm not interested in chemical EDUCATION. However,
> when Art Ellis got into the demonstrations, it really got my attention. I
> find demonstrations to be much more effective in making me (and I
> believe other students as well) more interested in the subject matter being
> taught....I must admit, if I had known that it would have been that
> fascinating, I would have gone whether or not I was getting extra credit
> (but might as well since it's available <grin>).

It is obvious from these students' comments that they love to learn
through seeing chemical demonstrations. This encouraged me to do more
chemistry demonstrations during class time. Instead of just doing them
myself, I involved students in demonstrations, so that other students
would identify with their fellow students, while they started to develop a
discourse community. For instance, when we studied how equilibrium
constants depend on temperature, students helped me do the class
demonstration of how the equilibrium process between room
temperature, brown NO_2 gas is temperature dependent as it dimerizes to
colorless N_2O_4 gas at liquid nitrogen temperatures.

Near the end of the semester in a chemistry departmental seminar,
we had another chemist who showed a demonstration of Raman
spectroscopy. A number of my students from Honors Chemistry attended
this seminar. Samuel responded:

> When I walked in about five minutes late I thought that [the speaker] was
> trying to sell me something instead of teach me about chemistry. His talk
> was about Raman Spectroscopy. It was kind of interesting to me since we
> just did a two-week laboratory on it. His equipment was, however, a bit
> more advanced than ours. He explained that it had "evolutionized" due to
> many improvements in its design. It has a multichannel and energy
> through-put advantage, which improved it sixteen-fold. It also had better
> sensitivity and more flexibility. The Raman gives information in a more

rapid and robust manner. It was really neat to see someone actually give examples of what they were talking about, as he did with his computer. I can't see me purchasing one of these fascinating pieces of technology in the near future due to its price of $85,000.

I concluded that students learn and remember better when they actually get to see experiments and see how the equipment performs. For me, it is worth the additional effort of preparing and presenting experiments of this kind in chemistry classrooms. Getting feedback from my students while I was teaching them chemistry encouraged me to use more demonstrations and to involve them in helping me do and explain the experiments during "lecture" time.

Gaining Feedback on Learning Strategies

Midway through the semester I also used e-mail to get some feedback from students about various learning strategies employed in the course:

Sarah commented:

> For the most part, I learn best from lectures, but the learning groups are a nice balance with the lectures because it checks what I have learned by making me practice with other people. The only thing that I find difficult about the groups is when we have assignments to do together outside of class because it is very hard to get four people in the same place at the same time.

Wayne commented:

> As to your inquiry about relative efficacy of learning strategies, although I'm no cognitive psychologist, I feel that group learning provides somewhat more interaction than straight lecture; however, impact of demonstration laboratories is directly related to pertinence and novelty of subject, so that each can only be judged on its own merit. I really like it when you bring in outside material relevant to what we're studying.

This feedback from students during the middle of the semester helped me shape how I utilized learning groups during the remainder of the term. I utilized more demonstrations than I had originally planned and tried to make demonstrations relevant to students. I learned of students' interests from their e-mail messages. Although I did not find a way during that semester to improve group interaction, later, when Tobin developed the "Connecting Communities of Learners" website (Tobin, Chapter 13 in this book), I realized the importance of finding a way for students to share their learning in groups on a website. During the semester I encouraged them to work in their groups using e-mail, since all students had access to e-mail.

Learning from Student Questionnaire

At the end of the semester I asked the students to provide anonymous feedback related to opportunities for them to attend chemistry seminars and to communicate with me by e-mail. The following protocol was used to initiate a response from the students:

> Did you participate in attending departmental/divisional seminars, the ChemiNole Poster session, the National High Magnetic Field Laboratory open house, or listening to chemistry videotapes for extra credit? If so, what did you learn? Was it worth the time you spent, including the time it took you to correspond with the teacher by e-mail? Would you recommend it be continued in the future? Were the teacher's comments to you in response e-mail constructive and/or helpful? Did participation in these activities change how you view science?

Some of the student responses to this set of questions are organized in categories that follow in Table 1.

Table 1: Comments on anonymous questionnaire at end of semester

Broadens scope of chemistry
• "Yes, some [seminars] were good, some I had no clues on what they were about—they were too advanced for beginning chemistry students. I did not really learn as much as I would have liked to, but did learn some. It was worth the time, it broadens your scope of what is really out there and being done in research. It should definitely be continued but maybe not for so much extra credit. It helped me see other aspects of chemistry (i.e., physical and biochemistry)."
• "Yes, I learned how chemistry is applied in the real world in all the seminars. It is definitely worth the time, and I would recommend future use. Your responses were neat, and I thought very considerate,

Continued on next page

Table 1 (Continued)

considering the volume I'm sure you had to respond to. These activities did change my view of science from something you just teach to something you actually use to make the world a better place."

• "I attended several departmental seminars, the ChemiNole, and the National High Magnetic Field Laboratory open house. The struggle to find free time was more than recompensed for by the holistic overview of applied science. They weren't merely educational opportunities but often were good entertainment."

• "I thought that the different chemistry seminars were great. I think it should be continued. It gave you an idea of the chemistry that took place outside of the classroom. It allowed you to get a closer look at certain aspects of chemistry. I definitely think it was worth my time, even the time spent relating what I learned to the chemistry I already knew and discussing through e-mail with the teacher."

Seminars not applicable to our classroom

• "I attended several of the extra-credit seminars. While I learned a great deal, none of it was applicable to my studies in this class. It was certainly worth the time, but I felt like I should have spent more time on the class material."

Seminars hard to understand

• "I really didn't understand most of the seminars that I attended, but by asking…, I was able to figure out some of them. I do believe this should be continued."

• "The departmental seminars were definitely worth the time. Yes, they were often about very difficult concepts, but it helps to hear something that is so complex. Also it may have proved useful when deciding whether or not to major in chemistry."

The students' comments are organized into three groups: those in which students thought attending seminars broadened their scope of understanding, several that felt that the seminars were too complex for their current understanding, and one who said that seminars did not pertain to this classroom. Some keywords in their writing, "broadens your scope," "real world," "holistic," give me hope that this was a start in the right direction. I was struck by one student saying, "These activities did change my view of science from something you just teach to something you actually use to make the world a better place."

Also, there were other thoughts from my students, such as, "none of it was applicable," "often about very difficult concepts," and "I really didn't understand most of the seminars," that make me realize that this attempt at action research in my classroom had some successes, but it definitely could be improved with more action research in another classroom during another semester. Overall, I think that students learned and understood more chemistry from attending optional extra-credit seminars, and it helped them see the utility of what we did in class because they would remember hearing about some of the ideas from class in the chemistry seminars.

In other courses such as Physical Science for Teachers and Science, Technology & Society, I have changed the method of communication, so that students read each other's writing on content (instead of the student just sending the writing to me to respond to), so the students learn how to critically review each other's work (Tobin, Chapter 13 in this book) and get ideas on what more they might do to improve their own work by reading other students' innovations.

Conclusion

It was additional work for me to advertise seminars to students, read their e-mail messages, and correspond back to each one of them, in addition to the usual work of organizing and teaching a class. Therefore, it was gratifying to read what Samuel said, "It is neat how all this stuff is starting to come together." Additional work for me is worthwhile if students learn more and develop a deeper interest in the field of chemistry. Also I learn more about teaching by conducting action research in my own classroom.

Communicating with my students about chemistry seminars and other opportunities to learn chemistry opened up a new dimension to my teaching and allowed me to learn about my students' learning through action research. I focused on encouraging students to learn chemistry while experiencing the culture of science in professional seminars. I utilized e-mail as a way to communicate with my students, to let them know of upcoming seminars, and to allow them to correspond with me about what they learned (or did not learn). I provided my chemistry colleagues with a collated sample of some of e-mail responses from students (using pseudonyms), giving them an idea of how our students react to such settings, thereby encouraging them to dialogue with their students. Thus, I experienced my "triple point" by merging my three professorial responsibilities, that is, teaching, research and service, so that my three roles interconverted, allowing me to tap into the tremendous energy where different domains intersect, as at a triple point in a phase diagram.

My students and I developed the beginning of a discourse community in which we utilized the language of science. I provided students with an opportunity to learn chemistry and to reflect on their learning. Students often went to the same seminar, so they had a chance to interact with each other after the seminar. Students heard speakers and other members of the scientific community discuss science in seminars and at poster sessions. Students witnessed a "science as argument" atmosphere and members of the audience grilling speakers with questions on results of research, as reported.

Attending seminars provided a way for students to make connections to their prior knowledge, with what we were learning in chemistry class and laboratory, and with what was said at seminars or poster sessions. Students learned not only chemistry content but also about the culture of science and the process of student learning. Students had opportunities to share with me their own interests in chemistry.

Sometimes it was hard for me to comment on a seminar because I had not attended it. When there were questions from physical chemistry seminars, the teaching assistant from the laboratory part of the same course helped some students understand the science content.

Since the time of this original study I have used a course website called Connecting Communities of Learners (Tobin, Chapter 13 in this book), so students can read each other's comments and have an opportunity to peer evaluate each other's responses. It also allows students to have virtual meeting rooms where student groups can meet electronically. I have tried using this website in a course called Science, Technology, and Society that I have taught several times, with students posting critical reviews of articles from the Science Times section of Tuesday's *New York Times* newspaper. Each student self-assessed his/her own critical review while peer reviewing two other students' critical reviews. In this way we become a true community of learners, co-participating in the process, not only between teacher and each student, but also among students (Gilmer & Alli, 1998).

Two students summarized what attending seminars meant to them:

Sarah commented:

> As a first-year university student, I would have had no idea what departmental seminars are or when they take place if it weren't for you telling us about them and encouraging us to attend them. Thank you for telling me about the seminar on science education. Since I am considering teaching, I think it will a good thing for me to see. I definitely think that the seminars are a good idea because they do talk about very complex, specific things, but it exposes us to them and that is important.

Laura said:

> I think the best thing about attending seminars is that it shows us science in practice. The seminars (especially ChemiNole) made me realize the great amount of study and research that goes on...around the world. On the news or in mass media one doesn't normally hear about many current experiments or discoveries.

The students also gave me feedback on how to improve their learning in class. Getting feedback from students both during and at the end of the semester helped me address their needs and helped me reflect on whether or not to continue providing opportunities for students to learn science content and the culture of science from attending seminars. This was the first semester I had such an e-mail-oriented discourse with students about chemistry.

Hearing what Sarah said about her need to hear lectures to help her learn chemistry indicates to me that I need to ensure that my future teaching addresses explicitly the issue of students' established epistemologies of (learning) practice, so that students are better able to appreciate the need for adopting my theoretical framework (i.e., constructivism and co-participation) as referents for their learning and, if they plan to become teachers like Sarah, for the learning of their future students.

Near the end of the semester, I conducted a member check (Guba & Lincoln, 1989) with my students and sent them a draft copy of a paper I had presented at a regional science education conference (Gilmer, 1996). Students made suggestions that I incorporated into this chapter.

Milton commented:

> I actually read your paper and found it interesting. The best part about e-mail is that it provides a nonthreatening venue for communication. For instance, I dread telephones, but I am an excellent writer. I find e-mail allows me time to refine my thoughts without having to lick stamps, which is another activity I dread.

Bringing research, teaching, and service together at a triple point has become my mantra. It is energizing to try new things in teaching, to see how they fit or do not fit within educational and scientific theory, and to encourage others to experience the same. The dream metaphor of triple point enabled me to bring my research, teaching, and service together and allowed me to tap into the energy where these worlds meet. I encourage other science content faculty to focus on a problem in their university classroom and conduct action research to address a particular issue of concern and interest.

Penny J. Gilmer
Department of Chemistry and Biochemistry
Florida State University, Tallahassee, Florida, USA

Postscript

Perhaps the triple, triple point goes beyond my own life, not only to other's lives, but also to the creation of the universe. Dr. Andrei Linde, a cosmologist at Stanford University, has speculated that the beginning of the universe may have been due to a phase change. "In this view, the Big Bang is more like a transformation, like the melting of ice to become water, than a birth. 'Maybe the universe is immortal. Maybe it just changes phase. Is it nothing? Is it a phase transition? These are very close to religious questions' (Overbye, 2001)."

On sheer speculation and a belief that we can learn through the symbols in our dreams, perhaps the universe started at the intersection of a triple point, or perhaps a triple, triple point. The Big Bang could have happened when the conditions slipped out of the dynamic equilibrium due to some statistical perturbation, leading to a rapid, phenomenal expansion. Substances at their triple point are opalescent and beautiful, so the result of the expansion would also be beautiful.

EDITORS

METALOGUE

Summoning the Energy for Transformative Teaching

KT: Penny's chapter brought back a rush of memories for me. Some are very sad because she speaks of a time when she came close to death, and I experienced those moments of great trauma. I should point out that in those moments she also was a great role model for many other people who were afflicted with cancer. Her courage gave them the faith to live on.

PG: Writing this chapter was a catharsis for me as I recovered from cancer. I wrote about it and how I became empowered as I conducted my first action research study, as I thought it might give hope and courage to others. Approximately 25% of us will have cancer sometime in our life, and most of us know at least one person close to us who has had cancer and perhaps already died from it. My recovery came during the semester that I took Ken Tobin's graduate course in evaluation, during the same semester that I was being considered for promotion to full professor. Hearing Peter Taylor's presentation in Ken Tobin's Evaluation class triggered my triple point dream. The energy from the dream still surges through me. When I really concentrate on it, I can feel the energy of the universe flow through me, entering at my toes, funneling through my body, and passing out through my head. I feel charged from the energy of the universe. I feel that my recovery was a bit like the feeling that was depicted in the movie *Contact* (based on Carl Sagan's book), when the woman astronaut passed through an enacted cosmological "worm-hole" as she passed from one world into another. That is what it felt like for me as I recovered from cancer. It was a rough, bumpy journey, and I was not sure I would come out at the other end. I will never forget it. Also I'll never forget what the other side was like, and it gives me great compassion for those who are sick, especially those with cancer.

Transforming Teaching and Acting Strategically

KT: This dream business is not for everyone. I dream too but my dreams are harder to see as metaphors with a great deal of significance. However, I must say that Penny's dream has provided her with an object for thinking about her life and making sense of the many struggles she has had as a woman in a community with few women participants. There are two aspects of these struggles that I will address here. The first is her struggle for promotion. My experience has been that universities do not encourage scientists to engage in research in science education. One of my colleagues, a geologist, was once called into the dean's office and told that if he had time to do research in science education, he had time to do more research in geology and that is what he should do. Another in the same department was instructed by his department chair not to participate in a project to improve the quality of science teaching and learning in the university. Disincentives to engage in research in science education are sufficient to make many wary of collaborative research with colleagues in colleges of education, many of whom are held in low regard by scientists.

PG: I took a big chance to get involved in science education before I was a full professor of chemistry. What I did to be sure that my scholarly work in science education would "count" toward my research is to ask my senior colleagues who evaluated me each year to put it in writing that it would "count," if it was published. That letter from my senior colleagues came in very handy when my department voted to promote me but the science area and college committees voted me down. In the end, the university committee voted in my favor, and I feel certain that the letter stating that my science education research would count made a difference. Other science faculty who are considering getting involved in educational research might consider doing something similar for their own situation.

KT: My second point relates to the *persistence* Penny has shown in attempting to improve her teaching of chemistry even in the face of strong opposition from some of her colleagues who would prefer her to prepare the students for subsequent courses in chemistry in a more conventional way. Her emphasis on understanding and engaging students in the discourse of chemistry appears to be seen as watering down the concern for knowing valued chemistry content. From my way of looking at education, Penny is providing opportunities that will have long-term payoff for her students. She is assisting them to join the discourse community of chemistry.

PG: Ken Tobin is right that I am persistent, almost doggedly so. However, I realize that cultures are slow to change, and I need to use the education discourse with my biochemistry and chemistry colleagues to encourage them to become part of the *discourse community* in science education. At a recent annual departmental chemical education seminar, we had a packed audience, and Ken Goldsby (Chapter 7) told our speaker, Loretta Jones from the University of Northern Colorado, that all of us have an interest in education, faculty and graduate students alike. Therefore, education does matter to my colleagues and our students. As I engage my colleagues, I realize that it takes time and active reflection to change. It took me years to change, and I still am as I learn more and reflect more deeply. There were some catalytic events like the triple point dream that got me over a high energy barrier and into the next realm. Scientists are trained to be skeptical, and that is our nature. Therefore, it should not surprise me that we are skeptical of educational research in an area in which scientists are just learning the theoretical underpinnings and the discourse of science education.

PT: Penny's story of self-transformation highlights the powerful agency of the individual for reconstructing not only her own private world but also her professional world. A professional world is populated by others and (uneven) power relationships tend to be the "glue" that holds it all together, particularly in educational institutions. But it is just too easy for us to construct ourselves as powerless individuals buffeted by institutional circumstance. Such a philosophy of despair (cynicism, perhaps?) serves to maintain the momentum of history that fuels cultural myths of the naturalness and privileged status of objectivism and technical rationality. These myths can rob the unwary of their agency as reality co-constructors.

Penny's story gives us deep insight into how metaphor can be used with nonrational (but not irrational) ways of thinking, such as listening with respect to our unconscious mind, to find the energy, enthusiasm, and values for transforming our professional worlds. What I find fascinating is that Penny has tapped into imagery emanating from her unconscious mind and given rational meaning to it via the linguistic tool of metaphor.   The power of metaphor is that it is readily com municable to others and, in this case, the triple point metaphor speaks authentically to chemists (using icons of chemistry) about the possibility and desirability of transforming their epistemologies of teaching practice. For the adventurous, it also invites critical self-reflection on one's sense of professional (and personal?) identity, a hallmark of the reflective practitioner.

Enacting the Dream

KT: It would be remiss of me if I didn't mention that the triple point metaphor also applies to Penny's efforts to become a *co-learner, co-teacher, and co-researcher* of science and science education.

PG: In this effort to be at the triple point of co-learner, co-teacher, and co-researcher, I am now in the midst of working toward a second doctorate in science education at Curtin University of Technology in Perth, Western Australia. I am doing this by distance education while I am professor of chemistry at Florida State University. I had taken four graduate level courses at Florida State as a special student, and then transferred those hours to Curtin. I spent a one-semester sabbatical at Curtin, for my one semester in residence on campus as a graduate student. I have two co-major professors: Peter Taylor and Ken Tobin, my two co-editors of this book. Co-editing this book, following up the authors' references, and really reading the chapters critically has been a powerful way for me to learn. I collected my data for my thesis in a Web-enhanced biochemistry course that I taught in a radically different way than by traditional lecture—using collaborative groups to develop and present websites to the rest of the class. It is a journey.

PT: Using constructivism as a referent for designing our teaching in ways that promote students' co-constructive learning is but one half of a good story. The other half is taking constructivism as a referent for evaluating the efficacy of our innovative teaching strategies. How do we know that we have improved students' learning? How can we obtain valid evidence, and what standards should underpin the validity of our knowledge claims? These questions are about ways of evaluating the relationship between (transformative) teaching and learning. They are questions that can be addressed successfully by science education research. In adopting the intersecting roles of teacher, learner, and researcher, Penny is learning to design a well-warranted interpretive research approach for inquiring into the efficacy of her innovative (Web-based) science teaching. It is this eclectic professional practice that exemplifies the advantage to professors of science of adopting a serious career-long interest in the practice of science education.

BIBLIOGRAPHY

Ashkin, A., & Dziedzic, J. M. (1985). Observations of radiation-pressure trapping of particles by alternating light beams. *Phys. Rev. Lett. 54*, 1245–1258.

Ashkin, A., Dziedzic, J. M., Bjokholm, & Chiu (1987). Optical trapping and manipulation of viruses and bacteria. *Science 235*, 1517–1520.

Atkins, P. (1998). *Physical chemistry* (6th ed.). New York: Freeman & Co.

Berliner, D. F. (1998, April). Introduction to action research and its relationship to instructional change and reform. Paper presented at the annual meeting of the National Association for Research in Science Teaching, San Diego, CA.

Bruffee, K. A. (1993). *Collaborative learning: Higher education, interdependence, and the authority of knowledge.* Baltimore, MD: The Johns Hopkins University Press.

Collins, A., & Spiegel, S. A. (1995). So you want to do action research? In S. A. Spiegel, A. Collins, & J. Lappert (Eds.), *Action research: Perspectives from teachers' classrooms* (pp. 117–128). Tallahassee, FL: SouthEastern Regional Vision for Education (SERVE).

Coppola, B. P., Ege, S. N., & Lawton, R. B. (1997). The University of Michigan undergraduate chemistry curriculum. 2. Instructional strategies and assessment. *Journal of Chemical Education 74*, 84–94.

Dougherty, R. C. (1997). Grade/study-performance contracts, enhanced communication, cooperative learning, and student performance in undergraduate organic chemistry. *Journal of Chemical Education 74*, 722–726.

Driver, R., Squires, A., Rushworth, P., & Wood-Robinson, V. (1994). *Making sense of secondary science: Research into children's ideas.* London: Routledge.

Ege, S. N., and Coppola, B. P., & Lawton, R. G. (1997). The University of Michigan undergraduate chemistry curriculum. 1. Philosophy, curriculum, and the nature of change. *Journal of Chemical Education 74*, 74–83.

Ellis, A. B. (1995, March). Treating students and industry as customers. *Chemtech*, 15–21.

Erickson, F. (1998). Qualitative research methods for science education. In B. J. Fraser & K. G. Tobin (Eds.), *The International Handbook of Science Education* (pp. 1155–1173). Dordrecht, The Netherlands: Kluwer Academic Publishers.

Gallagher, J. J. (1991). Uses of interpretive research in science education. In J. J. Gallagher (Ed.), *Interpretive research in science education (Monograph #4)* (pp. 5–17). Manhattan, KS: National Association for Research in Science Teaching.

Gardner, M. B., & Ayres, D. L. (Eds.). (1998). *Journeys of transformation: A statewide effort by mathematics and science professors to improve student understanding.* College Park, MD: Maryland Collaborative for Teacher Preparation.

Gilmer, P. J. (1996). Networking with chemistry honors students. Paper presented at the annual meeting of the Southeastern Association for the Education of Teachers of Science, Smyrna, GA.

Gilmer, P. J., & Alli, P. (1998). Action experiments: Are students learning physical science? In S. R. Steinberg & J. L. Kincheloe (Eds.), *Students as researchers: Creating classrooms that matter* (pp. 199–211). London: Falmer Press.

Gilmer, P. J., Grogan, A., & Siegel, S. (1996). Contextual learning for premedical students. In J. A. Chambers (Ed.), *Selected papers from the 7th national conference on college teaching and learning* (pp. 79–89). Jacksonville, FL: Florida Community College of Jacksonville.

Glynn, S. M., & Muth, K. D. (1994). Reading and writing to learn science: Achieving scientific literacy. *Journal of Research in Science Teaching 31*, 1057–1073.

Guba, E. G., & Lincoln, Y. S. (1989). *Fourth generation evaluation.* Newbury Park, CA: Sage Publications.

Johnson, D. W., Johnson, R. T., & Smith, K. A. (1991). Cooperative learning. Increasing college faculty instructional productivity. *ASHE-ERIC Higher Education Report No. 4.* Washington, DC: The George Washington University, School of Education and Human Development.

Kemmis, S. (1983). Action research. In T. Husen & T. Postlethwaite (Eds.), *International encyclopedia of education: Research and studies.* Oxford: Pergamon.

Kemmis, S., & McTaggart, R. (1988). *The action research planner* (3rd ed.). [1st ed., 1981]. Victoria: Deakin University Press.

Kury, P. G. (Gilmer), & McConnell, H. M. (1975). Regulation of membrane flexibility in human erythrocytes. *Biochemistry 14*, 2798–2803.

Lakoff, G., & Johnson, M. (1980). *Metaphors we live by.* Chicago: University of Chicago Press.

Lemke, J. L. (1995). *Textual politics: Discourse and social dynamics.* Bristol, PA: Taylor & Francis Inc.

Lewin, K. (1946). Action research and minority problems. *Journal of Social Issues 2*, 34–46.

Lyons, L., & Millar, S. B. (1995). *You do teach atoms, don't you? A case study in breaking science curriculum gridlock.* Madison: University of Wisconsin, The Lead Center.

McDonald, J. B., & Gilmer, P. J. (Eds.). (1997). *Science in the elementary school classroom: Portraits of action research* (Monograph). Tallahassee, FL: SouthEastern Regional Vision for Education (SERVE).

Muire, C., Nazarian, M. J., & Gilmer, P. J. (1999). Web-based technology in a constructivist community of learners. *British Journal of Educational Technology 30*, 65–68.

National Research Council (1997). *Science teaching reconsidered: A handbook.* Committee on Undergraduate Science Teaching. Washington, DC: National Academy Press.

Overbye, D. (2001). Before the Big Bang, there was...What? *New York Times*, p. D1-D2, 22 May.

Oxtoby, D. W., Nachtrieb, N. H., & Freeman, W. A. (1994). *Chemistry: Science of change* (pp. 219–223). Chicago: Saunders College Publishing.

Spiegel, S. A., Collins, A., & Gilmer, P. J. (1995). Science for early adolescence teachers (Science FEAT): A program for research and learning. *Journal of Science Teacher Education 6*, 165–174.

Spiegel, S. A., Collins, A., & Lappert, J. (Eds.). (1997). *Action research: Perspectives from teachers' classrooms* (Monograph). Tallahassee, FL: SouthEastern Regional Vision for Education (SERVE).

Taylor, P. C. (1996). College teaching of science and mathematics in Florida: A preliminary foray into the field. In R. Elmesky, C. Muire, N. Griffiths, P. Taylor, & K. Tobin, *A statewide evaluation of mathematics and science courses in Florida community colleges* (pp. 117–135). Tallahassee, FL: Florida State University Program in Science Education in collaboration with the Florida Department of Education and the Division of Community Colleges.

Tobin, K. (1995). Learning from the stories of science teachers. In A. Haley-Oliphant (Ed.), *This year in school science* (pp. 161–180). Washington, DC: A.M.S.

———. (1997). The teaching and learning of elementary science. In G. D. Phye (Ed.), *Handbook of academic learning: Construction of knowledge* (pp. 369–403). Orlando, FL: Academic Press.

Williams, M. C. (1996). *A self-study of teaching reform in a university business computing course: "...it all went wrong..."* Doctoral thesis, Perth, Western Australia: Curtin University of Technology.

Notes on Contributors

Abdullah O. Abbas is Assistant Professor of Science Education at Sanaa University in Yemen. He received his B.S. in Education at Sanaa University, his M.S. in Health Education from Indiana University, and his Ph.D. in Science Education from Florida State University. He has taught in elementary, middle, and high schools and in the University in Yemen. His research interest focuses on faculty change in the sciences in higher education.

> *"...myths are socially constructed and within a community of practice are accepted as referents for action. They define what is normal practice and might be used to justify customs."* Kenneth Tobin commenting in an e-mail, reacting to Peter Taylor's seminar at Florida State University.

Craig Bowen is Senior Research Assistant of Academic Affairs at Towson University. He completed his B.S. in Chemistry at Emory University, M.S. degrees in Chemistry and Educational Psychology at Purdue University, and a Ph.D. in Science Education at Florida State University. In 1999 he was awarded the Early Career Research Award from the National Association for Research in Science Teaching. His research continues to focus on development and validation of new means of assessing chemistry learning and various aspects of the chemistry learning environment.

> *Education is an immensely social activity that involves interaction among groups with different perspectives on teaching and learning.*

Carol Briscoe is Associate Professor in the College of Professional Studies at the University of West Florida. She received her B.S., M.S. and Ph.D. from Florida State University. Dr. Briscoe taught science in the public schools (grades 5–12) for twenty years and has been at the university for nine years. Research interests include teacher thinking and change, prospective teachers' understanding of problem solving in mathematics and science, and alternative assessment.

We are the visionaries of action; we are inspired with change. We think the past preserves itself in the future of itself, the way Isaac Newton is changed and still preserved in Albert Einstein. We are the culture of living change. J. Bronowski in "The Abacus and the Rose: A New Dialogue in To World Systems" in *Science and Human Values* (1965).

Sabitra Brush, Associate Professor of Chemistry at Armstrong Atlantic State University, received her B.S. in Chemistry from the University of the West Indies and her M.S. in Analytical Chemistry and Ph.D. in Science/Chemical Education at Florida State University. She has taught high school chemistry and physics, and now she teaches general chemistry courses to chemistry majors, chemistry for nurses and graduate science education, and chemistry courses for teachers. Her research involves designing and teaching graduate summer courses, as well as monthly staff development workshops to upgrade K-12 science teacher education.

There can be no understanding without reflection. Students need to carry out reflection for themselves, and teachers cannot be the mechanisms for transfer.

Margarita Cuervo is Professor of Mathematics at Miami-Dade Community College, Kendall campus, where she has been teaching for over twenty-five years. She was born and raised in Cuba until the eleventh grade. She earned a Bachelor in Business Administration, *magna cum laude*, from the University of Puerto Rico. She received her Master of Science in Mathematics from New York University and holds a Doctor of Education degree from University of Miami. Margarita was awarded an Endowed Teaching Chair, the highest achievement in faculty excellence at Miami-Dade Community College. She is married and has one son.

Your most precious gift to another is your time and attention.

Ben Cunningham, a Christian Brother, currently director of the two-year Diploma Course in Addiction Counseling in Ireland, received his research doctorate in education at the University of Bath, England. Over a twenty-year span he has taught at both primary and secondary school level, and he has been both a secondary school principal and counselor. Since 1990 he has been involved in teaching teachers and trainee counselors, using action research as his medium.

> *I am interested in my own self-study as I ask questions of the kind, "How do I improve what I am doing?" My "I" in that question exists as a living contradiction in holding values (particularly freedom and love!) and experiencing their denial at the same time as asking the question. Because I experience my "I" as a contradiction I am motivated to improve what I am doing. The descriptions and explanations of the consequent learning that I create constitute what I call my living educational theory.*

Penny J. Gilmer, Professor of Chemistry at Florida State University, received her B.A. in Chemistry from Douglass College, M.A. in Organic Chemistry from Bryn Mawr College, and her Ph.D. in Biochemistry from University of California, Berkeley. Currently, she is completing a second doctorate in Science Education from Curtin University of Technology. Penny's research interests include equity in science education, action research, science research experiences for teachers, and teacher change in K-16. Penny is a Fellow of the American Association for the Advancement of Science. Penny's husband is a chemist too, and they enjoy their teenage daughter and son.

> *Michael Crichton's quote, "Life finds a way," from* Jurassic Park *strikes a bell. The life force is strong. Life finds a way for some to survive, from the smallest viruses to humans to larger plants and animals. My living through a tornado and surviving cancer have given me energy and motivation to do what I feel I need to do in life.*

Kenneth A. Goldsby is Distinguished Teaching Professor of Chemistry and Director of the University Honors Program at Florida State University. Ken received his B.A. in Chemistry and Mathematical Science from Rice University and his Ph.D. in Inorganic Chemistry from the University of North Carolina at Chapel Hill. His research focuses on redox reactions of transition metal complexes and self-assembled monolayers based on these complexes. He has worked on several projects with faculty in Florida State University's Program in Science Education. Ken's wife is a pharmacist, and they have three daughters.

One of my jobs as a teacher is to help students find something that makes the course material interesting or relevant or important to them. I cannot transfer my interest or my sense of relevance to other students. They have to construct, if you allow me to borrow the word, that for themselves.

Noelle Griffiths is currently teaching chemistry at Benson High School in Omaha, Nebraska. She received her bachelor's degree in biology and master's degree in science education at Florida State University. After her first year of teaching at Florida High School in Tallahassee she won the Seed Pearl Award for Science Teaching. She moved with her husband to Nebraska, where she teaches. Her experiences include interdisciplinary teaching and curriculum development. Noelle has worked with Upward Bound to encourage minority involvement in the sciences.

Thou hast only to follow the wall far enough and there will be a door in it.
from *The Door in the Wall*, Marguerite de Angeli, 1949.

Rosalind Humerick is a chemistry professor at St. Johns River Community College in Florida. She received her bachelor's degree in Management and Chemical Sciences from the University of Manchester, United Kingdom, and her doctorate in Inorganic Chemistry from Auburn University. Rosalind has been included in *Who's Who among American Teachers* twice in the 1990s. She is married and has one son.

I treasure those precious moments when faces light up and mouths bubble with enthusiasm.

Susan Mattson is teaching high school biology in a high school in Tallahassee, Florids. She received her undergraduate degree in Biology from the University of California, Berkeley and her master's degree in Biology and Ph.D. in Science Education from Florida State University. Her experiences include curriculum development in the sciences and professional development for teachers. She has developed videotapes with the Harvard-Smithsonian Center for *Case Studies in Science Education*.

I believe that sometimes, to make an abstract point you have to create an image. And that image can serve as a place of departure for thinking and acting.

Hedy Moscovici is Assistant Professor in Teacher Education at California State University-Dominguez Hills. Hedy, born and raised in socialist Romania until the twelfth grade, received her bachelor's and master's degrees in Biology and Microbiology in Israel and her Ph.D. in Science Education at Florida State University. She has also served on the biology faculty at Western Washington University for three years and a science consultant. Her research interest focuses on the analysis of the power relationships during scientific inquiries. She's married and has three children.

While my parents, family and friends provided me with roots and wings, I spend my life jumping on the continuum between these two biologically incompatible infinities.

Wolff-Michael Roth, Landsdown Chair in Applied Cognitive Sciences at University of Victoria, received his doctorate in Science Education at the University of Southern Mississippi. He has taught for ten years in middle or high schools and eight at the university. Michael has received many prestigious awards, such as the Early Career Research Award in 1993 from the National Association for Research in Science Teaching (NARST) and the Kappa Delta Pi Research Award in 1994 from the American Educational Research Association (AERA).

Participation in practice becomes the most important aspect of teaching and learning; and learning becomes changing participation in a changing practice. Teaching then becomes a situation in which teachers provide for settings in which students do, co-participating with others of heterogeneous competencies in whatever they are supposed to become competent in.

Kathryn Scantlebury is Associate Professor in the Department of Chemistry and Biochemistry at the University of Delaware, where she is the Secondary Science Education Coordinator for the College of Arts and Science. Kate's research focuses on gender and ethnicity equity issues in science education, teacher education, and systemic reform.

Feminist philosophy is the underpinning of my research, teaching, and personal values.

Peter C. Taylor is a Senior Lecturer in Science and Mathematics Education at Curtin University of Technology in Western Australia. For many years Peter's teaching, research, and research supervision have been shaped by his passionate concern for making teaching and learning a life-enriching and transformative experience. To this end he is an ardent innovator, employing a range of referents (constructivism, postmodernism, spirituality, alternative genres, action research) in the design of his own professional practice, a practice which is continually transforming in response to his dedication to self-study.

Through the storying of science and mathematics we encounter the mysteries of the world and are enchanted——as teachers are we not morally obliged to take our students with us on this fabulous journey?

Kenneth Tobin is Professor of Science Education at the University of Pennsylvania. He is involved in research on teacher learning and reforms in urban schools and has a strong interest in relating sociocultural factors to the teaching and learning of science. He has published extensively on science education and most recently on learning to teach science in urban high schools. Tobin is a former president of the National Association for Research in Science Teaching, currently serves as North American editor of Learning Environments Research, and is Chair-elect for the Education group within the American Association for the Advancement of Science.

In all events and phenomena are opportunities to learn and teach. Do both in equal proportions.

Harold B. White III, a Professor of Biochemistry who has been teaching at the University of Delaware since 1971, began using problem-based learning in his classes in 1993 as a way to get students to go beyond superficial understanding of content. Hal is particularly interested in exploiting the rich interdisciplinary nature of research articles as a source of problems suitable for problem-based instruction.

Sometimes, when I get all wrapped up with my teaching, I have to remind myself that none of the most important lessons I have learned in life arose in the classroom. They all emerged unexpectedly out of interpersonal situations beyond my control that I couldn't avoid.

Mark Campbell Williams, a Ph.D. in Science Education from Curtin University of Technology, is now an undergraduate lecturer in the School of Management Information Systems at Edith Cowan University. He has worked in the educational arena in tertiary, secondary, and technical spheres. Mark has been involved in phenomenological, heuristic, autobiographical, and life-history case studies of people within management and information systems and the ubiquitous area of electronic commerce. Mark has inaugurated the Electronic Commerce Club—a Web club for senior management to share experiences in a virtual "Edwardian Oxbridge" Web atmosphere.

"There's a blaze of light in every word, it doesn't matter which are heard, the holy, or the broken, Alleluia."

—Leonard Cohen.

Index

Studies in the Postmodern Theory of Education

General Editors
Joe L. Kincheloe & Shirley R. Steinberg

Counterpoints publishes the most compelling and imaginative books being written in education today. Grounded on the theoretical advances in criticalism, feminism, and postmodernism in the last two decades of the twentieth century, Counterpoints engages the meaning of these innovations in various forms of educational expression. Committed to the proposition that theoretical literature should be accessible to a variety of audiences, the series insists that its authors avoid esoteric and jargonistic languages that transform educational scholarship into an elite discourse for the initiated. Scholarly work matters only to the degree it affects consciousness and practice at multiple sites. Counterpoints' editorial policy is based on these principles and the ability of scholars to break new ground, to open new conversations, to go where educators have never gone before.

For additional information about this series or for the submission of manuscripts, please contact:

Joe L. Kincheloe & Shirley R. Steinberg
c/o Peter Lang Publishing, Inc.
275 Seventh Avenue, 28th floor
New York, New York 10001

To order other books in this series, please contact our Customer Service Department:

(800) 770-LANG (within the U.S.)
(212) 647-7706 (outside the U.S.)
(212) 647-7707 FAX

Or browse online by series:

www.peterlangusa.com